SOUND PROPAGATION THROUGH THE STOCHASTIC OCEAN

The ocean is opaque to electromagnetic radiation and transparent to low-frequency sound. So acoustical methodologies are an important tool for sensing the undersea world. Stochastic sound-speed fluctuations in the ocean, such as those caused by internal waves, result in a progressive randomization of acoustic signals as they traverse the ocean environment. This signal randomization imposes a limit to the effectiveness of ocean acoustic remote sensing, navigation, and communication. At the same time, signal fluctuations can be used as an important probe to better understand stochastic ocean processes.

Sound Propagation through the Stochastic Ocean provides a comprehensive treatment of developments in the field of statistical ocean acoustics over the last thirty-five years. This book will be of fundamental interest to oceanographers, marine biologists, geophysicists, engineers, applied mathematicians, and physicists. Key discoveries in topics such as internal waves, ray chaos, Feynman path integrals, and mode transport theory are addressed with illustrations from ocean observations. The topics are presented at an approachable level for advanced students and seasoned researchers alike.

JOHN A. COLOSI is a professor of oceanography at the Naval Postgraduate School in Monterey, California. He is a Fellow of the Acoustical Society of America (ASA) and was the recipient of the 2011 ASA Medwin Prize in Acoustical Oceanography and the 2001 UK Institute of Acoustics A.B. Wood Medal. He has twice been a Cecil H. and Ida M. Green Scholar at the Scripps Institution of Oceanography.

SOUND PROPAGATION THROUGH THE STOCHASTIC OCEAN

JOHN A. COLOSI

Naval Postgraduate School

CAMBRIDGE
UNIVERSITY PRESS

One Liberty Plaza, 20th Floor, New York, NY 10006, USA

Cambridge University Press is part of the University of Cambridge.

It furthers the University's mission by disseminating knowledge in the pursuit of education, learning, and research at the highest international levels of excellence.

www.cambridge.org
Information on this title: www.cambridge.org/9781107072343

First published 2016

Printed in the United States of America by Sheridan Books, Inc.

A catalog record for this publication is available from the British Library.

Library of Congress Cataloging-in-Publication Data
Names: Colosi, John A.
Title: Sound propagation through the stochastic ocean / John A. Colosi, Naval Postgraduate School.
Description: New York NY : Cambridge University Press, 2016. | Includes bibliographical references and index.
Identifiers: LCCN 2016008185 | ISBN 9781107072343 (hardback : alk. paper)
Subjects: LCSH: Underwater acoustics. | Sound–Transmission. | Oceanography.
Classification: LCC QC242.2.C646 2016 | DDC 620.2/5–dc23
LC record available at http://lccn.loc.gov/2016008185

ISBN 978-1-107-07234-3 Hardback

Dedicated to my father, my mother,
my wife Denise, and our children Julia, Anna, Paul, and Luke

Contents

Foreword

Modern ocean acoustics began with the discovery of the ocean sound channel on April 3, 1944. Ewing and Worzell sailed out of Woods Hole Harbor aboard the Saluda and lowered a hydrophone to 1,300 m to listen to 2-kg charges at up to 1,000-km ranges. They reported a signature so sharp that it would be "impossible for the most unskilled observer to miss it" (not true). They also predicted that some day it would be possible to transmit over 10,000-km ranges (true). Later that year, the Russian acoustician Brekhovskyikh accidentally recorded a signal that he attributed correctly to transmission through a waveguide. In both countries the work was part of a classified effort in Anti-Submarine Warfare (ASW); Ewing and Brekhovskyikh did not learn of each other's work for years to come.

The following three decades saw intensive international efforts in ASW and associated acoustic problems. Progress was limited, not so much by a lack of understanding of acoustics but instead by a lack of understanding of underlying ocean physics.

The very essence of ocean acoustics is its inherent variability. Phases and amplitudes vary in a complex manner. Fadeouts are the rule rather than the exception. Early on, this inherent variability was attributed to isotropic homogeneous ocean turbulence; ocean turbulence is neither isotropic nor homogeneous. *Sound Transmission through a Fluctuating Ocean*, published by Flatté and co-workers in 1979, suffered from a lack of understanding of the ocean environment and limited observations.

At this very time, oceanography was going through the mesoscale revolution. For more than a century we had ignored the decisive role played by ocean weather. In fact, the ocean has a weather; storms are called mesoscale eddies. They have typical scales of 100 km and 100 days (compared to 1000 km and 3 days in the atmosphere). Single ships sailing at 12 knots and never repeating a station are incapable of resolving ocean weather. Carl Wunsch and I had spent a sabbatical year at Cambridge University trying to come up with an observational strategy that could. We ended up with "Ocean Acoustic Tomography," the use of sound transmission as a tool for learning about the oceans rather than observing the oceans as a tool for learning about sound transmission.

Since the publication of *Sound Transmission through a Fluctuating Ocean*, there has been tremendous progress in observational techniques and analyses of the propagation of sound through the ocean. The author has significantly contributed to both; this book pays proper attention to both.

The time to review a subject is when it is being rapidly developed rather than when it is being put to bed. This is the right time for publication!

Walter H. Munk

Preface

This book describes the phase and amplitude fluctuations of acoustic signals that travel in the ocean acoustic waveguide with little interaction with the ocean boundaries. The book is a vastly expanded update of the 1979 text *Sound Transmission through the Fluctuating Ocean*, edited by Stan Flatté with contributions from Roger Dashen, Walter Munk, Ken Watson, and Fred Zachariasen. The focus here is on physical concepts and methodologies that over the last three decades have proven to be most useful.

The original text was written when a basic grasp of the subject was just emerging from three important lines of research. First, a few state-of-the-art field efforts utilizing controlled electronic sources (mostly in the kilohertz range) and a few hydrophone receivers on stable or navigated platforms were providing reliable observations of phase and amplitude statistics for single-frequency (CW) transmissions and most importantly for fixed acoustic paths: Fadeouts or scintillations and rapid phase jumps were seen to be all too common. Interestingly at this stage, classified military systems vastly exceeded the capabilities available to civilian investigators, one example being the remarkable SOund SUrveillance System (SOSUS). Second, from work in the late 1960s and early 1970s, oceanographers and acousticians had come to the realization that sound-speed fluctuations in the ocean were both anisotropic and inhomogeneous and that the ocean internal-wave field, described in some approximation by the recently developed Garrett-Munk (GM) internal-wave spectrum, was a significant contributor to this variability. Lastly, a new theoretical tool, the Feynman path integral method, allowed one to calculate several acoustic field statistics for both CW and fixed path and account for the anisotropic and inhomogeneous internal-wave field as well as the important fact that sound interacting with this random field is deterministically refracted by the ocean waveguide. These breakthroughs well justified the writing of the classic 1979 text; however, in the last thirty-five years the field has come a long way.

Using technology that was developed for Ocean Acoustic Tomography, experiments are now being conducted using wide-aperture vertical receiving arrays that are precisely navigated and time-synchronized. The analogy to astrophysics would be the difference between gazing at the heavens using the 200-in. Mt. Palomar telescope compared to the present-day observations on the 10-m, 36-element Keck telescope on the Hawaiian Mauna Kea volcanic peak. Further, astronomers used to study only the visible spectrum, but now virtually the whole electromagnetic spectrum is analyzed. Similarly, ocean acousticians are now able to transmit sound to study a wide acoustic spectrum of the ocean sound field from 20 Hz to 20 kHz. This capability has been enabled by energy-efficient broadband electronic sources, which are particularly challenging to build in the low-frequency end of the spectrum. And lastly, observational efforts have been bolstered by the tremendous increase in data-processing power supplied by modern computers and new signal-processing methodologies, enabling, for example, the processing of acoustic normal-mode signals and statistics that complement fixed-path observations.

From the theoretical side we have also come a long way. While the path integral is still a power tool, which provides profound insight into wave propagation, we now know that the expansion about the deterministic ray, which is needed to solve the path integral, has its fundamental limitations due to ray instabilities, termed ray chaos, as was demonstrated by the landmark SLICE89 experiment that first utilized a large vertical aperture-receiving array. Ray chaos means that a ray propagating through ocean sound-speed perturbations is expected to diverge exponentially from its neighbors and that exponential divergence range is a stochastic quantity varying from roughly 50 to 500 km. New insight, also from advances in ray theory, allowed researchers to appreciate the effects of the background sound-speed profile in dictating the sensitivity of acoustic fields to perturbations and to realize that acoustic energy is scattered primarily in the direction along the wave front rather than in the direction of propagation. Finally, new transport theory methods based on normal modes have shown to be extremely accurate in predicting several acoustic moments, and this method shows some promise in "solving" the ocean fluctuation problem given an accurate model of the sound-speed spectrum and background ocean profiles.

Now on the matter of stochastic ocean sound-speed structure, many new measurements have been made, but it has been found that the GM internal-wave model is indeed a first-order description of internal waves in many regions of the ocean, including the continental shelf. Of course there are deviations in the spectrum in places such as the Arctic and near abrupt topography, but acoustic models utilizing the GM spectrum have given the best results when comparing to observations. Ongoing work related to ocean spiciness, that is, ocean sound-speed structure carried along constant density surfaces and vortical internal-wave motions, offers some hope of better characterizing the entire stochastic ocean sound-speed field, but no models of these processes yet exist to be utilized in the same way that the GM model is used.

Thus the objective of this book is to provide the reader with an up-to-date view of (1) observations, (2) ocean sound-speed fluctuations, and (3) ocean wave propagation through random media theory. The approach was to be as pedagogical as possible for a monograph on the state of the art in the field. This means we have incorporated toy models into the discussion to help illustrate basic concepts without the burden of complex calculation or mathematics. The hope is that this approach will make the text as useful for graduate students and postdocs as it will be for experts in the field. The original goal was to also include a series of chapters on applications such as internal-wave tomography, internal-wave effects on large-scale acoustic remote sensing, transport theory applied to ocean mixed layer propagation, and combined effects of stochastic internal waves and shallow-water nonlinear internal-wave packets. In the end it was determined that the text would simply be too big and so it was decided to leave these topics for some other volume.

The writing of a book is a significant task that can rarely be accomplished alone. This book was enabled by many friends and colleagues. Foremost among them is Dr. Peter Worcester, who has been my mentor, colleague, and friend for nearly twenty-five years and who painstakingly went over nearly every line of the manuscript and provided sage advice on the content and organization. Any flaws in the manuscript are directly attributable to my stubbornness and not to his attempted guidance. Special thanks go to Mike Brown, Bruce Cornuelle, Bill Kuperman, and Rob Pinkel for looking over specific chapters. Tim Duda, Jim Lynch, and Allan Pierce critiqued the entire book and provided many helpful comments. The figures were artfully done by Jennifer Matthews of the Scripps Institution of Oceanography (SIO). Funding for the book was provided by the Office of Naval Research, sabbatical funds from the Naval Postgraduate School, and the Cecil H. and Ida M. Green Foundation at the SIO Institute of Geophysics and Planetary Physics (IGPP). Warm hospitality was provided by the SIO IGPP, and the University of California, Santa Cruz departments of Physics, Earth and Planetary Sciences, and Ocean Sciences. Special thanks to Walter and Mary Munk for opening up their fabulous La Jolla home for me (with its access to one of the best surfing beaches in California, Blacks Beach) during my summer 2014 visit.

The book would never have been possible without the teaching, inspiration, and friendship of my thesis advisor Stan Flatté. Stan and I had started the book in the early 2000s, but after his death in 2007, work came to a halt. At this point I did not know if I had it in me to single-handedly put the last thirty years of progress in the field into some semblance of perspective. But with unrelenting encouragement from Stan's wife Renelde and from other colleagues, I dedicated my 2014 sabbatical year to the task. Contrary to my feelings going into the work, I found writing my first book to be quite enjoyable, and so my hope is that the reader comes away with a true sense of passion for the topic. The words of Roger Revelle (1974) capture this spirit when he said:

From 1955 to 1961, I experienced the fierce joys of helping found a new university. As with most things one does for the first time,
Making Love,
Getting a PhD,
Becoming a Father,
this task was done with more enthusiasm than knowledge.

Finally, the reader will undoubtedly find multiple errors in the manuscript and passages in which there may be some controversy. Please pass along any correspondence on these matters to John Colosi, Department of Oceanography, Naval Postgraduate School, 833 Dyer Road, Monterey, California 93943, or jacolosi@nps.edu.

Notation

Notational ambiguities and breaks in convention are the rule, not the exception, in interdisciplinary subjects such as this. While there is no perfect solution to this problem, a notation has been chosen that is the least of all possible evils. When a break in notational convention is perpetrated, the event is noted in the text or in a footnote. The notation f, $\mathbf{f}$, $\mathbf{F}$, $\mathbf{F}^T$ refers to scalar, vector, matrix, and matrix transpose representations of the quantity f. When individual elements of a vector or matrix are referenced, the notation is f_n or F_{mn} for the case of an $M \times N$ matrix.

Oceanographic and Internal-Wave Variables

$\mathbf{r} = (x, y, z)$	Spatial coordinate with z upward from the ocean surface
t, τ	Geophysical time, time delay
ϑ	Azimuthal angle
σ	Internal-wave frequency
$\kappa = (k, l, m)$	Internal-wave wavenumber
j, J	Internal-wave mode number, maximum mode number
$\kappa_j, \kappa_h = \sqrt{k^2 + l^2}$	Internal-wave modal horizontal wavenumber, horizontal wavenumber
$\hat{\kappa}_j = \pi j f / (N_0 B)$	Roll-off wavenumber in the GM spectrum
$\psi_j(z)$	Internal-wave normal mode functions
$\mathbf{u} = (u, v, w)$	Fluid velocity
ζ	Vertical displacement
Ω	Angular velocity of the earth
$f = 2\Omega \sin(\text{latitude})$	Coriolis parameter
$c, \bar{c}(z)$	Sound speed, background sound-speed profile
γ_a	Adiabatic sound-speed gradient
c_0	A typical sea water sound speed (usually 1,500 m/s)
$U(z) = \bar{c}(z) / c_0$	Fractional background sound speed

T, S	Temperature, salinity
$\rho, \rho_0(z)$	Density, background density profile
σ_p	Potential density referenced to pressure p
θ	Potential temperature
N	Buoyancy frequency
$N_0 B = \int_0^D N(z)dz$	WKB scaling parameter
$(dc/dz)_p$	Gradient of potential sound speed
$\delta c,\ \mu = \delta c/c_0,\ \mu_u = u/c_0$	Sound-speed perturbation, fractional sound-speed perturbation, Mach number
F_ξ, F_E, F_u, F_μ	Spectra of displacement, energy, horizontal current, and fractional sound speed
L_x, L_y, L_z	Internal-wave coherence lengths
T_{iw}	Internal-wave coherence time
$\hat{L}_x, \hat{L}_y, \hat{L}_z$	Internal-wave coherence length with the perpendicular wavenumber constraint
$\hat{T}_{iw}$	Internal-wave coherence time with the perpendicular wavenumber constraint

Acoustic Variables

ω	Acoustic frequency
$p(\mathbf{r}, t)$	Acoustic pressure field
$\mathbf{k} = (k_x, k_y, k_z)$	Acoustic wavenumber
$l_n = k_n + i\alpha_n$	Acoustic modal horizontal wavenumber
$\phi_n(z)$	Acoustic normal mode functions
λ, T_p	Acoustic wavelength, wave period
$q_0 = \omega/c_0$	A typical sea water acoustic wavenumber
γ	Fractional bandwidth, bandwidth divided by center frequency
Θ	Acoustic phase
τ	Signal travel-time fluctuation
R, D	Propagation range, water depth
Γ, θ_r	Acoustic ray path, local ray angle with respect to the horizontal
z_0, y_0	Acoustic coherence lengths in the vertical and transverse directions
t_0	Acoustic coherence time
L_p	Internal-wave coherence length in the direction of the ray

Φ	RMS phase fluctuation in the geometric limit
Λ	Diffraction parameter
a, I, χ, ι	Acoustic amplitude, intensity, natural log of amplitude, natural log of intensity
a_n	Acoustic normal mode amplitude
α, β	Ray stability parameter, waveguide invariant
ν	Lyapunov exponent
I, ϑ	Ray action, angle variables
$R^{\pm}, T^{\pm}$	Ray upper/lower horizontal loop distance and travel time
R_L, T_L	Ray double loop distance and travel time
$K_R = 2\pi/R_L$	Ray double loop spatial frequency
W	Wave energy density
TL	Transmission loss
H	Hamiltonian function
L	Lagrangian function
S_a	Acoustic path length
R_{fz}, R_{fy}	Vertical, transverse Fresnel zones
R_{mr}	Vertical spread of microray bundles
M	Ray Maslov number
ξ	Ray tube function in the parabolic approximation
ρ_{mn}	Mode coupling matrix
P	A probability density function
$\int D(path)$	Feynman path integral
i	$\sqrt{-1}$

PART I
Introduction and Prerequisites

1

Sound Propagation through the Stochastic Ocean

1.1 Introduction and Historical Background

In the undersea world the sound wave rules supreme, as is evidenced by the highly evolved acoustic physiology and sophisticated auditory processing capability of fish and marine mammals. Humans are relative latecomers to the world of undersea sound. Leonardo da Vinci (1483) developed a device to listen to approaching ships, writing

If you let your ship stop, and dip the end of a long blowpipe in the water and hold the other end to your ear, then you can hear ships which are very far distant from you.

There are also reports that for centuries Inuit whalers have been using acoustic methods to localize their prey. By placing the butt of a dipped oar against one's jawbone, the underwater vibrations of the vocalizing whales could be sensed.

Curiously the giants of classical physics (e.g., Newton, Euler, Lagrange, Laplace, Helmholtz) seem to have paid little attention to the subject, instead devoting their energies in acoustics toward musical problems. Even the great acoustical authority, Lord Rayleigh, makes scarce mention of underwater sound in his masterpiece publications *The Theory of Sound*, Parts 1 and 2, in 1887 and 1896. Indeed, the speed of sound in fresh water was not measured until 1826 (Colladon and Sturm, 1827), and the first crude measurements in sea water came roughly a century later (Wood, 1930).[1]

Developments in ocean acoustics seem to be inseparably tied to matters of military importance, the first of which was the problem of knowing the seafloor adequately so as to avoid vessel grounding. This was the genesis of the acoustic fathometer (echo sounder) invented shortly after the turn of the twentieth century, which was also in great favor with the European royals whose fleets suffered many losses due to grounding. The first measurements of the speed of sound in sea water were largely motivated by echo sounding and sound ranging.

---

[1] The first accurate, empirical equation of state for sound speed in sea water as a function of temperature, salinity, and pressure was developed by Wilson (1960).

The addition of the submarine to the naval arsenal provided a particularly strong catalyst for advancement in the twentieth century. Though there were some significant developments during and after World War I (Wood, 1930), the independent discovery of the ocean sound channel by both US and Soviet scientists toward the end of World War II brought the field to an entirely new level (Ewing and Worzel, 1948; see also the discussion of the Soviet discovery in Munk et al. 1995). Shortly after the discovery of the sound channel, a string of experiments revealed time and again remarkable detections of acoustic signals at extremely long ranges. Ewing and Worzel (1948) remark on their receptions at ranges of up to 1450 km:

> the end of the sound channel transmission was so sharp that it was impossible for the most unskilled observer to miss it.

Included in this list of long-range detections is a remarkable global-scale experiment in which 300-lb charges detonated off Perth Australia were detected halfway around the world in Bermuda (Munk and Forbes, 1989) (Figure 1.1).[2]

But, as the focus turned from simple detection of explosive events on single hydrophones or small receiver arrays to more challenging uses of sound for anti-submarine warfare (ASW) and other applications, the variability of acoustic fields became more evident and troublesome. The unskilled observation and interpretation of the transmission finale did not apply to other features of the transmission arrival pattern. Signal dropouts and rapid phase jumps were seen to occur at an alarming rate, presenting serious challenges in the development of a number of applications, including remote sensing, communication, and navigation. Eckart and Carhart (1950) state:

> If sound of constant intensity and frequency is transmitted through the sea from one ship and received on another at some fixed distance, the intensity of the signal received from one second to the next will not be constant; it fluctuates, often by a factor of ten. Indeed, the presence of fluctuation is perhaps the most constant characteristic of sound in the sea!

However, the primary motivation for this book is the inescapable fact that the ocean is an exceptional medium in which to utilize sound for many applications, but a knowledge of sound propagation through the stochastic ocean is often necessary to realize these applications. The problems of signals, fluctuations, and noise are all too common in many fields of science and engineering (a few examples from atmospheric optics, astrophysics, and seismology are Wheelon, 2003; Andrews and Phillips, 2005; Sato et al., 2012). The general branch of study that includes sound propagation through the stochastic ocean is often referred to as wave propagation through random media (WPRM).[3]

---

[2] This demonstration of global-scale acoustics was repeated in the 1990s' Heard Island feasibility test (Munk et al., 1994).

[3] The field of ocean acoustic WPRM may be compared to other fields such as electromagnetic propagation in the atmosphere, astrophysics, and seismology. In this comparison it is important to realize that there are significant

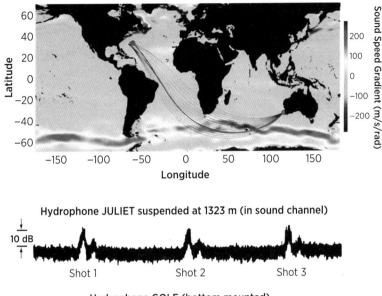

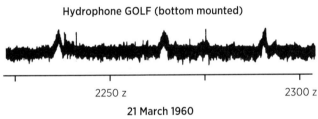

Figure 1.1. The 1960 Perth to Bermuda experiment. The upper panel shows refracted geodesic acoustic paths from the source to the receiver. Superimposed on the map is the horizontal sound-speed gradient at the depth of the sound-channel axis. The lower panel shows the acoustic receptions at two different receivers.
Source: Munk and Forbes (1989); Dushaw (2008).

It should be stated here that the topic of sound propagation through the stochastic ocean is quite broad in scope, including the areas of random boundary and biological scattering, scattering by bubbles, particulates, and seafloor inhomogeneities, nonlinear wave scattering, as well as high-frequency (tens of kilohertz) scattering. The focus of this monograph is on ocean sound (1) that is primarily trapped in the sound channel with little influence from the boundaries save attenuation and (2) with frequencies below 10 kHz, because these are

civilian as well as military operational systems that provide a catalyst for the development of these fields. In ocean acoustics, things are quite different. Presently there is a light civilian acoustical presence in the ocean, and operational systems are largely associated with restricted military concerns. The difficulty of maintaining a civilian presence in the ocean has many root causes including issues with marine mammals, but a major impediment comes from the powerful navies of the world that are understandably concerned with the stealth of their undersea forces. Recently developed applications of passive acoustics and cabled systems for research are a positive step forward, but these facilities are still operated under military oversight.

the acoustic frequencies that can travel significant distances. In these cases it has been shown by numerous experiments that stochastic internal-wave-induced sound-speed fluctuations are the primary source of acoustic variability.

1.1.1 Ocean Dynamics and Fluctuations

While many of the oceanographic causes of acoustic fluctuations were correctly identified shortly after World War II (Eckart and Carhart, 1950), it took some time for the community of scientists studying long-range propagation in the ocean sound channel to make the connection between specific ocean processes and acoustic scales of variability. The classical view of the ocean was a dominant, large-scale, geostrophically balanced, general circulation pattern with small perturbations due to ocean "weather," but the situation is in fact the other way around. In the 1960s and 1970s it became evident that the ocean has energetic weather patterns termed eddies that in many cases swamp the general circulation (Gould et al., 1974; MODE-Group et al., 1978). In addition, physical oceanographers were discovering that the ocean has a rich spectrum of small-scale structure associated with internal waves. This appreciation was spurred by the surprising observation that thermometers or current meters separated by 100 m vertically or several kilometers horizontally showed fluctuations that were uncorrelated (see references in Garrett and Munk, 1972).[4]

Given this situation, Hamilton's 1961–1963 time series of precisely located and timed explosions sending impulses from near Antigua to the islands of Eleuthera and Bermuda were all the more remarkable, showing 200 ms travel-time fluctuations associated with the energetic North Atlantic eddy field (Hamilton, 1977) (Figure 1.2; Table 1.1). These observations, which were not sampled finely enough in time to reveal internal-wave effects, may in fact have been the first measurements demonstrating the existence of the ocean eddy field.

Closely following Hamilton's work, between 1962 and 1973, a series of experiments were carried out by the University of Michigan and the University of Miami (termed MIMI) in the Florida Straits and along an acoustic path between Eleuthera and Bermuda. MIMI utilized newly developed controlled electronic sources transmitting single-frequency continuous-wave ([CW]) and broadband signals at a high enough rate to resolve internal-wave effects. These measurements allowed for the first time a quantification of the detailed time scales of acoustic variability (Figure 1.3; Table 1.1). Not surprisingly, the acoustic time scales had a

---

[4] Many observationalists did not trust these measurements and concluded that there were issues with the equipment. During this time Eckman was looking for observational evidence for the spiraling vertical current that bears his name. After a decade of frustration, he eventually concluded that his current meters were working properly, but that the spirals were swamped by internal-wave noise.

Table 1.1. *Significant deep-water ocean acoustic fluctuation experiments pre-1980*

Year(s)	Experiment	Institutions[a]	Specifics/Comments
1944	Ewing/Worzel	CU/WHOI/ Navy	4-lb TNT shots and single hydrophone receiver both at 1220 m depth. Signals received at several ranges up to 1450 km. First demonstration of the deep sound channel (Ewing and Worzel, 1948).
1961–1963	SCAVE	LGO/CU	27-month time series of sound axis explosions, Antigua to Bermuda and to Eleuthera, at ranges roughly 2000 km. Demonstration of travel-time fluctuations due to eddies (Hamilton, 1977).
1963–1973	MIMI	UMic/UMia	400 and 800 Hz CW and pulse transmission across the Florida straits. Range up to 65 km. Demonstration of phase stability, amplitude instability, and tidal fluctuations (Steinberg and Birdsall, 1966; Kennedy, 1969).
1972–1973	MIMI	UMic/UMia	400 Hz CW transmissions Eleuthera to Bermuda and mid-station. Ranges of 550 and 1250 km. Demonstration of the fully saturated regime and tidal fluctuations (Nichols and Young, 1968; Dyson et al., 1976).
1972	Cobb	UW	4- and 8-kHz pulse transmissions over a 17.2-km near-axial ray path. Demonstration of unsaturated and partially saturated regimes (Ewart, 1976).

(*continued*)

Table 1.1. *(cont.)*

Year(s)	Experiment	Institutions[a]	Specifics/Comments
1975	AFAR	NUWC	400-, 1000-, and 4700-Hz pulse transmissions over a 35-km near-axial path. Demonstration of partially and fully saturated regimes, and time coherence (Reynolds et al., 1985; Flatté et al., 1987b).
1977	MATE	UW	2-, 4-, 8-, and 13-kHz pulse transmissions over an 18-km near axial path and a steep upper ocean path. First experiment with detailed oceanographic observations. Demonstrated all three propagation regimes and presented phase and intensity frequency spectra (Ewart and Reynolds, 1984; Macaskill and Ewart, 1996).
1978	Bermuda	SIO	220 Hz pulse transmissions to 900-km range. Several pulse and CW statistics demonstrated (Spiesberger and Worcester, 1981).
1978	SD	SIO	2-kHz pulse transmissions to a range of 23 km, resolving 4 acoustic paths. Several pulse and CW statistics demonstrated (Worcester et al., 1981).

[a]CU, Columbia University; LGO, Lamont Geological Observatory; NUWC, Naval Underwater Systems Center; UMic, University of Michigan; UMia, University of Miami; SIO, Scripps Institution of Oceanography; UW, University of Washington; WHOI, Woods Hole Oceanographic Institution.

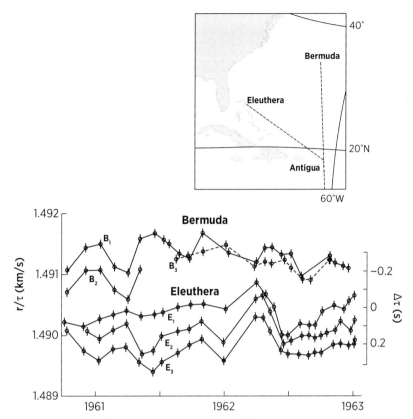

Figure 1.2. The 1961 Sound Channel Axis Velocity Experiment (SCAVE) in which precisely timed and located explosions off Antigua were received at Eleuthera and Bermuda (upper). The lower panel shows time series of the travel time of the pulse final cut-off at different receivers. The three receivers at Eleuthera show similar variations having traveled through similar ocean eddies. The records at Bermuda and Eleuthera are not similar since the paths see a different eddy field.
Source: Hamilton (1977); Munk et al. (1995).

close correspondence to those of the ocean, including eddies, tides, internal waves, and surface gravity waves (Nichols and Young, 1968; Dyson et al., 1976).[5]

With experimental results giving reliable statistical descriptions of phase and amplitude fluctuations, efforts to develop a physical understanding of the ocean acoustic scattering mechanisms got underway. This early theoretical work on sound-field fluctuations was not conducted entirely in earnest but relied heavily on ideas from other fields, such as optical and radio wave propagation through the

---

[5] The SCAVE and MIMI efforts were associated with the development of SOund SUrveillance System (SOSUS) whose purpose to locate and track submarines took on new urgency with the emergence of the Soviet nuclear submarine in the 1960s.

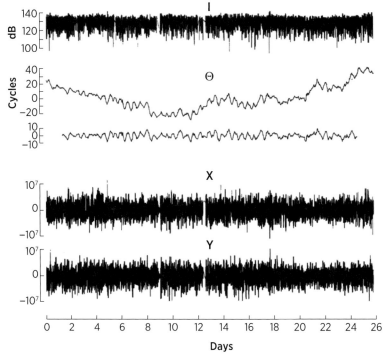

Figure 1.3. A 26-day record of the acoustic field for the 1972, single-frequency, 406-Hz MIMI transmissions from Eleuthera to Bermuda. The upper panel shows received intensity and the middle panel shows unwrapped phase (raw and high-pass filtered). Lower panel shows the real (X) and imaginary (Y) parts of the demodulated acoustic field.
Source: Dyson et al. (1976).

wind advected, homogeneous and isotropic, index of refraction fluctuations of the turbulent atmosphere described by the Kolmogorov model (Tatarskii, 1971).

A key breakthrough occurred when Garrett and Munk identified internal waves as a significant source of ocean sound-speed fluctuations and found that the internal-wave spectrum has a surprisingly universal form (Garrett and Munk, 1972). Internal waves induce sound-speed perturbations by vertical displacement, ζ, of the background sound-speed structure, that is,

$$\delta c(\mathbf{r},t) = \left(\frac{dc(z)}{dz}\right)_p \zeta(\mathbf{r},t), \tag{1.1}$$

where the vertical gradient here is the total gradient minus the adiabatic gradient (termed the potential gradient), because of the adiabatic nature of the displacements. Figure 1.4 shows an example of contemporary measurements of internal-wave displacements on a vertical mooring together with their frequency spectrum. Internal-wave-induced sound-speed fluctuations are (1) inhomogeneous in depth, with larger fluctuations in the main thermocline and smaller ones at

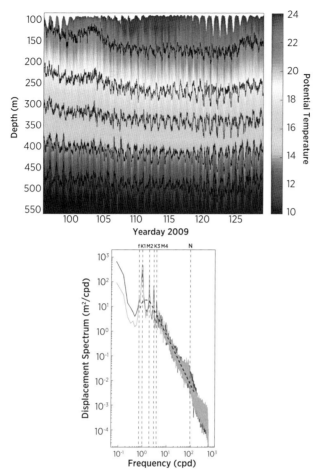

Figure 1.4. (Top) A month-long time/depth series of temperature fluctuations observed in the Philippine Sea. Superimposed on the temperature fluctuations are vertical displacements of a few isopycnals, showing random internal waves, internal tides, and eddies. (Bottom) Frequency spectra of the displacements at different depths (blue and green) and the GM internal-wave spectrum (dash). Tidal peaks are marked, and the inertial and buoyancy frequencies are labeled f and N. *Source*: Colosi et al. (2013b).

depth, (2) anisotropic in the vertical/horizontal plane with horizontal and vertical correlation lengths of roughly 10 km and 0.1 km, respectively, and (3) the waves have their own intrinsic space/time dependence imposed through the dispersion relation. The nature of sound-speed structure in the ocean is quite different from the homogeneous and isotropic turbulence models borrowed from the early days of atmospheric WPRM.

Even with the Garrett-Munk (GM) spectrum in hand, a significant obstacle, which is still faced today, is accurately accounting for the fact that the sound-speed fluctuations are superimposed on a background sound channel, which causes the sound waves to follow curved (refracted) trajectories through the fluctuations. From the full-wave perspective, the channel imposes a normal mode structure and the fluctuations induce random mode coupling. It has been only recently discovered that the background sound-speed profile, which varies geographically due to the climatological structure of the ocean, has a profound effect on the acoustic fields' sensitivity to sound-speed fluctuations from internal waves and other perturbations.

1.1.2 Theory of Sound Propagation

In the 1970s four important lines of research allowed the field to make significant advances. First, there was the description of the ocean sound-speed fluctuations provided by the GM internal-wave spectrum. Second was the advent of controlled broadband electronic sources and precisely timed and located receivers that allowed high-quality observations of acoustic fluctuations. Third, there was new theoretical work using the Born and Rytov approximations, covering the regime of weak fluctuations, and the Feynman path integral technique, covering the regime of strong fluctuations. This body of theory was significant because it was able to simultaneously account for all the unique aspects of ocean acoustic WPRM, that is propagation in a deterministic background waveguide with inhomogeneous and anisotropic wavelike fluctuations.[6] In addition to the more ray-like approaches noted here, there were also significant advances in the area of stochastic coupled mode theory (Dozier and Tappert, 1978a). However, the mode results were not fully appreciated until a few decades later.

Lastly, the fourth important advancement that occurred in the 1970s was the emergence of a new method of numerical simulation termed the parabolic equation (PE) method (Tappert and Hardin, 1973; Tappert, 1974).[7] The PE method, which is based on a forward marching algorithm using the Fast Fourier Transform (FFT), provided a means for computing acoustic fields with significantly greater speed and accuracy than was available through ray and normal mode methods. In addition, the method could easily handle range-dependent environments such as those caused by internal waves, thus opening up the field of Monte Carlo numerical simulation (Flatté and Tappert, 1975).

During this period, theoretical and observational development focused on acoustic fluctuations of resolved arrivals for specific paths that could be isolated using broadband transmissions. For long-range propagation in the ocean waveguide, these

---

[6] The development of the Born/Rytov and Feynman path integral methods for sound propagation through the stochastic internal-wave field was largely completed during the JASON summer studies of 1974–1976.

[7] The PE method and many other critical ASW advancements resulted from a collaboration of academics, naval officers, and industry who operated within the Long Range Acoustic Propagation Program (LRAPP) managed by the Navy Director of Antisubmarine Warfare Programs (OP-095). The LRAPP group was formed in 1964 and continued its work until the fall of the Soviet Union.

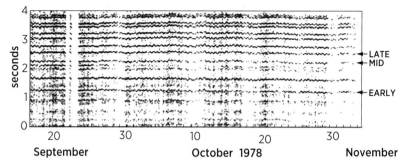

Figure 1.5. The 1978, 220-Hz, 900-km Bermuda transmissions showing multiple arrivals that are stable over several months. Each arrival has its own specific fluctuation statistics because the signals have followed different ray paths that sample distinct regions of the water column.
Source: Spiesberger and Worcester (1981).

specific paths result from the evolution of an initial spherical wave front to a highly folded double accordion pattern. Therefore, a receiver at a given depth records multiple pulses (arrivals) from a single-impulse transmission, and these arrivals are associated with specific paths through the ocean (Figure 1.5). Acoustic fluctuation statistics depend on the arrival considered because some arrivals are associated with paths that cycle steeply through the waveguide (perhaps hitting the ocean surface) while others are trapped near the deep sound-channel axis. This behavior can be contrasted with the single-frequency MIMI transmissions (Figure 1.3), where the phase and intensity records are a sum of all the paths simultaneously.

From the theoretical development for weak fluctuations and strong scattering, there emerged the important notion of propagation regimes, termed unsaturated, partially saturated, and fully saturated. Here the critical physical concept is that acoustic fluctuations occur through microray interference and large-scale modulation (Flatté et al., 1979). In the unsaturated regime, perturbative approaches like Born and Rytov theory apply. In this case there is only one ray path, whose ray tube is modulated by the internal-wave fluctuations. Two ways of quantifying this variability are the scintillation index (SI), defined as the normalized intensity variance

$$SI = \frac{\langle I^2 \rangle - \langle I \rangle^2}{\langle I \rangle^2},$$ (1.2)

and the log-intensity ($\iota = \ln I$) variance

$$\sigma_\iota^2 = \langle \iota^2 \rangle - \langle \iota \rangle^2.$$ (1.3)

The scintillation index is sensitive to high-intensity fluctuations, while the log-intensity variance is sensitive to fading due to the logarithmic distortion of low intensities. In the unsaturated regime one typically has $SI < 0.6$ and $\sigma_\iota^2 < (4 \text{ dB})^2$.

As the scattering gets stronger, new ray paths in the neighborhood of the unperturbed path are generated by the sound-speed fluctuations, leading to microray interference. In the partially saturated regime, the microrays are strongly correlated by the large-scale internal-wave structure and intense focusing can result, leading to *SI* values significantly more than 1 and rms log-intensity as large as 10 dB. The familiar bright bands of sunlight focused on the bottom of a swimming pool by random surface waves and the twinkling of stars are both associated with the partially saturated regime. In the fully saturated regime, the acoustic field is the sum of many uncorrelated microrays, which by the central limit theorem leads to Gaussian statistics. In the fully saturated regime $SI \simeq 1$ and $\sigma_{\iota}^2 \simeq (5.6 \text{ dB})^2$.

This body of work was able to bring theory and observation into relatively good agreement (within factors of 2) for short-range, high-frequency experiments such as the Cobb, AFAR, MATE, and SD (see Table 1.1).[8] The 1979 publication of the first text on the subject, *Sound Transmission through the Fluctuating Ocean* (Flatté et al., 1979), was largely motivated by these significant steps forward.

Quite a bit of time has passed since the publication of this original text. The purpose of the present text is to provide an updated treatment of the subject, emphasizing concepts and methodologies that, over time, have proven to be most useful. The text is also intended to underscore the multitude of gaps in our understanding. As such, the following sections give an overview of the progress that has been made in the last three decades, and a summary of the most pressing issues in the field.

1.2 Three Decades of Development: Observations

The last three decades of ocean acoustic observations related to fluctuations from stochastic internal waves involve both deep and shallow-water environments, mainly in mid-latitudes although there are two examples of polar experiments (Tables 1.2 and 1.3). The primary focus of the field, however, has been on deep-water propagation, where boundary and other effects are absent or minimal. In shallow water a stochastic field of internal waves also exists, along with several other important physical, biological, and geological processes of acoustical significance.

1.2.1 Deep Water

In the late 1970s and early 1980s the field of ocean acoustic tomography was developing (Munk et al., 1995), and the two fields have been closely linked ever

---

[8] The body of theory has also been successfully applied in the Arctic (AATE), as well as in some short- and mid-range, low-frequency experiments such as the AET (87 km range) and PhilSea09 (107 km range) (see Table 1.2).

Table 1.2. *Significant deep-water ocean acoustic fluctuation experiments post-1980*

Year(s)	Experiment	Institutions[a]	Specifics/Comments
1983	RTE83	SIO/WHOI	400-Hz reciprocal pulse transmissions at a range of 305 km. Demonstration of internal-wave current effects on travel time (Stoughton et al., 1986).
1985	AATE	UW	2-, 4-, 8-, and 16-kHz transmissions to 6 km range under ice in the Beaufort Sea. Demonstrated Arctic internal waves are weak and have a different spectrum. Weak fluctuation theory works for phase and amplitude. (Ewart and Reynolds, 1993)
1989	SLICE	SIO/UW	250-Hz pulse transmissions to a 3-km vertical receiving array at 1000-km range. Demonstration of shadow zone scattering, early wave front stability, and microray interference (fracturing) along the front (Duda et al., 1992).
1994	AET	SIO/UW	75-Hz pulse transmissions to 700 m vertical receiving arrays at 3250- and 87-km range. Long-range demonstration of shadow zone scattering, early wave front stability, and microray interference. Short range shows utility of weak fluctuation theory for phase and amplitude (Colosi et al., 1999, 2009).
1995–1956	ATOC	SIO/UW	75-Hz pulse transmissions to 1400 m vertical receiving arrays at 3500- and 5200-km range, and receptions at Navy SOSUS stations. Quantification of mode statistics, high angle path shadow zone scattering, and finale statistics and coherence (Dushaw et al., 1999; Colosi et al., 2005; Wage et al., 2005).
1996	AST	SIO/UW	Similar analysis to ATOC yet at frequencies of 28 and 84 Hz (Colosi et al., 2005; Wage et al., 2005).

(continued)

Table 1.2. *(cont.)*

Year(s)	Experiment	Institutions[a]	Specifics/Comments
1998–1999	NPAL	SIO/UW	75-Hz pulse transmissions to a billboard receiving array at 3800 km range. Quantification of horizontal coherence. Troublesome bottom interaction (Voronovich et al., 2005).
2004	LOAPEX	UW/SIO	75-Hz pulse transmissions for axial and shallow source depths received on a 1400-m vertical array for ranges from 50 to 3200 km. Quantification of range evolution of mode statistics. (Chandrayadula et al., 2013)
2004–2005	SpiceEx	SIO	250-Hz pulse transmissions to a 1400-m vertical array at 500- and 100-km ranges. Quantification of high angle path shadow zone scattering. (Van Uffelen et al., 2009)
2009	PhilSea09	SIO/UW NPS	250-Hz pulse transmissions to 185 km range, and 285-Hz pulse transmissions to 107 km range. Extensive oceanographic observations.
2010–2011	PhilSea10	SIO/UW NPS	250-Hz pulse transmissions on a pentagonal tomography array of 6 transceiver moorings and a large-aperture vertical array.

[a]NPS, Naval Postgraduate School; UMic, University of Michigan; UMia, University of Miami; SIO, Scripps Institution of Oceanography; UW, University of Washington; WHOI, Woods Hole Oceanographic Institution.

Table 1.3. *Significant shallow-water ocean acoustic fluctuation experiments and results related to scattering by the stochastic internal-wave field.*

Year(s)	Experiment	Institutions[a]	Specifics/Comments
1983	Yellow Sea	IAAS	Explosive and electronic transmissions to ranges of 20–30 km, showing 10–30 dB fluctuations
1992	Barents Sea	WHOI/NPS/ SAIC	224-Hz pulse transmissions to 25-km range, and extensive oceanographic observations. Ray and mode travel-time statistics (Lynch et al., 1996).
1995	SWARM	WHOI/NPS/ JH/NRL/UD	400- and 224-Hz pulse transmissions to 42-km range, and extensive oceanographic observations. Mode pulse statistics and coherence. Null result for exponential growth of scintillation (Apel et al., 1997).
1996	Shelfbreak Primer	WHOI/HU/ NPS	400-Hz pulse transmissions to 42- and 60-km range, and extensive oceanographic observations. Quantification of scintillation and intensity PDF (Fredricks et al., 2005).
2001	ASIAEX	WHOI/NPS/ UD	400-Hz pulse transmissions to 21- and 31-km range, in the along- and across-shelf directions. Extensive oceanographic observations. Quantification of scintillation and nonstationarity (Duda et al., 2004).
2006	SW06	WHOI/UD	85- to 450-Hz pulse transmissions to 19- and 30-km range, in the along- and across-shelf directions. Extensive oceanographic observations. Quantification of horizontal coherence and nonstationarity (Duda et al., 2012).

[a]IAAS, Institute of Acoustics, Academia Sinica, HU, Harvard University; UD, University of Delaware; NPS, Naval Postgraduate School; NRL, Naval Research Laboratory; JH, Johns Hopkins University; SAIC, Science Applications International Corporation; WHOI, Woods Hole Oceanographic Institution.

since. Ocean tomography requires broadband transmissions to resolve individual arrivals and propagation paths free of complex bottom interaction. Here precisely timed and navigated sources and receivers allow measurement of acoustical fields (primarily travel time) that reflect changes in the ocean. Using inverse methods, observed travel times of the arrivals can be used to estimate the large-scale intervening ocean temperature structure or, better yet, the travel times can be assimilated into an ocean circulation model (Munk et al., 1995). These data are also what is needed to study acoustic fluctuations that are not corrupted by platform motion or complex bottom interaction. The knowledge gained concerning fluctuations, in turn, helps better inform the forward tomographic problem.[9]

Furthermore, in the 1990s the development of precisely navigated and timed, large-aperture vertical line arrays (VLAs) benefited both tomographic measurements and acoustic fluctuation studies. During the seminal SLICE89 experiment where broadband 250-Hz pulses were transmitted to a 3-km-long VLA at 1000-km range, the first detailed observations of a *time front* were obtained (Duda et al., 1992) (see Figure 8.1). A time front is defined as the depth/time pattern of a wave front as it sweeps by a vertical array of receivers at fixed range. These arrays provide many benefits to tomographic observing systems including (1) improved signal-to-noise ratios for long-range propagation; (2) improved ray identification and ray arrival angle resolution (Worcester et al., 1999); (3) observation and identification of diffracted arrivals, that is, arrivals in a caustic shadow zone (Brown, 1982); (4) improved horizontal tomographic resolution (Cornuelle and Howe, 1987); and (5) observation of acoustic normal mode arrivals whose travel times can be used to increase tomographic vertical resolution (Munk et al., 1995).

Figure 1.6 shows a time front observed with a water column spanning VLA in the 2010 Philippine Sea experiment. More modest vertical arrays were also used in the basin scale Acoustic Thermometry of Ocean Climate (ATOC) experiments (700- and 1400-m VLAs) and the North Pacific Acoustic Laboratory (NPAL) experiments (700- and 1400-m VLAs), which are discussed at length in later chapters (Table 1.2). Acoustic energy that arrives early is associated with acoustic paths that cycle with high angles (relative to the horizontal) through the waveguide, while the late arriving energy is associated with small-angle propagation. The vertical array observations in Figure 1.6 reveal several important aspects of acoustic fluctuations caused by the stochastic internal-wave field that have been observed in one degree or another in all experiments.

---

[9] As with any relationship, there were some downsides too. Joint experiments had to accommodate both the objective of tomographically analyzing the slowly varying large-scale ocean and at the same time resolve the rapid fluctuations caused by internal waves. With limited battery power for transmissions and limited data storage capability, compromises had to be struck. In addition, interest in oceanographic questions such as climate change brought transmission ranges to basin scales, where the notion of a controlled internal-wave experiment is nonexistent.

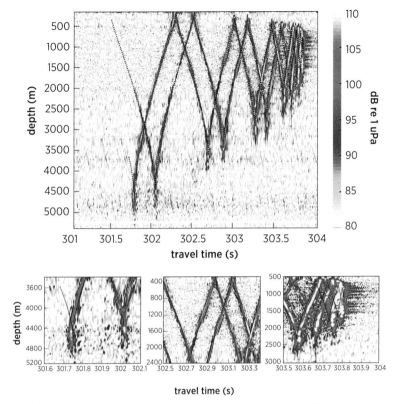

Figure 1.6. A time front at 450-km range observed during the 2010 Philippine Sea experiment using a watercolumn spanning vertical array. The pulse transmissions from an on-axis source had a 250-Hz center frequency and 50-Hz bandwidth. Superimposed on the time front with a dashed curve is a ray prediction computed using climatology (Antonov et al., 2006; Locarnini et al., 2006). The ray prediction is delayed by 82 ms to better line up with the observations. The upper panel shows the entire arrival pattern and the lower panels show expanded views of the early deep caustics, the microray interference pattern along the time front branches, and the arrival finale. (Courtesy M. A. Dzieciuch)

First, the time extent of the arrival pattern is limited due to lossy boundary interactions. In this case the duration of the arrival is a little over 2 seconds and the duration generally increases with increasing propagation range. Acoustic energy with path angles larger than roughly $\pm 15°$ has multiple interactions with the surface and/or bottom at this range and would arrive at the VLA at early travel times. The fact that this energy is virtually absent in the observations indicates how strongly dissipative these boundary interactions are. The observed acoustic fields are dominated by the ocean sound-speed and current fields and not by the boundaries.[10]

---

[10] Internal-wave currents generate weaker acoustic scattering than internal-wave-induced sound-speed perturbations. The effects of currents are, however, observable via travel-time fluctuations when reciprocal transitions are carried out. Sum travel times isolate the sound-speed effect, while difference travel times give the current effect.

Another important feature of the observed time front is that the early part of the arrival pattern, where time-resolved wave front branches are seen, qualitatively matches the ray theory prediction, thereby revealing an important stability in the pattern. It has been discovered that due to small-angle local scattering by internal waves, the cumulative scattering of sound is primarily along the wave front rather than across it (Flatté and Colosi, 2008). This along-wave front scattering leads to an interference pattern *along* the observed time front, which is evident as the pattern of fade-outs and intensification in Figure 1.6. Associated with this narrow interference pattern is an apparent fracturing along the front and the appearance of closely spaced multiple arrivals (micro-multipath) for each branch (Colosi et al., 2001; Dzieciuch, 2014). The fracturing of the front is associated with new acoustic paths that are generated by the sound-speed fluctuations, an effect termed microray generation. The generation of microrays is a result of ray instability called ray chaos (more about it later).

In the region near the end of the arrival pattern, which is composed of acoustic energy propagating at small angles relative to the horizontal, the field has a complex interference pattern that can be interpreted using ray ideas (Beron-Vera et al., 2003) or mode coupling (Wage et al., 2005; Chandrayadula et al., 2013). The scattering in this region leads to a strong blurring or outright loss of the time front branches that are seen so clearly in the early part of the arrival pattern.[11] The strong interference in this region compared to the early arriving region means that the statistical properties of the signal are strongly anisotropic. Different approaches are, therefore, needed to analyze these regions. Path-based theories, such as the path integral and Born theory, are generally used for the time-resolved early arrivals, and mode-based theories, such as transport theory, are used for the late arrivals.

Lastly, internal-wave scattering is seen to fundamentally change the nature of acoustic shadow zones. In particular, shadow zones are observed to become increasingly ensonified beyond the cusp of the unperturbed ray caustics for both early and late arriving energy (Figure 1.6, top panel). For the late arrivals, this shadow zone effect was first observed in the SLICE89 experiment (Duda et al., 1992), which was the first field effort to utilize a large aperture (3000 m) vertical array (see Figure 8.1). The discovery of shadow-zone scattering in the early part of the arrival occurred when Navy SOSUS receivers were utilized in the ATOC network in an effort to quantify climate change and variability (Dushaw et al., 1999) (Figure 1.7). Work by Colosi et al. (1994) and Van Uffelen et al. (2009) unambiguously identified these effects with scattering from random internal waves.

---

[11] The blurring of the time front finale is seen more clearly in longer range acoustic transmissions. See, for example, Figures 5.10 and 8.1.

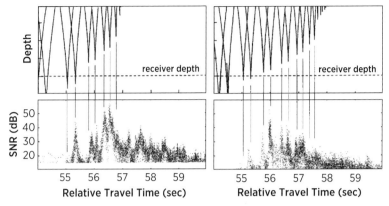

Figure 1.7. Broadband, 75-Hz transmissions from Kauai and Pioneer seamount to North Pacific Navy SOSUS stations (see Figure 7.3). The propagation range in both cases is roughly 2000 km. The upper panels show predicted time fronts based on climatology, and the lower dot plots reveal the average arrival pattern of the acoustic signals. Significant acoustic energy is observed below the cusps of the time front in the predicted shadow zones.
Source: Dushaw et al. (1999).

1.2.2 Emergence of Shallow-Water Acoustics

Largely motivated by the 1990s' appearance of quiet diesel-electric submarines patrolling Western Pacific continental shelves and slopes, there emerged a significant thrust in many countries to understand littoral or shallow-water acoustics. Here many geological, biological, and physical oceanographic processes combine in concert to cause significant acoustic variability, as was revealed by the early measurements of Zhou et al. (1991) in the Yellow Sea demonstrating 25-dB signal fluctuations with timescales of minutes to hours.

Subsequent observational programs in shallow water (see Table 1.3) established the importance of internal waves in this problem, including nonlinear internal tides, nonlinear internal-wave packets or solitons, and stochastic linear internal waves (Apel et al., 1997, 2007). Because of complex boundary interactions, attenuation, and the thinness of the waveguide relative to the acoustic wavelength, the method of normal modes has many advantages over other methodologies. To date, measurements of mode and full-field statistics have focused largely on the problem of anisotropic mode coupling and horizontal refraction behavior caused by nonlinear internal waves that advance up the continental shelf as broken fronts (Figure 1.8).

Figure 1.9 shows 200- and 400-Hz broadband transmissions in the along- and across-shore directions to an L-shaped receiving array (L-array) on the New Jersey continental shelf (Table 1.3; SW06 experiment). The L-array is a combination of a vertical array and a bottom-mounted horizontal array. The signals begin with an abrupt strong first arrival followed by a gradual fade. This structure is

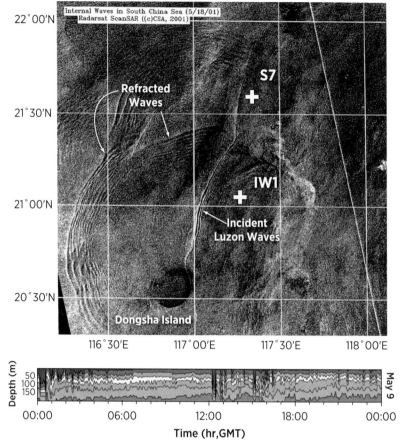

Figure 1.8. Nonlinear internal waves observed in the South China Sea, as a part of ASIAEX. Upper: Radar image of South China Sea nonlinear internal waves from May 18, 2001. Lower: A 24-hour thermistor record from May 9, 2001 in the South China Sea showing the vertical structure and semidiurnal time separation of nonlinear internal-wave packets. Temperature contours are in increments of 2°C starting at 14°C (blue). Temperature perturbations from the stochastic internal-wave field are also evident.
Source: Ramp et al. (2004).

present because shallow-water geometric signal dispersion is the opposite of that in deep water. Small horizontal angle energy arrives first, followed by higher angle energy that has ever-increasing lossy boundary interactions. Even though the signal bandwidth is rather large, ±50 Hz, no time front branches are evident. The signal is in many ways similar to the late arriving energy shown in Figure 1.6. A vertical normal mode structure is sometimes evident, however, particularly at the initial onset of the transmission. Modes one and two can be discerned by their telltale single and double lobe vertical structure. Another noteworthy feature of these observations is that the signal sometimes shows strong coherence in the horizontal

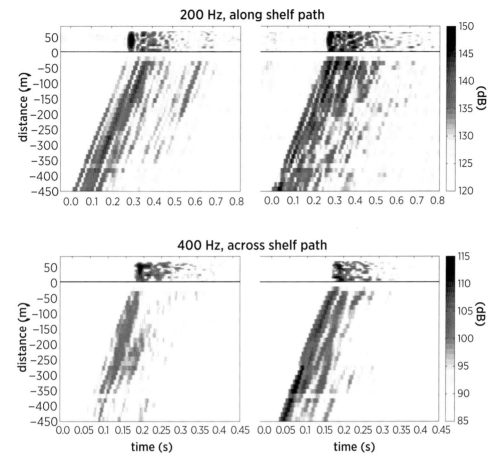

Figure 1.9. Examples of broadband 200 (upper) and 400 (lower) Hz transmissions to an L-array on the New Jersey continental shelf. The upper section of the plot shows the vertical aperture and the lower section shows the horizontal aperture along the seafloor. The 200-Hz transmissions were over a 30-km propagation path along the continental shelf, while the 400-Hz transmissions were over a 19-km path going across the shelf. (Courtesy T. F. Duda)

direction. Periods of high coherence are associated with the absence of nonlinear internal waves, while periods of low coherence are associated with nonlinear internal-wave packets whose crests are aligned with the acoustic transmission path (Duda et al., 2012).

While initial interest in the stochastic shallow-water internal-wave field focused on the theoretical possibility of exponential growth of the scintillation index (Creamer, 1996; see Chapter 8), the topic was largely ignored after a null result in the SWARM experiment (Table 1.3). Recently, however, interest in the role of stochastic internal-wave-induced scattering in shallow water has been revived by the observation that acoustic fluctuations on paths that are roughly perpendicular to nonlinear internal–wave fronts are statistically stable, while paths parallel to the fronts show strong episodic fluctuations associated with the nonlinear waves (Duda et al., 2012). There is also interest in how the stochastic internal-wave field modifies acoustic interactions with nonlinear waves. It has been demonstrated numerically that stochastic waves make acoustic scattering by nonlinear tides and solitons less coherent (Colosi et al., 2011). Shallow-water environments, therefore, are a region where much remains to be learned about the acoustical effects of stochastic internal waves.

1.2.3 Ocean Sound-Speed Spectrum

Many areas of WPRM have difficulty adequately characterizing the random medium, and ocean acoustics is no exception. Here legitimate questions have been raised concerning observed deviations from the GM internal-wave spectral model, and there is a considerable spread of opinions concerning how important these deviations are for predicting acoustic fluctuations. Furthermore, there are other sources of stochastic sound-speed structure in the ocean that have been poorly observed and studied. Among these are intrusive thermohaline fine structure, often termed spice,[12] and there is a class of internal-wave motions at zero frequency termed vortical modes. Spicy sound-speed structure exists due to the mixing of water masses with differing temperature and salinity. Hot and salty water (high spice) can have the same density as cold and fresh water (low spice). Along surfaces of constant density, therefore, temperature and salinity can vary, sometimes significantly. Regions of strong spice include ocean fronts, shallow water, and the surface mixed layer. Density-compensating temperature and salinity anomalies produce sound-speed anomalies because sound speed increases with both temperature and salinity. Vortical motions create sound-speed anomalies in the same way as other internal waves, that is, by vertical displacement of fluid parcels. Generation and dissipation mechanisms, as well as space–time scales of spice and vortical modes, are poorly understood and almost certainly do not have universal behavior (Ferarri and Rudnick, 2000; Pinkel, 2014). Spice and vortical motions are briefly discussed in Chapter 3.

In the 1980s and 1990s a series of experiments were carried out to better quantify the internal-wave spectrum and to come to a dynamical understanding of the processes that make the spectrum appear to be rather universal (Pinkel, 1984; Sherman and Pinkel, 1991). Unlike the Kolmogorov turbulence spectrum (Tatarskii, 1971) and the Philipps saturated surface-wave spectrum (Phillips, 1977) that have clear dynamical underpinnings, the cause of the near-universality of the GM spectrum has been difficult to determine. The issue is not a lack of

---

[12] The term spice was coined by Walter Munk in the early 1980s and is used commonly but not universally (Munk, 1981).

possible mechanisms but an overabundance of them that are difficult to disentangle observationally (Munk, 1981; Müller et al., 1986; Pinkel, 2008; Polzin and Lvov, 2011). The sources and sinks of internal-wave energy are also poorly understood (Wunsch, 1976).

Absent a dynamical understanding of the spectrum, observational efforts have revealed deviations that can occur from the canonical GM model, including (1) variations in energy by factors of 2, (2) variations in spectral slope by factors of 10–20 percent (the canonical model has a slope of −2 in both frequency and vertical wavenumber), (3) nonseparability of the spectrum in terms of frequency and mode number (primarily at low frequency), and (4) variation of the energy in the lowest internal-wave modes, which is controlled by a GM modal bandwidth parameter called j_*. These deviations make it desirable to have observations of the internal-wave spectrum in conjunction with acoustic transmission experiments, an example of which is shown in Figure 1.4 for the Philippine Sea. In some cases, parameters in the GM model can be adjusted based on the observations, and fluctuation calculations can be carried out with these adjusted spectra. The detailed acoustical consequences of these deviations from the GM model are not understood at present.

Regions where the GM spectral model is known to be in serious error are the Arctic and near abrupt topographic features like seamounts, canyons, and continental slopes.

1.3 Three Decades of Development: Theory

Spurred on by the intense observational advances afforded by close collaboration with the ocean acoustic tomography effort and ever-increasing computational power, significant theoretical advances occurred in the last three decades (Table 1.4).

It cannot be stated enough that much of the theoretical progress was due to the development of Monte Carlo numerical simulation techniques, which allowed exploration of a diverse range of propagation regimes and environments. Presently, Monte Carlo simulations using parabolic equation, ray, and normal mode propagation methods are rather routine and often accompany experimental results or are used for theoretical validation. In addition, with the remarkable increases in computational power over the years, some Monte Carlo simulations are fast enough to be legitimate competitors to reduced physics theories that carry out the ensemble averaging analytically but nonetheless require extensive numerical evaluations.

Reduced physics models are, of course, still essential because a basic understanding of scattering physics is a prerequisite for interpreting ocean observations, as well as for formulating useful Monte Carlo studies. The theoretical results also

Table 1.4. *Significant developments in ocean acoustic fluctuation theory*

Year	Topic	Reference	Comments
1972	Sound-Speed Fluctuations	Garrett and Munk, 1972	Development of the internal-wave spectrum
1973	Numerical Simulation	Tappert and Hardin, 1973	Parabolic equation method
1975	Monte Carlo Simulation	Flatté and Tappert, 1975	First CW simulations
1976	Weak Fluctuation Theory	Munk and Zachariasen, 1976	Predictions of variances of log-intensity and phase
1979	Path Integrals	Dashen, 1979	Framework of WPRM path integrals with no waveguide
1979	Wave Regimes/Path Integrals	Flatté et al., 1979	Overview of the state of the art
1979	Mode Coupling	Dozier and Tappert, 1978a,b	Mode transport theory
1981	Internal-Wave Currents	Munk et al., 1981	Vertical momentum flux
1981	Sound-Speed Fluctuations	Munk, 1981	Refinement of the internal-wave spectrum
1982	Moment Equations	Uscinski, 1982	Solution for the fourth moment
1985	Path Integrals	Dashen et al., 1985	Mutual coherence functions
1986	Path Integrals/Moment Equations	Codona et al., 1986a,b	Comparison of techniques
1986	PDF of Intensity	Ewart and Percival, 1986	Generalized gamma distribution
1987	Path Integrals	Flatté et al., 1987a	Intensity moments
1987	Ray Scattering	Cornuelle and Howe, 1987	Loop resonance
1992	Ray Chaos	Smith et al., 1992a,b	Chaos and mesoscale fluctuations
1994	Monte Carlo Simulation	Colosi et al., 1994	First broadband simulations

Year	Topic	Reference	Description
1996	Fourth Moment Equations	Macaskill and Ewart, 1996	Intensity spectrum
1996	Mode Coupling	Creamer, 1996	Transport theory with attenuation
1996	Ray Chaos	Tappert and Tang, 1996	Chaos and eigenrays
1997	Ray Chaos	Simmen et al., 1997	Full-wave manifestations of chaos
1999	Markov Approximation	Colosi et al., 1999; Henyey and Ewart, 2006	Path integral/high-angle rays
2002	Sound-Speed Fluctuations	Levine, 2002	Refinement of the internal-wave spectrum
2003	Ray Chaos	Beron-Vera and Brown, 2003	Introduction of the α parameter
2003	Ray Chaos	Brown et al., 2003; Beron-Vera et al., 2003	Chaos and internal waves
2004	Microray Kinemetics	Colosi and Baggeroer, 2004	Approach to saturation
2008	Ray Scattering	Flatté and Colosi, 2008	Scattering along and across fronts.
2009	Mode Coupling	Colosi and Morozov, 2009	Transport theory cross-mode coherence
2011	Mode Coupling	Colosi et al., 2011	Transport theory shallow water
2012	Sound-Speed Fluctuations	Colosi et al., 2012	Shallow-water internal-wave spectrum
2013	Mode Coupling	Colosi et al., 2013a	Transport theory time and space coherence

provide critical information on the relevant oceanographic space and time scales needed to accurately predict any given acoustic observable. Advances in wave propagation theory through the stochastic internal-wave field occurred in many directions over the last three decades, including ray methods (Chapter 5), weak scattering Born and Rytov theory (Chapter 6), Feynman path integral methods (Chapter 7), and coupled mode transport theory (Chapter 8). Moment equation methods that were introduced to the ocean acoustics field by Uscinski (1982) have also seen some success (Macaskill and Ewart, 1996). However, because the path integral method is slightly more general than the moment equation approach and the two are equivalent when the Markov approximation is made (Codona et al., 1986b), the moment equations are not discussed in this text. A review of this topic is given by Flatté (1983).

1.3.1 Ray Theory

The most useful applications of ray theory have been to acoustic observables primarily related to phase or travel time. Indeed, results from weak fluctuation theory and path integrals show that many ray results for phase have a quite broad regime of applicability. This is because phase is sensitive to the largest scales of the internal-wave field, and the phase behavior is much less sensitive to interference effects compared to intensity. Therefore, estimates of phase variance and phase correlation functions using ray theory have been found to be quite successful, and they form necessary ingredients for several path integral theory expressions. In addition, phase effects due to internal-wave currents that are isolated using reciprocal acoustic transmissions have also been investigated.

From a conceptual standpoint, the most compelling development in our understanding of acoustic wave propagation in the ocean was the realization that ray trajectories in an ocean with random internal waves[13] are primarily unstable and exhibit exponential sensitivity to initial conditions, an effect termed *ray chaos* (Beron-Vera et al., 2003; Brown et al., 2003). The nonlinear ocean ray equations have a Hamiltonian form given by (Brown et al., 2003)

$$\frac{d\mathbf{r}}{dt} = \frac{\partial \omega}{\partial \mathbf{k}} = \mathbf{c_g}, \tag{1.4}$$

$$\frac{d\mathbf{k}}{dt} = -\frac{\partial \omega}{\partial \mathbf{r}}, \tag{1.5}$$

where $\mathbf{c_g}$ is the group velocity and the dispersion relation,

$$\omega(\mathbf{k}, \mathbf{r}) = (\bar{c}(z) + \delta c(\mathbf{r}))|\mathbf{k}|, \tag{1.6}$$

---

[13] Other oceanographic processes that perturb the sound speed can also give rise to chaotic ray behavior. These include eddies, surface waves, and bottom roughness.

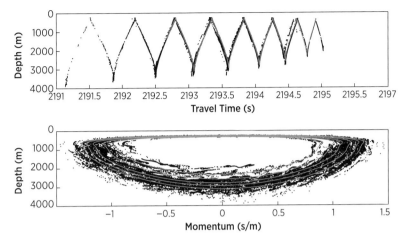

Figure 1.10. A ray simulation for the AET experiment of a section of the time front with (black) and without (gray) internal-wave sound-speed perturbations. The time front is shown in the upper panel and the Lagrangian manifold is shown in the lower panel. Each point on the time front and on the Lagrangian manifold corresponds to a ray with a fixed launch angle integrated out to a 3250-km range.

plays the role of the Hamiltonian function. Here $\bar{c}(z)$ is the background ocean acoustic waveguide (assumed to be known), and δc is the internal-wave perturbation. A useful measure of the ray path instability is the finite-range Lyapunov exponent, which quantifies the exponential divergence of nearby ray paths and the exponential increase in microrays. This quantity is a stochastic variable and is estimated to have roughly a Gaussian distribution whose variance decreases as one over range, thereby reaching a single asymptotic value at long range. Values for the Lyapunov exponent in deep-water environments for ranges of order 1000 km are typically between 1/50 and 1/500 km^{-1}.

In the chaotic ray theory context, one can consider a simple two-dimensional phase space in which the conjugate variables are ray depth and vertical slowness (i.e., sine of the ray angle divided by the local sound speed), and the independent variable is range. An important entity derived from the phase space is the Lagrangian manifold (Figure 1.10), which maps the phase space variables for several initial conditions when integrated out to a fixed range. The time front can be considered to be a projection of the Lagrangian manifold. For acoustic fluctuations, phase space and Lagrangian manifold complexity play a large role. While details of the internal-wave spectrum are important, it has been found that the background sound-speed profile, $\bar{c}(z)$, has a strong influence as well. Deterministic shearing of the ray phase space by the sound channel was found to increase ray instability and affects many acoustic observables (Brown et al., 2005).

Ray theory has also been applied to show that acoustic scattering is strongly anisotropic in the sense that energy is scattered primarily along the wave front crests, rather than in the direction of propagation (Godin, 2007; Flatté and Colosi, 2008). Figure 1.10 shows that in spite of the chaotic Lagrangian manifold structure, the qualitative structure of the time front branches is preserved. This anisotropy has important consequences for acoustic fluctuations. An additional phenomenon associated with strong chaotic scattering along the wave front is wave front healing. That is, if a bathymetric feature, such as an island or seamount, takes energy out of some part of the wave front, chaotic ray scattering will quickly reensonify that section of the front as if the energy had never been lost.

It must be appreciated, however, that some portion of the chaotic ray phase space complexity is artificial since the finite value of the acoustic wavelength will introduce diffractive smoothing of this complexity. Ray theory itself does not provide useful guidance to this dilemma, and thus full-wave approaches are needed. This problem is particularly troublesome for amplitude fluctuations, which arc sensitive to the small-scale part of the internal-wave spectrum. In this regard ocean acoustics shares many of the same problems as the field of quantum chaos, which is concerned with the behavior of quantum systems that have underlying chaotic classical behavior (Giannoni et al., 1991; Casati and Chirikov, 1995).

Lastly and quite practically, ray instability has important consequences for popular theoretical methodologies, such as Born and Rytov approximations and path integrals, which utilize expansions about the deterministic ray path. Expansion about an unstable path, therefore, will have a restricted regime of validity for these approaches. The boundaries imposed by ray chaos are barely known at present.

1.3.2 Weak Fluctuation Theory

Developed by Munk and Zachariasen (1976), weak fluctuation theory introduced accommodations for diffraction that are necessary due to the small internal-wave scales in the ocean spectrum. This theory provided the first estimates of several acoustic observables for a specific ray path that were within a factor of two of the available observations. In this work the Rytov approximation was used.[14] At that time, the Rytov theory was thought to be superior to the Born approximation because it appeared to account for multiple scattering, while the Born theory was a single scattering result. The apparent optimism about the Rytov approach is now known to be unfounded, and therefore the body of theory that the ocean acoustics community calls weak fluctuation theory may rightly be termed the Born approximation, with the attendant regime of validity associated with the Born series.

---

[14] Munk and Zachariasen use an approach they termed the super-eikonal approximation, but the results that were computed to first order are essentially the same as those of the Rytov and Born approximations to first order.

In accounting for the effects of the ocean waveguide, weak fluctuation theory uses an expansion about the unperturbed or deterministic ray path and the fact that internal waves induce weak small-angle forward scattering. This expansion is not unreasonable, because the unperturbed ray is a stationary path that to first approximation is not altered by sound-speed fluctuations (Munk et al., 1995). Furthermore, weak fluctuation theory is expected to apply only for short ranges at which chaotic ray effects have not fully emerged. A second important approximation associated with the theory, which significantly simplifies the expressions, is the so-called Markov or ray-tangent approximation, which gives an integral equation with a physically meaningful scattering kernel. In the ray-tangent approximation, the unperturbed ray path is assumed to be locally straight over an internal-wave correlation length, an approximation that can easily break down for high-angle ocean rays that have significant curvature, particularly at the important ray upper turning point. The extent to which these approximations degrade the accuracy of phase and log-amplitude statistical predictions is presently poorly studied, although it appears that the error in the predictions is at the factor of two level (Colosi et al., 2009). Weak fluctuation theory has successfully described experimental results for moments and spectra of phase (Tables 1.1 and 1.2). Two cases in which the theory accurately describes both phase and log-amplitude are the 1985 AATE Arctic experiment and the 1994, 87-km-range AET (Table 1.2, Figures 6.8, 6.9, and 6.10).

Weak fluctuation theory has also played a significant role in describing the boundary between unsaturated propagation and stronger scattering. Starting with weak fluctuation theory, one can approximately describe the regimes of unsaturated, partially saturated, and fully saturated propagation using two parameters termed Φ and Λ (Chapters 4 and 6). The parameter Φ is defined as the rms phase fluctuation along a ray path in the geometric limit and quantifies the scattering strength. Φ is often termed the *strength parameter*. The strength parameter increases with frequency, range, and internal-wave strength. To quantify diffraction, the Λ parameter is defined as a weighted average of the vertical Fresnel zone scale divided by the vertical correlation length of the internal waves. The Fresnel zone describes a diffractive region around a ray path in which nearby paths differ in phase by less than half a cycle. Because internal waves have small vertical scales, diffractive effects occur most significantly in the vertical. Λ is often termed the *diffraction parameter*. The diffraction parameter decreases with increasing frequency, increases with increasing range, and is independent of internal-wave energy. The three different propagation regimes correspond to different regions of the $\Lambda - \Phi$ diagram (Flatté et al., 1979; Colosi, 2015) as shown in Figure 1.11. The diagram provides a useful way to intercompare experiments or different frequencies, ranges, and environments (see Chapter 6, Figure 6.4).

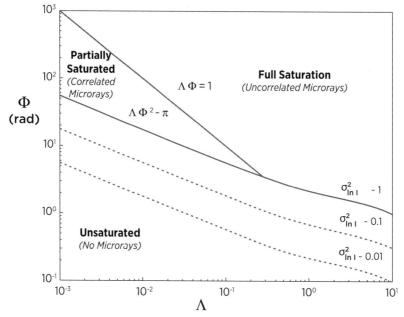

Figure 1.11.   The $\Lambda - \Phi$ diagram showing contours of log-intensity variance and approximate borders between unsaturated, partially saturated, and fully saturated propagation. An rms log-intensity value of 1 corresponds to 4.3 dB.

1.3.3 Path Integral Theory

Once precisely timed and located electronic sources and receivers were used in experiments, such as the mid-1970s' MATE and AFAR, it became evident that a new methodology was needed to handle strong scattering. In several cases the scintillation index was observed to be significantly above 1, and while phase statistics such as the phase spectrum were consistent with ray theory and weak fluctuation predictions, observables like the log-amplitude and intensity spectra showed dramatic departures from these theories. A quite successful tool to handle the strong scattering case was the path integral developed by Dashen, Flatté, and co-workers (Dashen, 1979; Flatté, 1983). It is important to point out too that many path integral results are also valid in the unsaturated and partially saturated regimes, so the technique is not strictly an asymptotic approach for full saturation.

A pleasing physical picture of acoustic scattering from internal waves emerges from the path integral formalism, where the acoustic pressure is written as a sum over all possible paths connecting the source and receiver (Feynman and Hibbs, 1965), that is,

$$p(\mathbf{r}) = N \int d(path) \exp\big(i\Theta(path)\big). \tag{1.7}$$

Here Θ is the accumulated phase over the path, and the factor N is a normalization. Rays are seen to be stationary phase points in this formalism. When the sound-speed fluctuations are zero, there is only one ray path and the acoustic field at the receiver is determined not only by the ray path but also by all other paths that are within a Fresnel zone of the ray. Paths outside the Fresnel zone are seen, by and large, to cancel out, and the paths within the Fresnel zone represent the effects of diffraction. When sound-speed fluctuations are added to the picture, in the initial weak fluctuation unsaturated regime, the path integral formalism tells us that there is still one dominant ray but that ray and its associated ray-tube (determined by the Fresnel zone) are perturbed by the sound-speed fluctuations. As the range increases in the stronger scattering regimes of partial and full saturation, one path is no longer dominant but there are several significant paths termed *micro-multipath*. In this regime, intensity fluctuations are caused by micro-multipath interference. The degree of fluctuation is determined by the spatial extent of the micro-multipaths relative to the Fresnel zone and the correlation lengths of the internal waves. Microrays that are within a Fresnel zone of the unperturbed ray will generally not make significant contributions to the fluctuations because they are smeared out by diffraction. In the case where the microrays extend beyond the Fresnel zone but are spread less than a correlation length of the internal waves, the conditions for partial saturation exist. In this case the large-scale internal-wave structure causes the microrays to be strongly correlated, resulting in a high probability of constructive interference or intense focusing. Lastly, if the microrays are spread beyond both the Fresnel zone and the internal-wave correlation lengths, the microrays become uncorrelated so that the acoustic field is an interference by many independent waves. This is the condition of full saturation.

Like weak fluctuation theory, the path integral method requires an expansion about the deterministic ray and the assumption of weak small-angle forward scattering. This physically means that the microrays cannot wander too far from the unperturbed ray, a requirement that can break down if the rays are strongly chaotic. In addition, to make analytic progress the expansion about the unperturbed ray is taken to second order only where standard techniques for Gaussian path integrals can be applied. Lastly, while not essential, the Markov or ray-tangent approximation is generally used to obtain physically interpretable results. The regimes for which these assumptions break down as a function of range, frequency, bandwidth, and environment are not understood.

The development of path integral theory for ocean acoustics led to predictions for many observables, including mutual coherence functions with space, time, and frequency separations; intensity correlation functions again as a function of space, time, and frequency separations; intensity variance; and pulse statistics such as wander, time spread, and time bias. Path integral theory is, therefore, the most mature body of work to handle ocean acoustic fluctuations.

The greatest successes for path integral theory were associated with the 35-km AFAR experiment in which pulses with frequencies of 410, 1010, and 4670 Hz were transmitted (Table 1.1). In AFAR the key questions were associated with the time behavior of the signals, such as the mutual coherence function (MCF) and the intensity correlation function (along with its Fourier transform the intensity frequency spectrum). Because AFAR was on the border between the partially saturated and fully saturated regimes, it took nearly a decade to develop the necessary theory to explain the observations. A key path integral result was that the MCF for space and time separations is given by the simple relation (Dashen et al., 1985)

$$\langle p(1)p^*(2)\rangle = \exp\left[-\frac{D(1,2)}{2}\right] \tag{1.8}$$

where $D(1,2)$ is the phase structure function for the separations (1) and (2) computed in the ray theory limit along the unperturbed ray path. Path integral predictions for the MCF matched the observations at all three frequencies to better than 10% (Reynolds et al., 1985; Figure 7.1).

Observations from both MATE and AFAR showed that while the phase spectra agreed well with predictions from weak fluctuation theory, the intensity spectra were markedly different with strong discrepancies, particularly at higher frequencies. Work by Flatté et al. (1987a) using path integral methods gave theoretical predictions that were in excellent agreement with the AFAR measurements (Flatté et al., 1987b; Figure 7.13). Similarly, good agreement for the intensity spectrum was obtained for the MATE experiment using moment equation methods (Macaskill and Ewart, 1996; Figure 7.14).

1.3.4 Coupled-Mode Transport Theory

Three important factors have led to interest in coupled-mode approaches originally introduced by Dozier and Tappert (1978a) in the late 1970s. First, in some cases observed acoustic signals do not lend themselves easily to the path integral notion of scattering about a deterministic ray path. For the sound fields observed in shadow zones, there is in fact no unperturbed ray. In the finale of the deep-water arrival pattern, there are no clear time front branches with which to identify an unperturbed path (Figure 1.6). Observed shallow-water acoustic fields also do not reveal easily identifiable paths (Figure 1.9). Second, path integral results are limited because of the inherent instability of the unperturbed ray and the ray-tangent approximation. Lastly, with the development of large-aperture mode-resolving VLAs, mode statistics can be observed directly and compared to theory. Coupled-mode approaches, therefore, have the potential to provide more accurate predictions in many circumstances, especially long ranges. The topic of modes is strongly relevant in acoustic tomography (Munk et al., 1995) and especially in shallow-water acoustics (Creamer, 1996).

In the mode picture the acoustic field for a single frequency can be written as an eigenmode expansion using the mode functions, $\phi_n(z)$, determined by the background sound-speed and density structure (i.e., the unperturbed modes in analogy to the unperturbed rays), giving

$$p(r, z; \omega) = \sum_n \frac{a_n(r)\,\phi_n(z)}{\sqrt{k_n r}}. \tag{1.9}$$

The stochastic mode amplitudes, $a_n(r)$, obey the one-way coupled equations in the weak, multiple forward scattering approximation (Dozier and Tappert, 1978a),

$$\frac{d\hat{a}_n}{dr} = -i \sum_{m=1}^{N} \rho_{mn}(r) e^{il_{mn}r} \hat{a}_m(r). \tag{1.10}$$

Here $\hat{a}_n(r) = a_n(r) e^{-il_n r}$ is the demodulated mode amplitude, the complex modal wavenumber is given by $l_n = k_n + i\alpha_n$, and $l_{mn} = l_m - l_n$. The matrix ρ_{mn} is a symmetric coupling matrix whose statistical properties are determined by the internal-wave spectrum. The complex wavenumber is necessary for handling shallow-water cases and the intrinsic attenuation in sea water. The evolution equation for the mode amplitudes is of the same form as the Schrödinger equation in the "interaction" representation (Van Kampen, 1981).

In mode representation, the statistical properties of the pressure field are determined by the statistical properties of the mode amplitudes $a_n(r)$. For example, the mean and mean square intensities depend on the moments $\langle a_n a_p^* \rangle$ and $\langle a_n a_p^* a_m a_q^* \rangle$, respectively. Using the Markov approximation, transport theories for the range evolution of second and fourth moments of the mode amplitudes have been successfully tested using Monte Carlo simulation for both deep-water and shallow-water environments (Colosi and Morozov, 2009; Colosi et al., 2011, 2013a). Transport theory methods do not suffer the high-angle propagation problems associated with the ray-tangent approximation, because the high-angle behavior is built in by virtue of the use of the unperturbed modes.[15] The only limitation of the technique appears to be the weak forward-scattering approximation that will break down at high frequency, a regime where mode calculations are generally impractical.

Transport theory methods have just started to be used in the interpretation of observations. In the LOAPEX (Table 1.2), the theory was shown to accurately describe the range evolution of mode statistics from deep-water, low-frequency, long-range transmissions (Chandrayadula et al., 2013; Figure 8.8). In particular,

---

[15] The Markov approximation is used in both transport and path integral theory, but the implementation in each case is quite different, as discussed in subsequent chapters.

the theory accurately predicts the transfer of energy in mode space from modes that initially have large energy to modes with initially small energy. The theory also accurately predicts the observed temporal coherence of modes (Figures 8.10 and 8.15).

In addition, coupled mode theory is poised to provide the first theoretical description of scattering into shadow zones. From the mode view, acoustic energy in a shadow zone is caused by higher order modes whose effective group speed has been slowed down by mode coupling. Because these higher order modes arrive at travel times associated with slower low modes in the unperturbed problem, the shadow zone becomes ensonified. Solution of this problem will require further theoretical analysis of cross-frequency behavior of coupled modes, a topic that is outlined in this text, but has not received any attention in the literature.

In shallow-water problems, transport theory describes the progressive ran-domization of modal phases leading to loss of cross-mode coherence. This decorrelation of the modes has important consequences for acoustic interaction with intense, but localized, nonlinear, shallow-water internal waves. Coupling caused by nonlinear internal waves depends on the relative phases of the modes when they interact with the nonlinear wave, and therefore because of stochastic internal-wave phase randomization, scattering by nonlinear wave packets can be much less coherent than estimates obtained where the nonlinear waves are considered alone (Colosi et al., 2011). Shallow-water transport theory has also been applied to the combined effects of random surface and internal gravity waves for propagation at kilohertz frequencies (Raghukumar and Colosi, 2015). Transport theory comparisons to shallow-water observations have not been carried out in any detail.

1.4 Where We Stand

The problem of sound propagation through the stochastic ocean is two-fold. One must establish the statistical variation of the ocean medium and then from this information observe and predict the statistical properties of acoustic signals. While Monte Carlo testing of acoustic scattering physics by stochastic sound-speed structure in the ocean is critical for identifying fundamental processes, the ultimate objective is, of course, the comparison with observations. In the past three decades a great deal of progress has been made in observations, theory, and characterizing the ocean sound-speed fluctuations, but concrete observational tests of fluctuation theories have been carried out only for a small subset of ocean acoustic conditions.

For ray-like theories in which a specific acoustic path is considered, results are primarily available for short ranges (100 km or less), but over a rather broad range of frequencies from 75 Hz to 16 kHz. In these cases, comparisons with path integral and weak fluctuation theory are quite good. A fundamental question is where these methods break down, perhaps due to ray chaos.

For the recently developed mode transport theory, the results are limited to a single experiment, LOAPEX, in which 75-Hz sound was transmitted to ranges of 50, 250, 500, 1000, 1600, 2300, and 3250 km. While the theory/observation comparisons for single-frequency second moments like mode energy and coherence were quite good, analyses for the higher moments and cross-frequency pulse propagation effects have yet to be done.

Thus presently there is strong uncertainty if the acoustic propagation regimes of unsaturated, partially saturated, and fully saturated as a function of frequency, bandwidth, range, and geographic location can be generally predicted and what information is required to do so. The knowledge of the degree of acoustic field randomization is, of course, a prerequisite for the use of sound in the ocean for remote sensing, communication, and navigation.

When the original text, *Sound Transmission through a Fluctuating Ocean*, was written in 1979, only a nascent understanding of the subject of sound propagation through the stochastic ocean had emerged. As of the writing of this book, one may say the field is in its adolescence, as the theoretical and observational foundations of the field have been strongly reinforced over the last three decades. With a future focus on hypothesis-driven data, theory, and model analysis, it is clear that many of the challenging questions of the field could be put to rest, bringing the field to full maturity. It is hoped that this monograph serves as a "lighthouse" for this future development of the field.

1.5 Utilizing This Book

This book is organized in two parts. Part I provides important prerequisites in acoustics (Chapter 2) and internal waves (Chapter 3) for those readers who need to refresh their knowledge of the subjects. These chapters are intended to be as self-contained as possible, but they are no substitute for introductory courses or texts in ocean acoustics and ocean waves. Also in Part I is a qualitative discussion of acoustic fluctuations (Chapter 4) that lays out the key physical principles of microray interference, resonances, mode coupling, and diffraction. These three chapters have served as a text (along with other supporting lecture material) for a two-quarter introductory graduate course in acoustics and fluctuations at the Naval Postgraduate School. While Chapters 2 and 3 could be skimmed over by a reader already familiar with the basics, Chapter 4 is essential for the interpretation of Part II of the book.

Part II is organized along the lines of propagation theories: ray theory (Chapter 5), weak fluctuation theory (Chapter 6), path integral methods (Chapter 7),

Table 1.5. *Summary of acoustic observables for single-path statistics and the*
methodologies best suited for prediction in different propagation regimes.

Observable	Regime	Best methodology	Chapter(s)
Mean Pressure	U/P/S	Path Integral	7
Phase/Travel-Time Variance	U/P/S	Ray Theory	4,5,6,7
Phase/Travel-Time Spectrum	U/P/S	Born Theory	6
Arrival Angle Variance	U/P/S	Ray Theory	4,5
Travel-Time Variance (chaos regime)	P/S	Ray theory	5
Travel-Time Bias (chaos regime)	P/S	Ray theory	5
Log-Amplitude Variance	U	Born Theory	4,6
Log-Amplitude Spectrum	U	Born Theory	6
Phase Log-Amplitude Correlation	U	Born Theory	6
Space/Time Coherence	U/P/S	Path Integral	4,7
Frequency Coherence	U/P/S	Path Integral	4,7
Pulse Time Wander	U/P/S	Path Integral	4,7
Pulse Time Spread	U/P/S	Path Integral	4,7
Pulse Time Bias	U/P/S	Path Integral	4,7
Intensity Covariance	U/P/S	Path Integral	7
Intensity Coherence/Spectra	U/P/S	Path Integral	7
Space/Time/Frequency			
Intensity Variance	U/P/S	Path integral	4,7

Propagation regimes are labeled by (U) unsaturated, (P) partially saturated, and (S) fully
saturated. Path integral and some ray theory results can be limited in their applicability
due to ray chaos and expansion about the deterministic ray. Single-frequency and normal
mode fluctuations are described quite well by coupled mode and hybrid transport theory
addressed in Chapter 8.

and mode transport theory (Chapter 8). Early versions of these chapters were
used for a semester-long advanced graduate acoustics course in the Massachusetts
Institute of Technology and Woods Hole Oceanographic Institution (MIT/WHOI)
joint program in applied ocean physics and engineering. Every effort was made to
make these chapters stand alone, but in some cases a review of material in Part I
will be required.

While the chapters provide a strong foundation for the reader in the different
approaches to wave propagation, further study using the references herein will
be required to master many of the techniques. The subject of this monograph
presents many challenges to the student or the researcher because of the diverse
skill sets needed to make progress. Among these skills are a knowledge of (1) the
methods of mathematical physics, especially the difficult topics of Feynman path
integrals and Hamiltonian/nonlinear dynamics; (2) probability and statistics; (3)
signal processing; and (4) oceanography.

Lastly, because the text is organized by propagation theory, the reader concerned with a specific acoustic observable will find relevant material scattered through a number of chapters. Here the use of the index can provide useful guidance, and Table 1.5 points to specific chapters that will provide the best treatment of any particular acoustic observable.

2

Acoustical Prerequisites

2.1 Introduction

The difficulty of the subject of sound propagation through the stochastic ocean owes much of its complexity to the fact that sound transmission in the ocean occurs in a waveguide. Sound waves are continuously refracted and diffracted by the sound channel and the waves are bounded on the edges by the sea floor and surface. Therefore, this chapter provides an overview of the basic ocean acoustics concepts of guided propagation that are needed as a foundation to understand transmission through the fluctuating ocean waveguide. The key propagation methodologies to be covered in this chapter include ray theory, the Born approximation, Feynman path integrals, and the method of normal modes. Most of the background information provided here will involve sound propagation concepts in a vertically stratified ocean, but an effort is made to present the fundamental range-dependent equations for each propagation methodology and to briefly discuss implications for acoustic scattering.

Those readers with a solid background in underwater acoustics may not find anything particularly new in this chapter, but studying the following material is encouraged to foster a familiarity with the notation and broad approach of the monograph. By necessity this is an abridged treatment of the subject of underwater acoustics, and the reader is referred to numerous quality texts dedicated to this subject (see, for example, Brekhovskikh and Lysanov, 1991; Frisk, 1994; Jensen et al., 1994; Munk et al., 1995; Katznelson et al., 2012).

2.2 Fundamental Equations of Hydrodynamics

For ocean sound propagation the fundamental hydrodynamic equations are the continuity equation (conservation of mass), the momentum equation (Newton's Law), and the adiabatic equation of state (pressure as a function of density and entropy). These are written

$$\frac{\partial \rho'}{\partial t} = -\boldsymbol{\nabla} \cdot (\rho' \mathbf{u}'), \tag{2.1}$$

$$\rho' \left(\frac{\partial \mathbf{u}'}{\partial t} + \mathbf{u}' \cdot \boldsymbol{\nabla} \mathbf{u}' \right) = -\boldsymbol{\nabla} p', \tag{2.2}$$

$$p' = p'(\rho', s = \text{const.}), \tag{2.3}$$

where p' is the pressure, ρ' is the density, $\mathbf{u}'$ is the fluid velocity, and in the equation of state changes in pressure and density involve no heat conduction, that is, entropy (s) following a water parcel is constant (e.g., $\partial s/\partial t + \mathbf{u}' \cdot \boldsymbol{\nabla} s = 0$). Because most ocean acoustic problems involve propagation over many wavelengths, nonlinear effects are small. Further assuming a fluid otherwise at rest and expanding the pressure, density and entropy fields about a time independent state $p_0(\mathbf{r})$, $\rho_0(\mathbf{r})$, and $s_0(\mathbf{r})$ the following linearized acoustic equations are obtained:

$$\frac{\partial \rho}{\partial t} = -\rho_0 \boldsymbol{\nabla} \cdot \mathbf{u}, \tag{2.4}$$

$$\rho_0 \frac{\partial \mathbf{u}}{\partial t} = -\boldsymbol{\nabla} p, \tag{2.5}$$

$$p = \rho c^2, \tag{2.6}$$

where the acoustic quantities are now p, ρ, $\mathbf{u}$, and the speed of sound is given by the thermodynamic relation $c^{-2} = (\partial \rho_0/\partial p_0)_{s=\text{const}}$. A little manipulation of these equations leads to the acoustic wave equation (Pierce, 1994):

$$\frac{1}{c^2}\frac{\partial^2 p}{\partial t^2} = \rho_0 \boldsymbol{\nabla} \cdot \left(\frac{1}{\rho_0} \boldsymbol{\nabla} p \right). \tag{2.7}$$

Here the sound speed and density fields are dictated by the ocean fields of potential temperature θ, salinity S, and pressure p_0. The fluctuation of these ocean fields include the effects of internal waves (more later). In this derivation the time dependence of the ocean has been ignored. Because the speed of sound ($\sim$1500 m/s) is so much larger than the typical speeds of ocean processes (order cm/s) this "frozen medium" approximation is well justified. Taking into account background currents $\mathbf{U}$, the acoustic wave equation becomes (Pierce, 1994)

$$\frac{1}{c^2}\frac{D^2 p}{Dt^2} = \rho_0 \boldsymbol{\nabla} \left(\frac{1}{\rho_0} \boldsymbol{\nabla} p \right), \tag{2.8}$$

where $D/Dt = \partial/\partial t + \mathbf{U} \cdot \boldsymbol{\nabla}$, and it is understood that this advection current may include the effects of internal waves. A comparison between the magnitude of internal wave terms in Eq. 2.8, that is $\delta c/c_0$, $|\mathbf{U}|/c_0$, and $\delta \rho/\rho$, reveals that the flow and density field terms are respectively one and two orders of magnitude smaller than the sound-speed term (Flatté et al., 1979). Thus the density effect is completely ignored in the treatment of sound propagation and the current effect is generally considered only when treating reciprocal transmissions where signals are sent in opposite directions thereby canceling the large sound-speed term and leaving the smaller current effect.

2.2.1 Parabolic Wave Equation

The parabolic approximation has played an important role in the development of the subject of sound propagation through the stochastic ocean not only through the area of numerical simulation (Flatté and Tappert, 1975; Colosi et al., 1994) but also in the theoretical development associated with Feynman path integrals (Flatté et al., 1979). Thus some time is spent on this topic here and further development is given in Chapter 7.

   If density variations and background flow fields are ignored then the standard wave equation is obtained, that is, $\partial^2 p/\partial t^2 = c^2 \nabla^2 p$. Examining a single-frequency ω such that $p = \Psi e^{-i\omega t}$, the wave equation becomes the Helmholtz equation:

$$\nabla^2 \Psi + \frac{\omega^2}{c^2} \Psi = 0. \tag{2.9}$$

Physically the parabolic approximation involves examining waves that move primarily in one direction, that is, in a small range of angles centered around a primary direction. Mathematically this preferred direction is chosen to be along the x-axis where it is useful to define $\Psi(\mathbf{r}) = A(\mathbf{r})e^{iq_0 x}$, with $q_0 = \omega/c_0$ a reference wavenumber. Plugging this form into the Helmholtz equation then gives

$$\frac{\partial^2 A}{\partial x^2} + \frac{\partial^2 A}{\partial y^2} + \frac{\partial^2 A}{\partial z^2} + 2iq_0 \frac{\partial A}{\partial x} + \left(\frac{\omega^2}{c^2} - q_0^2\right)A = 0. \tag{2.10}$$

Qualitatively, considering the amplitude function A to be a slowly varying function of x compared to the phase function $e^{iq_0 x}$, then this gives $\partial^2 A/\partial x^2 << 2iq_0 \partial A/\partial x$ with the result that

$$\frac{i}{q_0} \frac{\partial A}{\partial x} = -\frac{1}{2q_0^2}\left(\frac{\partial^2 A}{\partial y^2} + \frac{\partial^2 A}{\partial z^2}\right) + V(\mathbf{r})A, \tag{2.11}$$

where $V(\mathbf{r}) = (1 - c_0^2/c^2(\mathbf{r}))/2$. Equation 2.11 is the parabolic equation. It has the same form as the Schrödinger wave equation in which the x-coordinate is substituted for the time coordinate, and $q_0 = 1/\hbar$ ($\hbar$ is Planck's constant) for a unit mass particle (Flatté, 1986). The analogy between quantum and acoustic systems has been quite fruitful because many of the tools that have been developed for quantum systems can be directly applied to underwater acoustics (e.g., Feynman path integrals, time independent and dependent perturbation theory, and classical and quantum chaos). However with the correspondence between q_0 and Planck's constant, it is apparent that there is no quantum analogy to transient acoustic fields because these are characterized by a distribution of wavenumbers q_0.

The derivation can be a bit more rigorous, providing some additional insight. Using operator notation, define

$$P = \frac{\partial}{\partial x}, \quad \text{and} \quad Q = \left(1 - 2V(\mathbf{r}) + \frac{1}{q_0^2}\left(\frac{\partial^2}{\partial y^2} + \frac{\partial^2}{\partial z^2}\right)\right)^{1/2}, \quad (2.12)$$

and using these expressions, Eq. 2.10 becomes

$$\left(P + iq_0(1 - Q)\right)\left(P + iq_0(1 + Q)\right)A - iq_0[PQ - QP]A = 0. \quad (2.13)$$

Here the first term corresponds to the product of an outgoing $(1 - Q)$ and incoming wave $(1 + Q)$. Further, the last term is identically zero for $U = U(z)$, and for $(1/\omega)dc/dx << 1$ the term is negligible. Thus only keeping the outgoing wave, the one-way wave equation is obtained (Jensen et al., 1994):

$$\frac{\partial A}{\partial x} = iq_0\left[\left(1 - 2V(\mathbf{r}) + \frac{1}{q_0^2}\left(\frac{\partial^2}{\partial y^2} + \frac{\partial^2}{\partial z^2}\right)\right)^{1/2} - 1\right]A. \quad (2.14)$$

The parabolic equation results from Taylor expanding such that $Q = (1 + \epsilon)^{1/2} \simeq 1 + \epsilon/2$ (i.e., small-angle approximation) giving Eq. 2.11 as before. Higher order, wide-angle approximations to Eq. 2.14 are quite commonly used in ocean acoustic numerical analysis and are discussed in detail in other textbooks (see Jensen et al., 1994).

2.3 Character of the Oceanic Acoustic Waveguide

As previously mentioned, one of the most striking aspects of sound propagation in the ocean is the effect of the vertical acoustic waveguide. Sound waves propagate due to the compressibility of sea water, and in the ocean, like in the atmosphere, this is an adiabatic not isothermal process, meaning there is no heat conduction during the passage of the wave. The thermodynamic relation gives the sound-speed equation,

$$\frac{1}{c^2} = \left(\frac{\partial \rho_0(p_0, \theta, S)}{\partial p_0}\right), \quad (2.15)$$

where $\rho_0(p_0, \theta, S)$ is the adiabatic equation of state, that is density as a function of pressure, p_0, potential temperature, θ, and salinity S. Potential temperature of a parcel of water is defined as the temperature that parcel would acquire if it were brought adiabatically from some in situ pressure p and temperature T to a reference pressure p_{ref} (Gill, 1982). In oceanographic cases the reference pressure is typically taken to be at the ocean surface, $p_{ref} = 0$ decibars.

A simple empirical equation for sound speed due to Medwin and Clay (1997) in terms of in situ temperature (T in °C), depth (z in m), and salinity (S in ppth) is

$$c(T,S,z) = 1449.2 + 4.6T - 0.055T^2 + 0.00029T^3$$
$$+ (1.34 - 0.01T)(S - 35) + 0.016z. \tag{2.16}$$

Sound speed is seen to increase with increases in all three of the dependent variables, but the sensitivity to salinity is rather weak compared to the others. Sound-speed equations used in practice are similar to Eq. 2.16 except the polynomial expansions include many more terms (e.g., DelGrosso, 1974; MacKenzie, 1981).

The vertical gradient of sound speed is the largest gradient in the ocean and it has three terms, namely

$$\frac{dc}{dz} = \left(\frac{\partial c}{\partial \theta}\right)_{p_0,S} \frac{d\theta}{dz} + \left(\frac{\partial c}{\partial S}\right)_{p_0,\theta} \frac{dS}{dz} + \left(\frac{\partial c}{\partial p_0}\right)_{\theta,S} \frac{dp_0}{dz}. \tag{2.17}$$

The first two terms constitute the potential sound-speed gradient because a parcel of water is being followed at fixed pressure. These terms are critical for modulating internal-wave-induced sound-speed fluctuations because of the adiabaticity of internal-wave displacements (see Section 2.3.2). The potential sound-speed gradient is then written

$$\left(\frac{dc}{dz}\right)_p = \left(\frac{\partial c}{\partial \theta}\right)_{p_0,S} \frac{d\theta}{dz} + \left(\frac{\partial c}{\partial S}\right)_{p_0,\theta} \frac{dS}{dz}. \tag{2.18}$$

The third term is called the adiabatic gradient because this is the change in sound speed that would be obtained with fixed θ and S, and a change in pressure with no heat conduction. The adiabatic sound-speed gradient is thus written

$$\left(\frac{dc}{dz}\right)_a = \left(\frac{\partial c}{\partial p_0}\right)_{\theta,S} \frac{dp_0}{dz} = -\rho_0 g\left(\frac{\partial c}{\partial p_0}\right)_{\theta,S} = -c\gamma_a, \tag{2.19}$$

where the hydrostatic equation $dp_0/dz = -\rho_0 g$ has been used. In the deep ocean where the temperature and salinity gradients are small, the sound-speed profile will be largely adiabatic. The quantity γ_a defined in the last line of Eq. 2.19 will be used extensively and has a value of 0.0113 km^{-1} with little variability, except in high latitudes.

A closely related quantity which is important for both acoustics and internal waves is the vertical gradient of density, which is written

$$\frac{d\rho_0}{dz} = \left(\frac{\partial \rho_0}{\partial \theta}\right)_{p_0,S} \frac{d\theta}{dz} + \left(\frac{\partial \rho_0}{\partial S}\right)_{p_0,\theta} \frac{dS}{dz} + \left(\frac{\partial \rho_0}{\partial p_0}\right)_{\theta,S} \frac{dp_0}{dz}. \tag{2.20}$$

As in Eq. 2.17 the first two terms represent the potential gradient of density and the third term is the adiabatic gradient, which can be written using Eq. 2.15:

$$\left(\frac{d\rho_0}{dz}\right)_a = -\frac{\rho_0 g}{c^2}. \tag{2.21}$$

Because parcels of water displaced vertically in the ocean do so adiabatically it is the potential gradient of density that establishes the static stability of the water column (Gill, 1982). This stability quantity is conventionally expressed in terms of the buoyancy or Brunt-Väisälä frequency, whose square is written

$$N^2 = -\frac{g}{\rho}\left[\left(\frac{\partial\rho_0}{\partial\theta}\right)_{p_0,S}\frac{d\theta}{dz} + \left(\frac{\partial\rho_0}{\partial S}\right)_{p_0,\theta}\frac{dS}{dz}\right] = -\frac{g}{\rho}\left[\frac{d\rho_0}{dz} + \frac{g\rho_0}{c^2}\right], \tag{2.22}$$

where $\rho = \rho(p, T, S)$ is the in situ density. The water column will be statically stable if $N^2 > 0$. The buoyancy frequency squared is proportional to the potential gradient of density, which is the actual density gradient minus the adiabatic gradient.

2.3.1 Canonical Sound-Speed Profiles

Empirical numerical relations are used to compute sound speed from profiles of temperature, salinity, and depth. However, there exists a useful canonical relation (Munk, 1974). For the case in which there is approximately a linear temperature–salinity relation over all depth,

$$\frac{d\theta}{dz} = \frac{1}{a}\frac{dS}{dz}, \tag{2.23}$$

then the total sound-speed gradient can be related to the ocean stratification, namely

$$\frac{dc}{dz} = c\left(GN^2(z) - \gamma_a\right). \tag{2.24}$$

Here the constant G is given by

$$G = \frac{\frac{1}{c}\left(\frac{\partial c}{\partial\theta} + a\frac{\partial c}{\partial S}\right)}{-\frac{g}{\rho_0}\left(\frac{\partial\rho_0}{\partial\theta} + a\frac{\partial\rho_0}{\partial S}\right)}, \tag{2.25}$$

and it is expressible in terms of "typical" coefficients of thermal expansion, haline contraction, and thermal and saline effects on sound speed (Munk, 1974). Some of these values are given by

$$\frac{1}{c}\frac{\partial c}{\partial\theta} \simeq 2.0\times10^{-3}\,^\circ\mathrm{C}^{-1}, \frac{1}{\rho_0}\frac{\partial\rho_0}{\partial\theta} \simeq -0.25\times10^{-3}\,^\circ\mathrm{C}^{-1},$$

$$\frac{1}{c}\frac{\partial c}{\partial S} \simeq 0.74\times10^{-3}\,\mathrm{PSU}^{-1}, \frac{1}{\rho_0}\frac{\partial\rho_0}{\partial S} \simeq 0.75\times10^{-3}\,\mathrm{PSU}^{-1}. \tag{2.26}$$

The value of a for the linear temperature–salinity equation can be derived from the Turner Number, Tu, which gives the relative contributions of salinity and potential temperature to water column stability (Turner, 1979), that is

$$Tu = -\frac{\frac{1}{\rho_0}\frac{\partial \rho_0}{\partial S}\frac{dS}{dz}}{\frac{1}{\rho_0}\frac{\partial \rho_0}{\partial \theta}\frac{d\theta}{dz}}. \tag{2.27}$$

For North Pacific Intermediate waters it is found $Tu \simeq -0.3$, and near Bermuda at shallow depth $Tu \simeq 0.8$ (Munk, 1974). Mid-latitude values of G range from 1 to 3.

If one further assumes an exponentially stratified ocean, that is $N(z) = N_0 \exp(z/B)$, then Eq. 2.24 can be solved yielding

$$c(z) = c_A\left[1 + \frac{\gamma_a B}{2}\left(e^{2(z-z_A)/B} - \frac{2(z-z_A)}{B} - 1\right)\right], \tag{2.28}$$

where c_A is the sound speed on the channel axis z_A. Equation 2.28 is called the canonical sound-speed profile, and values typical of a mid-latitude location are $c_A = 1500$ m/s and $B = z_A{=}1000$ m. The Munk profile will be used extensively in this monograph, for example, calculations in deep-water, mid-latitude environments.

In high latitudes the thermocline essentially disappears, leaving a nearly uniform temperature water column. In the absence of thermal effects the high-latitude ocean stratification is often supported by a halocline in the upper few hundred meters of the water. Thus in this instance there is some utility in a canonical polar profile (Munk et al., 1995) which is given by

$$c(z) = c_A\left[1 + \gamma_a z + \frac{\delta c}{c_A}(1 - e^{-z/\hat{B}})\right]. \tag{2.29}$$

Here c_A is the sound speed at the surface, and in addition to the constant adiabatic sound-speed gradient (γ_a), the third term gives a nonadiabatic sound speed contribution in the upper ocean. Typical values are $c_A \simeq 1440$ m/s[1], $\delta c \simeq 15$ m/s and $\hat{B} \simeq 150$ m.

For shallow water there is quite a lot of variability in the background profile due to the multitude of processes that can come into play in this region. These include mixing events that occur at both the ocean boundaries, shelf break front meanders and front tidal advection, slope effects, and river outflow. However, a canonical form that can be fit readily to observations is of the form

$$c(z) = c_0 + \delta c\left(1 + \tanh\left(\frac{z - z_0}{\Delta}\right)\right) + \overline{(dc/dz)}z, \tag{2.30}$$

---

[1] This surface sound-speed value is consistent with Eq. 2.16, since typical upper ocean Arctic temperatures are less than $0°C$.

where δc, z_0, and Δ physically represent the sound speed anomaly, the location, and width of the thermocline, and $\overline{dc/dz}$ represents a mean sound-speed gradient. This mean gradient is often related to the foot of the shelf break front pushing up onto the continental shelf.

2.3.2 Sound-Speed Fluctuations due to Internal Waves

The primary effect of internal waves on the ocean sound-speed field is due to distortions of the background vertical profile from vertical displacements of density surfaces, $\zeta(\mathbf{r},t)$. Adding vertical displacements to an otherwise stratified ocean gives

$$c(z + \zeta(\mathbf{r},t)) \simeq c(z) + \left(\frac{dc}{dz}\right)_p \zeta(\mathbf{r},t) + \cdots, \tag{2.31}$$

where the potential sound-speed gradient is important here (total sound-speed gradient minus the adiabatic gradient) because of the adiabaticity of the internal-wave displacements, that is, no heat conduction occurs when an internal wave moves a water parcel vertically (Munk and Zachariasen, 1976). The second term in this Taylor expansion thus defines the internal-wave-induced sound-speed perturbation that will enter many calculations, namely,

$$\delta c(\mathbf{r},t) = \left(\frac{dc(z)}{dz}\right)_p \zeta(\mathbf{r},t). \tag{2.32}$$

It should be noted that in much of the literature, the potential sound-speed gradient component of Eq. 2.24 is often used in Eq. 2.32 to model internal-wave sound-speed perturbations, but a more proper treatment is to compute the potential sound-speed gradient directly from profiles of potential temperature and salinity.[2]

2.3.3 Example Profiles

Figures 2.1–2.3 show example vertical profiles of many acoustically relevant oceanographic quantities for mid-latitude, high-latitude, and continental shelf regions. For the mid-latitude case (Figure 2.1), the broad thermocline and relatively weak salinity variation give rise to a deep sound-channel axis at a depth somewhat less than 1000 m. This sound-speed profile is not unlike the Munk Canonical profile (Eq. 2.28), which is also plotted in the figure. The potential density and corresponding buoyancy frequency profiles are also not unlike the canonical exponential form. Lastly, the potential sound-speed gradient is seen to be largest in the upper ocean and essentially zero in the deep ocean. Because of Eq. 2.32 it is seen that the upper ocean will be the location where the largest sound-speed fluctuations from internal waves will occur.

---

[2] In practice the potential sound-speed gradient is computed using $(dc/dz)_p = dc(p_{ref}, \theta(z), S(z))/dz$, where one generally chooses $p_{ref} = 0$ decibars.

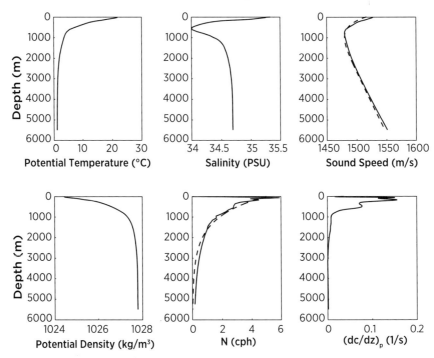

Figure 2.1. Examples of mid-latitude profiles of potential temperature, salinity, sound-speed, potential density, buoyancy frequency, and potential sound-speed gradient from the annual average World Ocean Atlas (Antonov et al., 2006; Locarnini et al., 2006). The profiles were taken from the location 30° N latitude, 150° W longitude. In the sound-speed panel the Munk canonical profile is plotted (dash, Eq. 2.28) for comparison, and in the buoyancy frequency panel an exponential profile $N(z) = 5\exp(z/1000)$ is plotted (dash) for comparison.

A typical high-latitude environment is depicted in Figure 2.2. Here the temperature is essentially uniform with depth, and there is a strong halocline in the upper ocean. The upper Arctic ocean is strongly freshened by significant river outflow. The resulting sound-speed profile is nearly linear (Eq. 2.29) with slope dictated by the adiabatic gradient. In the upper ocean there is a significant halocline effect on the sound-speed profile. The density and buoyancy frequency profiles reveal little stratification except in the halocline. The corresponding potential sound-speed gradient is small indeed relative to the mid-latitude case, meaning that an equal internal-wave displacement at mid-latitude and high latitude will result in a smaller sound-speed anomaly at high latitude. Internal waves are known to be quite weak in the high latitudes (see Chapter 3), so it is expected that acoustic fluctuations from these waves will be rather less important in this region. This situation may change as Arctic ice continues to diminish.

Figure 2.3 shows a typical summer continental shelf profile. The continental shelf has strong seasonal variability as storms can often mix the water column from top to bottom. In the example here there are both strong temperature and salinity

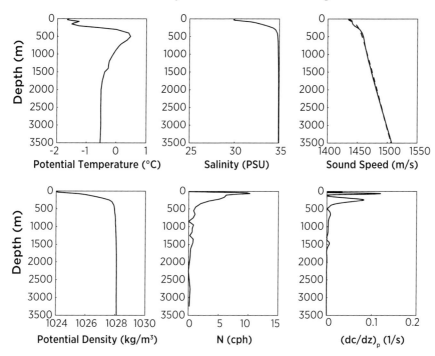

Figure 2.2. Profiles of potential temperature, salinity, sound speed, potential density, buoyancy frequency, and potential sound-speed gradient from the annual average World Ocean Atlas (Antonov et al., 2006; Locarnini et al., 2006). The profiles were taken from the location 80° N latitude, 150° W longitude. In the sound-speed panel a canonical Arctic profile (dash, Eq. 2.29) is shown for comparison.

gradients throughout the water column, resulting in a sound channel axis near 20-m depth. Here a canonical sound-speed profile based on Eq. 2.30 is plotted with the observation, where the parameters are: $c_0 = 1480$ m/s, $\delta c = 25$ m/s, $\Delta = 8$ m, $z_0 = 10$ m, and $\overline{dc/dz} = 0.3$ 1/s. The potential density profile varies rapidly though the strong thermocline, yielding extremely large values of the buoyancy frequency in the upper ocean. The potential sound-speed gradient is large in this case with peak values of order 100 and 10 times the high-latitude and mid-latitude values, respectively.

The interested reader can also refer to the ocean acoustic propagation atlas within the monograph Munk et al. (1995), which shows deep-water sound-speed profiles from around the world's oceans.

2.3.4 Attenuation of Sound

The attenuation of sound from seawater is generally quite small, and this fact is one of the most compelling ones in support of acoustic methodologies for remote sensing, navigation, and communication. But small is not zero, and much effort

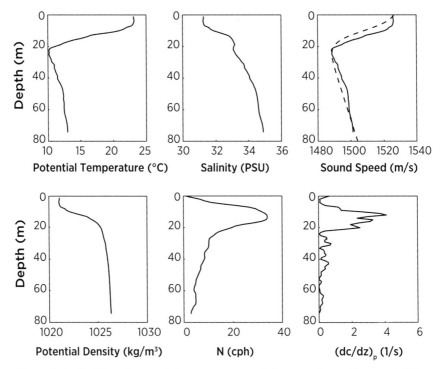

Figure 2.3. Profiles of potential temperature, salinity, sound speed, potential density, buoyancy frequency, and potential sound-speed gradient from measurements taken on the New Jersey continental shelf (Colosi et al., 2012). The profiles come from CTD data in the vicinity of 39° N latitude, 73° W longitude. A canonical sound-speed profile (dash) derived from Eq. 2.30 is shown with the observed sound-speed profile.

has been put toward formulating useful expressions for acoustic attenuation from seawater. Ocean acousticians choose to represent sound attenuation in many ways. The representation used here is to consider the acoustic wavenumber to have a small complex component, that is, the wavenumber is written $\omega/c + i\alpha$. Thus over one meter the relative change in acoustic amplitude is $e^{-\alpha(x+1)}/e^{-\alpha x} = e^{-\alpha}$. Written in units of dB/km the result is

$$\alpha'(\text{dB/km}) = 1000 \times (20\log_{10} e^{\alpha}) \simeq 8686\,\alpha. \qquad (2.33)$$

In this book the quantity α will generally be used for theoretical calculations, but tabulations of attenuation are generally given in terms of α', so the translation given above is quite practical.

A useful approximate expression for α' is (Thorpe, 1967; Fisher and Simmons, 1977; Urick, 1979)

$$\alpha' \simeq 3.3 \times 10^{-3} + 0.79A\frac{f^2}{(0.8)^2 + f^2} + \frac{36f^2}{5000 + f^2} + 3.0 \times 10^{-4}f^2 \quad \text{dB/km}, \quad (2.34)$$

Table 2.1. *Geoacoustic properties of common marine sediments and seafloors*

Bottom type	Density (kg/m^3)	Sound speed (m/s)	Attenuation (dB/λ)^a
Sand coarse	2034	1836	0.87
Sand medium	2014	1765	0.88
Sand fine	1962	1759	0.89
Sand very fine	1878	1709	1.05
Silty sand	1783	1658	1.13
Sandy silt	1769	1644	1.22
Silt	1740	1615	0.38
Sand-silt-clay	1575	1582	0.17
Clayey-silt	1489	1546	0.11
Silty-clay	1480	1517	0.08
Clay-mud	1312	1470	0.09
Chalk	2200	2400	0.2
Sandstone	2400	4350	0.1
Basalt	2550	4750	0.1
Granite	2650	5750	0.1
Limestone	2700	5350	0.1

^{a}The unit dB/λ is converted to dB/m by dividing by m/λ.

Values shown are representative of natural conditions, and variations will occur site to site. Values based on Clay and Medwin (1977), Hamilton (1980), Hamilton (1987), Jensen et al. (1994), Buckingham (2005), APL-UW (2008), and Ainslie (2010). (Courtesy G. Potty).

where f is the acoustic frequency in kilohertz. The ordering of the terms in this equation is significant, as going from left to right the terms apply to progressively higher acoustic frequency. The first term, which is perhaps the least understood of the four, is related to boundary losses due to leakage out off the deep sound channel. The second term, which dominates at frequencies between a few hundred and a few thousand Hz, is due to borate $(B(OH)_3)$ relaxation, which has a distinct pH dependence denoted by the coefficient A. Not much is known about the detailed geographical or seasonal variation of A (Lovett, 1980) but values for the Pacific (lower pH) and Atlantic (higher pH) Oceans are generally taken to be $A = 0.055$ and $A = 0.11$ respectively. The next term, which dominates from tens to hundreds of kHz, is associated with chemical relaxations due to magnesium sulfate $(MgSO_4)$, and the last term is due to the viscosity of seawater.

When sound interacts with the seabed, acoustic loss can go up markedly. The fields of geo- and seismo-acoustics, which treat these complex seabed interactions, are, much like the present monograph, concerned with wave propagation through a heterogeneous and stochastic medium. Rather than getting into these subjects, the idea here is to simply provide bulk properties of the seabed so as to enable shallow-water calculations of sound transmission through the stochastic internal-wave field. Typical geoacoustic parameters are summarized in Table 2.1.

2.4 Rays

Because of its strong geometric basis, ray theory applied to ocean acoustic propagation provides an extremely useful tool to understand the physics of sound scattering in the ocean (Flatté et al., 1979; Brown et al., 2003). In addition, ray theory provides the conceptual picture that is the foundation of many other approaches to sound scattering such as Born and Rytov theory, and Feynman path integrals. Here an overview of ray theory is presented that is meant to be brief but self contained. The emphasis will be on the Hamiltonian structure of the ray equations, and the topic of ray chaos will be introduced. Example calculations for a range-independent sound-speed profile, $c(z)$, will be presented and a qualitative discussion of internal-wave effects will be given.

2.4.1 Physical Picture and Basic Equations

There are a few ways to conceptualize and thus derive ray theoretical results. There is the variational approach that considers the ray to be an extremum of the travel time (Landau and Lifshitz, 1975; Gutzwiller, 1990), a methodology that is based on Fermat's principle (see Section 2.7.1). There is also the method of high-frequency asymptotic analysis of the wave equation (Frisk, 1994; Jensen et al., 1994), which has a pleasing rigor (see Section 2.4.2). Both of these methods will be presented in this chapter, but in this section a third method is put forward that is highly geometric in nature. Here the wave field of interest is considered to behave *locally* like a plane wave (Lighthill, 1978; Munk et al., 1995). Writing the local phase function as $\Theta(\mathbf{r}, t)$ then the plane wave approximation dictates that the vector wavenumber and scalar frequency are given by

$$\mathbf{k} = \nabla\Theta, \tag{2.35}$$

$$\omega = -\frac{\partial\Theta}{\partial t}, \tag{2.36}$$

where ω and $\mathbf{k}$ are related by the plane wave dispersion relation, that is $\omega = \omega(\mathbf{k}, \mathbf{r})$. Taking the divergence of Eq. 2.36 and using Eq. 2.35, the following equation is obtained

$$\frac{\partial\mathbf{k}}{\partial t} = -\left(\frac{\partial w}{\partial\mathbf{k}} \cdot \nabla\right)\mathbf{k} - \frac{\partial\omega}{\partial\mathbf{r}}. \tag{2.37}$$

Identifying $c_g = \partial w / \partial k$ as the group velocity and the total derivative in the direction of the wave as $d/dt = \partial/\partial t + c_g \cdot \nabla$ then from Eq. 2.37 it follows quite naturally that the wave energy will follow along ray paths given by the Hamiltonian-like equations (Lighthill, 1978)

$$\frac{d\mathbf{r}}{dt} = \frac{\partial \omega}{\partial \mathbf{k}} = \mathbf{c_g}, \tag{2.38}$$

$$\frac{d\mathbf{k}}{dt} = -\frac{\partial \omega}{\partial \mathbf{r}}. \tag{2.39}$$

Here the dispersion relation plays the role of the Hamiltonian function. Equation 2.38 simply states that wave energy moves along the ray at the group velocity, and Eq. 2.39 states that the direction of propagation will change due to spatial variations in the dispersion relation (refraction). Here six equations are given to solve for the six unknowns, that is, three components each of the ray position and the wave vector. Lastly, as a consequence of the Hamiltonian structure of these equations it is seen that ω is constant along the rays (e.g., $d\omega/dt = 0$). These equations are quite general and can be applied to any wave system. A few examples include light, seismic waves, surface gravity waves, internal waves, and Rossby waves.

At this point it is useful to tailor the discussion to the problem of ocean acoustic propagation, where the dispersion relation for an ocean otherwise at rest is given by

$$\omega(\mathbf{r},\mathbf{k}) = (k_x^2 + k_y^2 + k_z^2)^{1/2} c(\mathbf{r}). \tag{2.40}$$

For most ocean acoustic problems, only one-way wave propagation in a range-depth slice (x,z) need be considered. For propagation between fixed locations (e.g., sources and receivers), the convention is to use the range coordinate instead of the time coordinate as the independent variable. With these modifications the ray equations become

$$\frac{1}{\omega}\frac{dk_z}{dx} = -\frac{1}{(c^{-2} - k_z^2/\omega^2)^{1/2}} \frac{1}{c^3}\frac{\partial c}{\partial z}, \tag{2.41}$$

$$\frac{dz}{dx} = \frac{k_z}{\omega}\frac{1}{(c^{-2} - k_z^2/\omega^2)^{1/2}} = \tan\theta, \tag{2.42}$$

$$\frac{dT}{dx} = \frac{1}{c^2(c^{-2} - k_z^2/\omega^2)^{1/2}}, \tag{2.43}$$

where T is the acoustic travel time of the ray. It is evident that these equations are independent of acoustic frequency since $k_z \propto \omega$. Thus it is useful to define a frequency independent quantity $p_z = k_z/\omega = \sin\theta/c$ called the vertical ray slowness. Equations 2.41 and 2.42 with conjugate or phase space variables (z, p_z) are of

Hamiltonian form such that

$$\frac{dz}{dx} = \frac{\partial H}{\partial p_z}, \qquad \frac{dp_z}{dx} = -\frac{\partial H}{\partial z}, \qquad (2.44)$$

where the convention is to define $H = -k_x/\omega = -(c^{-2} - p_z^2)^{1/2} = -\cos\theta/c$. The third equation is given by

$$\frac{dT}{dx} = L = p_z\frac{dz}{dx} - H = \frac{1}{c^2(c^{-2} - p_z^2)^{1/2}} = \frac{\sec\theta}{c}, \qquad (2.45)$$

where L is the Lagrangian function. Equation 2.45 simply says that the ray moves along the ray path at the local sound speed, that is, $dT/ds = 1/c$ where $ds = (dx^2 + dz^2)^{1/2}$ is an infinitesimal ray arc length. In the parlance of Lagrangian dynamics and the variational approach to ray propagation (Fermat's principle), it is understood that the travel time of the ray path (e.g., the path integral of the Lagrangian function) is an extremum. These equations also indicate that the acoustic energy moving in the direction of the ray is perpendicular to the wave front (defined as a surface of constant phase). For a moving medium this is not the case (see Pierce, 1994; Munk et al., 1995; Ostashev and Wilson, 2015), but this situation will not be treated until Chapter 5.

In summary, these equations constitute a nonautonomous Hamiltonian system with one degree of freedom (Brown et al., 2003); z and p_z are canonically conjugate position and momentum variables, x is the independent variable, H is the Hamiltonian, L is the Lagrangian, and the travel time T corresponds to the action (kinetic energy minus potential energy) for a classical mechanics system. The system is nonautonomous because for the case of propagation through internal waves H will depend explicitly on the independent variable x. Ocean acoustics, as a Hamiltonian system, provides a unique view into the dynamics since the variational quantity that dictates the physics (e.g., travel time) is actually an observable quantity. This can be compared to the classical mechanics case where the action is not a direct observable (more later in Chapters 5 and 7).

The equations here are for the one-way Helmholtz equation but other Hamiltonian functions can be written down for approximations to the Helmholtz equation such as the parabolic approximation (Tappert and Brown, 1996; Brown et al., 2003). In the parabolic approximation a useful ray Hamiltonian function is given by

$$H = \frac{c_0 p_z^2}{2} + \frac{V(\mathbf{r})}{c_0}, \qquad (2.46)$$

where $p_z = \tan\theta/c_0$ and the function $V = (1 - c_0^2/c^2)/2$ where c_0 is a reference sound speed. Here the ray equations take the particularly simple form

$$\frac{dp_z}{dx} = -\frac{1}{c_0}\frac{\partial V}{\partial z}, \tag{2.47}$$

$$\frac{dz}{dx} = c_0 p_z, \tag{2.48}$$

$$\frac{dT}{dx} = \frac{1}{c_0}\left(1 + \frac{c_0^2 p_z^2}{2} - V\right). \tag{2.49}$$

Note that in the PE travel time equation an extra factor of $1/c_0$ has been included on the right-hand side and this term comes from the $e^{iq_0 x}$ term in the PE equation (Simmen et al., 1997). Also related to the travel time equation it is found that $dT/ds \neq 1/c$, that is, the ray does not travel at the local sound speed. This is an artifact of the small-angle approximation. It is also recognized that the reference sound speed c_0 scales out of of Eqs. 2.47 and 2.48, so often the definition $p_z = \tan\theta$ is used. In this case the Hamiltonian has the dimensionless form $H = p_z^2/2 + V$, and the Lagrangian function is $L = p_z^2/2 - V$, which yields a variational principle in which the ray is an extremum of the acoustic path length S_a or distance along the ray. This oddity is a result of introducing the scaling parameter c_0 into the parabolic equation so that time and length are interchangeable, that is, the ray travel time is related to the path length by $T = S_a/c_0$.

The ray equations as discussed so far (Eqs. 2.44 and 2.45) allow calculation of the wave phase as a function of position and time, that is the acoustic pressure field is given by

$$p(\mathbf{r}, t) = \sum_j a_j(\mathbf{r}, t)e^{i\omega(T_j(\mathbf{r})-t)}, \tag{2.50}$$

where the sum is over all rays that pass through the position $\mathbf{r}$. The only remaining piece of information that is needed to predict the pressure field is the ray amplitude a_j, which is discussed next.

2.4.2 Ray Theory: Asymptotic Analysis

While quite standard in many ocean acoustics textbooks (Jensen et al., 1994; Pierce, 1994), the asymptotic analysis that leads to the ray equations will be presented here so that the reader may see the geometric derivation discussed in the previous section in a different light. Here the ray amplitude equation will be derived for use in subsequent chapters. The starting point is the Helmholtz equation, where the Anzatz $p(\mathbf{r}, \omega) = a(\mathbf{r}, \omega)\exp(i\Theta(\mathbf{r}, \omega))$ is used. Plugging this expression into the Helmholtz equation yields equations for the real and imaginary parts that are given by

$$\nabla^2 a - a|\nabla\Theta|^2 + \frac{\omega^2}{c^2}a = 0, \tag{2.51}$$

$$2\nabla\Theta \cdot \nabla a + a\nabla^2\Theta = 0. \tag{2.52}$$

In the high-frequency limit one expects $\Theta \propto \omega$ and thus in Eq. 2.51 the last two terms are proportional to ω^2 while the first term has no frequency dependence. This yields the *Eikonal* equation written

$$|\nabla\Theta|^2 = \frac{\omega^2}{c^2}. \tag{2.53}$$

This equation defines the geometry of the ray paths which are lines perpendicular to the wave fronts for which Θ is a constant. The equation can also be interpreted in terms of Eqs. 2.35 and 2.36, thus giving the plane wave dispersion relation. Standard techniques, which will not be repeated here, can be used to render the Eikonal equation into a form suitable for computing ray trajectories (Jensen et al., 1994; Pierce, 1994) giving

$$\frac{d}{ds}\left(\frac{1}{c}\frac{d\mathbf{r}}{ds}\right) = -\frac{1}{c^2}\nabla c. \tag{2.54}$$

With some manipulation this second-order differential equation can be shown to be equivalent to the first-order Hamiltonian equations presented in the previous section.[3]

The ray amplitude equation can be interpreted geometrically. A unit vector in the direction of the ray can be written as $\mathbf{k}/q_0 = \nabla\Theta/q_0$, so the first term in Eq. 2.52 has the form $2q_0 da/ds$ where da/ds is the rate of change of the ray amplitude in the direction of or along the ray. So, the ray amplitude equation has the form

$$2q_0\frac{da}{ds} + a\nabla \cdot \mathbf{k} = 0, \tag{2.55}$$

which says that the rate of change of the ray amplitude is proportional to the divergence of the rays.[4] This interpretation follows the intuition that a diverging/converging bundle of rays will have decreasing/increasing amplitude. More about this matter follows in the next section.

2.4.3 Ray Amplitude and Stability

Ray theory gives us a means to understand the variation of acoustic amplitude through the ocean, and this issue is critically connected to the subject of ray path

---

[3] Also see the variational approach to ray propagation presented in Section 2.7.
[4] The factor of two q_0 in the first term comes from the fact that the ray amplitude is being solved for. In the next section where energy considerations are addressed (e.g., quadratic quantities) a simpler result follows.

stability. In the absence of any dissipation, it is well known that wave energy is conserved along ray bundles (Lighthill, 1978; Pierce, 1994; Brown et al., 2003). For wave energy density W and power flux $\mathbf{I} = W\mathbf{c_g}$ the energy conservation equation is written

$$\frac{\partial W}{\partial t} + \nabla \cdot \mathbf{I} = 0. \tag{2.56}$$

Expanding the divergence, this equation can be cast as a total time derivative along the ray giving $dW/dt = -W\nabla \cdot \mathbf{c_g}$. Hence the change in the local wave energy density is proportional to minus the divergence of the group velocity of a ray bundle, that is, the intuitive result is obtained that a diverging bundle has decreasing energy density while a converging bundle has increasing energy density. Equation 2.56, however, is not useful for computing ray amplitudes because evaluation of $\nabla \cdot \mathbf{c_g}$ requires knowledge of nearby ray trajectories. Instead an equation is sought that can be integrated together with the ray trajectory and travel time equations (Eqs. 2.44 and 2.45). This can be achieved by carrying out a traditional stability analysis of the ray equations (Tabor, 1989).

Consider the stability of a ray with respect to variations in the initial conditions. Using the chain rule, the resulting changes in the variables p_z and z are thus written

$$\delta p_z = \frac{\partial p_z}{\partial p_0}\bigg|_{z=z_0} \delta p_0 + \frac{\partial p_z}{\partial z_0}\bigg|_{p_z=p_0} \delta z_0, \qquad \delta z = \frac{\partial z}{\partial p_0}\bigg|_{z=z_0} \delta p_0 + \frac{\partial z}{\partial z_0}\bigg|_{p_z=p_0} \delta z_0, \tag{2.57}$$

where $(\delta p_0, \delta z_0)$ are the perturbations to the initial condition (p_0, z_0), and $(\delta p_z, \delta z)$ are the resultant perturbations to the ray. Writing this in matrix form, the result is

$$\begin{pmatrix} \delta p_z \\ \delta z \end{pmatrix} = \mathbf{Q} \begin{pmatrix} \delta p_0 \\ \delta z_0 \end{pmatrix}, \tag{2.58}$$

where the stability matrix $\mathbf{Q}$ is given by

$$\mathbf{Q} = \begin{pmatrix} q_{11} & q_{12} \\ q_{21} & q_{22} \end{pmatrix} = \begin{pmatrix} \frac{\partial p_z}{\partial p_0}\big|_{z_0} & \frac{\partial p_z}{\partial z_0}\big|_{p_0} \\ \frac{\partial z}{\partial p_0}\big|_{z_0} & \frac{\partial z}{\partial z_0}\big|_{p_0} \end{pmatrix}. \tag{2.59}$$

Now the question is how do the elements in the stability matrix evolve in range. As an example it is helpful to examine one of the elements, $\partial z/\partial p_0$. Taking the range derivative the result is

$$\frac{d}{dx}\frac{\partial z}{\partial p_0} = \frac{\partial}{\partial p_0}\frac{dz}{dx} = \frac{\partial^2 H}{\partial p_z^2}\frac{\partial p_z}{\partial p_0} + \frac{\partial^2 H}{\partial z \partial p_z}\frac{\partial z}{\partial p_0}, \tag{2.60}$$

where Hamilton's equations have been used. Considering all the elements in this way the stability equations become

$$\frac{d}{dx}\mathbf{Q} = \mathbf{KQ}, \qquad \mathbf{K} = \begin{pmatrix} -\frac{\partial^2 H}{\partial z \partial p_z} & -\frac{\partial^2 H}{\partial z^2} \\ \frac{\partial^2 H}{\partial p_z^2} & \frac{\partial^2 H}{\partial z \partial p_z} \end{pmatrix}, \tag{2.61}$$

where $\mathbf{K}$ is termed the propagator matrix and the initial condition for $\mathbf{Q}$ is the identity matrix. Physically the stability matrix tells us the variation in (z, p_z) at range x with respect to initial variations in the source depth and angle. For the case of a point source the variation of the source depth is irrelevant, and the critical term in the stability matrix is q_{21}, that is, the variation in the depth of the ray with respect to an initial variation in vertical ray slowness (i.e., variation in initial angle). There are also stability equations for travel time, which are given here for completeness and future use

$$\frac{d}{dx}\left(\frac{\partial T}{\partial p_0}\bigg|_{z_0}, \frac{\partial T}{\partial z_0}\bigg|_{p_0}\right) = p_z\left(\frac{\partial^2 H}{\partial p_z^2}, \frac{\partial^2 H}{\partial z \partial p_z}\right)\mathbf{Q}. \tag{2.62}$$

A simple geometric calculation for the ray amplitude at position (x, z) gives (Brown, 1994; Tappert and Tang, 1996)

$$a(x, z) = a_0 \frac{r_0}{\sqrt{x}} \frac{e^{-iM\pi/2}}{\sqrt{|H(x, z)|\,|q_{21}(x, z)|}}, \tag{2.63}$$

where $r_0 = 1$ m is the standard reference range and a_0 is the reference amplitude. For $a_0 = 1$, the transmission loss (TL) given by $20\log_{10}|a|$ is obtained. The ray amplitude equation can be better understood by looking at the small range behavior (e.g., $x \to 0$). In this limit the result is $q_{21}(x) \simeq x(\partial^2 H(0)/\partial p^2) = xc_s/\cos^3\theta_0$, and $H(0) = -\cos\theta_0/c_s$, where c_s and θ_0 are the sound speed and ray angle at the source. This yields $a(x, r) = a_0 r_0 \cos\theta_0/x = r_0/s$, where the slant range is $s = x/\cos\theta_0$. Thus the normalization yields spherical spreading at short range, and for $a_0 = 1$ the conventional TL of zero at the slant range of $r_0 = 1$ m is recovered.

The quantity M, termed the Maslov index, is the number of caustics that the ray has passed through. A ray has passed through a caustic when the stability matrix element q_{21} passes through zero, which means that two nearby rays have crossed one another. The crossing yields a π phase shift for q_{12}, and thus a $\pi/2$ phase shift in Eq. 2.63. Caustics are high-intensity regions of the wave field and are therefore of great interest. However, the troubling result is found that ray theory predicts an intensity singularity at caustics. There are methods to remove the singularities for the case in which the sound speed is only a function of depth or the range dependence is extremely weak (Brown, 1994), but these methods are not applicable to the problem of sound propagation through the stochastic internal-wave field because of the abundance of caustics created by the internal waves. While ray theory has this shortcoming with regard to caustics, nonetheless the theory does have the advantage of indicating precisely where the caustic regions are located.

The stability equations in the parabolic approximation are of great interest, and these results will be used extensively in the development of fluctuation theory using the method of path integrals. With the PE Hamiltonian (Eq. 2.46), the propagator matrix K has a simple form and writing $q_{21} = \xi$ the result is

$$\frac{d^2\xi}{dx^2} + V''\xi = 0, \tag{2.64}$$

where $V'' = \partial^2 V/\partial z^2$ is the second derivative along the ray path. Using the terminology of Flatté et al. (1979), Eq. 2.64 is called the ray-tube equation and the function ξ is called the ray tube function.[5] Here it is seen that ξ, and thus the ray amplitude, depends critically on the curvature of the sound-speed profile. For deterministic cases this means that the sound-speed field must be known precisely to compute the ray amplitude, and this result shows that acoustic amplitudes are most sensitive to small vertical scale structure in the ocean.

With knowledge of the amplitude the ray theory description of the pressure field is complete. The acoustic pressure for a given range and receiver depth is then a coherent sum over all the rays connecting the source and receiver: These paths are termed eigenrays. Assuming propagation free of attenuation this coherent sum can be written

$$p(x,z,t) = \sum_{j=1}^{N} a_j(x,z)\, E(T_j - t)\, e^{i\omega_c(T_j - t)}, \tag{2.65}$$

where T_j and a_j are the eigenray travel time and amplitude, N is the number of eigenrays, $E(T_j - t)$ is the unit amplitude pulse envelope, and ω_c is the carrier frequency. Depending on the signal bandwidth, which dictates the temporal width of the pulse envelope, the eigenrays may be interfering (e.g., overlapping in time) or they may form isolated arrivals.

2.4.4 Ray Chaos: Introduction

It has been well established that ray propagation through the stochastic internal-wave field results in exponential growth of the stability matrix elements (Beron-Vera et al., 2003; Brown et al., 2003). This exponential sensitivity to initial conditions is termed ray chaos, and it is central to the understanding of

[5] The ray-tube equation can also be derived directly from the PE ray equation, $d^2z/dx^2 + V' = 0$ (see Eqs. 2.47 and 2.48) by examining perturbations around a ray given by $z(x) = z_r(x) + \xi(x)$, where z_r is the ray trajectory. Plugging z into the ray equation and expanding around the ray path to first order in ξ, Eq. 2.64 is easily obtained.

ocean acoustic propagation. Ray chaos is also related to exponential sensitivity to perturbations in sound speed like those caused by internal waves. Of course it is not enough to simply say there is ray chaos; the growth rate of the instability must be estimated: this factor is called the Lyapunov exponent. The basic theory for the Lyapunov exponent is presented here and this subject is developed further in Chapter 5. There is a wealth of literature on dynamical systems stability and the interested reader is referred to the textbook by Tabor (1989) and the review article by Brown et al. (2003), which cover well the fundamentals. What follows is a brief self-contained overview.

The ray equations for an initial launch angle and depth will be stable or nonchaotic if there is a constant Υ along the ray trajectory such that

$$\frac{d\Upsilon}{dx} = \frac{\partial\Upsilon}{\partial z}\frac{dz}{dx} + \frac{\partial\Upsilon}{\partial p}\frac{dp}{dx} + \frac{\partial\Upsilon}{\partial x} = 0. \tag{2.66}$$

For $c = c(z)$, then $\Upsilon = H = -\cos(\theta)/c$ is a ray constant (i.e., the horizontal wavenumber) and all rays are stable or integrable; that is the equations of motion can be written in terms of an explicit integral or quadrature.[6] For ocean sound-speed structure that is generally a complicated function of both depth and range, like that caused by stochastic internal waves, there are no obvious ray constants, and extensive calculations show no indications of stable trajectories (Beron-Vera et al., 2003; Brown et al., 2003). Thus with a high degree of certainty all ray trajectories are assumed to be unstable. In this case the growth rate of the instability or the Lyapunov exponent is a critical piece of information. Furthermore, since there is a stochastic medium in the ocean, the Lyapunov exponent itself will be a stochastic variable whose moments and distribution function will be of critical interest.

An analysis of the stability matrix $\mathbf{Q}$ gives the key information. First, $\mathbf{Q}$ has some important properties. For Hamiltonian systems it can be shown that phase space "flow" $(dz/dx, dp_z/dx)$ satisfies the incompressibility condition,

$$\frac{\partial}{\partial z}\frac{dz}{dx} + \frac{\partial}{\partial p_z}\frac{dp_z}{dx} = 0, \tag{2.67}$$

and thus phase space volume is conserved regardless of the form of the Hamiltonian (Tabor, 1989). This is Liouville's theorem. So, $\mathbf{Q}$ can be viewed as a linear, canonical transformation and as such its determinant is unity. At finite range, x, the stability of a ray path is determined from the eigenvalues of $\mathbf{Q}$. It is well known that $\mathbf{Q}$ can be diagonalized by a linear, similarity transformation

$$\Lambda = \mathbf{L}\mathbf{Q}\mathbf{L}^{-1} \Rightarrow \begin{pmatrix} \lambda_1 = \lambda & 0 \\ 0 & \lambda_2 = \lambda^{-1} \end{pmatrix}. \tag{2.68}$$

---

[6] The constancy of $\cos(\theta)/c$ is termed Snell's law.

The last form applies to systems with a single degree of freedom because the determinant is unity. The similarity transformation is known to leave $\text{Tr}(\mathbf{Q})$ invariant, and it is found that $\text{Tr}(\mathbf{Q})$ is real. A well known result is (Tabor, 1989; Brown et al., 2003)

$$\text{For } |\text{Tr}(\mathbf{Q})| > 2, \lambda_1 \simeq \text{Tr}(\mathbf{Q}), \lambda_2 = 1/\lambda_1 \simeq 0, \text{Unstable Motion}, \tag{2.69}$$

$$\text{For } |\text{Tr}(\mathbf{Q})| < 2, \qquad \lambda_{1,2} = \exp(\pm i\alpha), \qquad \text{Stable Motion}. \tag{2.70}$$

Traditionally under strong instability one writes

$$|\text{Tr}(\mathbf{Q})| \simeq \lambda_1 \equiv \exp(vx), \tag{2.71}$$

which is the definition of the finite-range Lyapunov exponent, v, describing the local exponential divergence of nearby ray trajectories. Under stable conditions, namely when the sound speed is only a function of depth, nearby ray trajectories diverge in range by and large linearly. The largest Lyapunov exponent is expressed as

$$v_L = \lim_{x \to \infty} \frac{\ln(|\text{Tr}(\mathbf{Q})|)}{x}, \tag{2.72}$$

and in this limit all values of v approach v_L. The rays with nonzero Lyapunov exponent are termed chaotic rays because of their exponential sensitivity to initial conditions.

An interesting consequence of the exponential divergence of nearby ray trajectories is that the intensities of chaotic rays, which are proportional to $|\partial z/\partial p_0|^{-1}$, decay exponentially with range and therefore would be difficult to detect in experiments. On the other hand, energy conservation considerations suggest and numerical experiments show (Tappert and Tang, 1996) that under chaotic conditions the number of eigenrays increases exponentially with range and therefore collectively chaotic rays may be detectable in experiments.

This is where the subject will stand for now. The issue of ray stability will be further investigated in Chapter 5.

2.4.5 Ehrenfest Theorem

The significance of the ray picture is demonstrated by the important result of Ehrenfest (1927), who showed that the mean position and momentum of a quantum wavepacket follows the classical trajectory (see Merzbacher (1961) or Sakurai (1985)). In this section it will be shown that for acoustic propagation in the parabolic approximation the mean position and angle of propagation of the wavepacket in the vertical will follow the ray equations. This topic is relevant to the problem of sound propagation through the stochastic internal-wave field since chaotic ray dynamics are known to invalidate the correspondence between full wave– and ray-based representations after the waves have propagated over a distance termed the Ehrenfest range (Brown et al., 2003). For classical/quantum systems this breakdown is called the Ehrenfest time (Zaslavsky, 1980).

The mean vertical position of the wavefunction is defined as

$$\langle z \rangle = \int A^*(z,x) \, z \, A(z,x) \, dz. \tag{2.73}$$

Taking the x-derivative and using the parabolic equation the result is

$$\frac{d}{dx}\langle z \rangle = \int \left[\frac{\partial A^*(z,x)}{\partial x} z A(z,x) + A^*(z,x) z \frac{\partial A(z,x)}{\partial x} \right] dz,$$

$$= \frac{i}{2q_0} \int \left[A^* z \frac{\partial^2 A}{\partial z^2} - \frac{\partial^2 A^*}{\partial z^2} z A \right] dz. \tag{2.74}$$

The integrand of Eq. 2.74 has the form

$$\frac{\partial}{\partial z} \left[A^* z \frac{\partial A}{\partial z} + A^* A - \frac{\partial A^*}{\partial z} z A \right] - 2 A^* \frac{\partial A}{\partial z}. \tag{2.75}$$

Integrating Eq. 2.74, the first term is seen to be a total derivative, so it can be evaluated at the boundaries where A must be zero in order to have finite energy. Thus the result for the range evolution of the mean vertical position is

$$\frac{d}{dx}\langle z \rangle = \int A^* \left(\frac{-i}{q_0} \frac{\partial}{\partial z} \right) A \, dz. \tag{2.76}$$

Next it should be noted that the operator $p = (-i/q_0)\partial/\partial z$ appears in the parabolic equation such that

$$\frac{i}{q_0} \frac{\partial A}{\partial x} = \left(\frac{p^2}{2} + V(x,z) \right) A \equiv H_{op}(p,z,x) A,$$

where the Hamiltonian operator H_{op} has the same form as the ray Hamiltonian (Eq. 2.46). One may also view the momentum operator p as generating vertical translations, so that $pA \simeq k_z A/q_0 = A\tan\theta$; hence p is related to c_0 times the ray slowness p_z. Thus from Eq. 2.76 the final result is that the average depth position follows the ray trajectory equation (Eq. 2.48), that is,

$$\frac{d}{dx}\langle z \rangle = \langle p \rangle. \tag{2.77}$$

A similar (but tedious) calculation can be done for the range evolution of the canonical variable p, related to the wave angle. Here it is found that

$$\frac{d}{dx}\langle p\rangle = \int\left[\frac{\partial A^*(z,x)}{\partial x}pA(z,x) + A^*(z,x)p\frac{\partial A(z,x)}{\partial x}\right]dz,$$

$$= \int -A^*\frac{\partial V}{\partial z}A\,dz = -\left\langle\frac{\partial V}{\partial z}\right\rangle, \tag{2.78}$$

which is nearly the full wave equivalent of the ray equation Eq. 2.47. The only thing that limits us from making a perfect ray-to-wave analogy is that

$$\left\langle\frac{dV}{dz}\right\rangle \neq \frac{d}{d\langle z\rangle}V(\langle z\rangle). \tag{2.79}$$

For equality in Eq. 2.79 to hold then $V(z)$ must be a slowly varying function of z, relative to the vertical spread of the wave functions A. Physically this means that diffraction is weak.

2.4.6 Rays in a Range-Independent Ocean

In this section various ray-derived quantities are discussed for the case in which the sound speed is only a function of depth. Useful example calculations are provided to give important insight into ocean acoustic propagation, and they provide a benchmark from which to compare calculations involving internal-wave sound-speed perturbations. For simplicity and ease of reproduction by the reader, results are shown for the mid-latitude Munk canonical profile (Eq. 2.28) where parameters of $B = z_a = 1000$ m, and $c_A = 1500$ m/s are chosen. For this example a point source at the sound channel axis is considered and all the rays are nonsurface interacting. Examples for shallow-water propagation are not presented since the mode approach proves much more useful for that case. Properties of ray quantities in a range-independent profile have been discussed in many textbooks (e.g., see Munk et al., 1995), so only a brief account is given here.

Ray Paths

Figure 2.4 shows a collection of ray paths for the Munk sound-speed profile. In this case the Hamiltonian function is conserved (i.e., horizontal wavenumber is constant) meaning that $\cos\theta/c$ is a constant along a ray (i.e., Snell's law). Thus it is useful to define a quantity for identifying particular ray paths called the grazing angle, which is the absolute value of the ray angle at the sound-channel axis, namely $\cos(\theta_g) = c_{axis}/c_{turn}$ where c_{axis}/c_{turn} are the sound speed at the axis/ray turning point. A turning point is where the ray angle is zero, and the upper and lower turning point depths are labeled z^+ and z^- respectively. High grazing angle

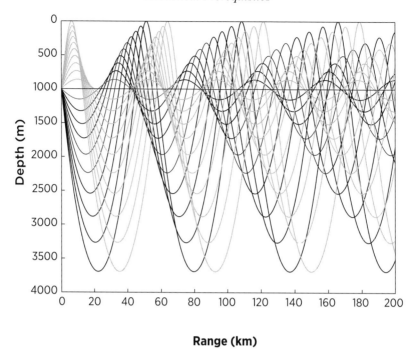

Range (km)

Figure 2.4. A selection of rays computed for the Munk canonical profile (Eq. 2.28). Initially upward/downward rays are colored gray/black.

rays have turning points quite distant from the sound-channel axis, while low grazing angle rays are confined vertically near the axis. In this particular case the grazing angles are between $0°$ and $12.6°$.

For a range-independent Munk profile, rays of a given grazing angle are periodic in range but asymmetrical above and below the axis. For this mid-latitude profile, the rays have a short upper loop that extends above the axis and a longer lower loop below the axis. The combination of the upper and lower loop is termed the ray double loop. For small grazing angles the upper and lower loops become symmetrical and the double loop appears more sinusoidal. Furthermore, as the grazing angle increases so does the horizontal double loop distance. The upper/lower horizontal loop distances $R^{\pm}$, and the travel time to traverse a loop $T^{\pm}$ can be easily derived from the ray equations with the result

$$R^{\pm} = 2 \int_{z_a}^{z^{\pm}} \frac{c^{-1}(z^{\pm})}{\sqrt{c^{-2}(z) - c^{-2}(z^{\pm})}} \, dz, \tag{2.80}$$

$$T^{\pm} = 2 \int_{z_a}^{z^{\pm}} \frac{c^{-2}(z)}{\sqrt{c^{-2}(z) - c^{-2}(z^{\pm})}} \, dz. \tag{2.81}$$

The double loop distance and time are $R_L = R^+ + R^-$ and $T_L = T^+ + T^-$. The ray double loop distance is important to scattering because of a ray loop resonance condition (Cornuelle and Howe, 1987; Munk et al., 1995) that makes rays most sensitive to ocean structure on the scale of R_L and its harmonics (more in Chapters 4 and 5). Another important quantity is the horizontal loop group velocity of a ray, defined as $c_g^\pm = R^\pm / T^\pm$. This quantity also varies as a function of grazing angle and upper/lower loop. At low grazing angles both the upper and lower loops have horizontal group speeds close to the axial sound speed, but as grazing angle increases, upper loops slow down while lower loops speed up. The horizontal double loop group speed R_L / T_L, however, increases with grazing angle, that is, the lower loop dominates. The properties that have been discussed for the mid-latitude/Munk profile also can help understand the polar and shallow-water cases because polar rays act like lower loops, and shallow-water cases act like upper loops. This dispersive behavior of the loops as a function of grazing angle has important consequences for the geometries of the wave front and time front which are discussed next.

Time Front

An important observational ray quantity is the time front, which is ray depth versus travel time for a fixed range (see Figure 2.5). This display shows how the wave front sweeps by a set of vertical receivers at fixed range. The accordion pattern seen in the display is formed by the folding of the wave front by the sound channel (loop dispersion), and the differences in wave speed from initially up-going (upper loop first) and down-going (lower loop first) rays. As previously discussed for the Munk profile, and general deep-water profiles, low grazing angle rays move down range more slowly than high grazing angle rays, so the time front is seen to spread out in depth from late to early arrival times.[7] The time front is seen to be composed of branches that terminate in depth with a cusp shape. The cusped regions are termed caustics where singularities exist. The branches are made up of rays with similar properties, namely all rays on a branch have the same number of ray turning points. Thus each branch can be labeled with a unique identifier (ID), which is the number of ray turning points and a plus/minus sign depending on whether the rays have an initial up/down angle at the source (up means the ray is pointing towards the ocean surface). Because early arriving branches are composed of high grazing angle rays with long double loop lengths they have fewer turning points (and thus a smaller ID number) compared to the low grazing angle branches. As range increases/decreases loop dispersion dictates that there will be more/fewer branches in the time front.

---

[7] For a polar profile the time front has this same pattern because in both cases the loop dispersion is dominated by the lower loops. In shallow water the low grazing angles are faster and thus the time front widens in depth from early to late travel times.

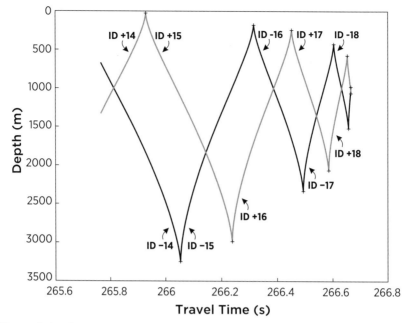

Figure 2.5. Time front for the Munk canonical profile for a range of 400 km. Initially upward/downward rays are colored gray/black. Caustics are marked with black crosses, and time front IDs are labeled.

Much of the progress that has been made in the subject of sound transmission through the stochastic internal-wave field is due to the ability to observe the evolution of time fronts over a large vertical aperture. When internal waves are included in the ray calculation, it has been discovered that the basic time front pattern for the early arrivals is not dramatically altered, although the caustic cusps are observed to vertically elongate, spreading into the shadow zones. These rays scatter primarily along the time front as opposed to across it due to small-angle scattering effects. For the late arrivals the ray scattering is strong enough to destroy the branch pattern, and significant vertical spreading of acoustic energy into the shadow zone is observed (more in Chapter 5).

Eigenray Plot: Caustics

Further information regarding the acoustic field is provided by an eigenray plot, which is a display of ray depth at the final range versus initial ray angle (see Figure 2.6). Eigenrays are identified in this plot by drawing a horizontal line at the receiver depth and noting the launch angles at which the eigenray curve intersects the receiver depth. The slope of this curve $\partial z/\partial \theta_0$ is proportional to the stability matrix quantity $q_{12} = \partial z/\partial p_0$ and thus gives us information about caustics and the ray amplitude. Caustics occur where the slope is zero.

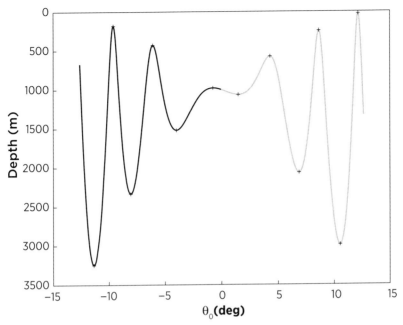

Figure 2.6. Eigenray plot showing ray depth versus initial angle for the Munk canonical profile at a range of 400 km. Caustics occur where the slope of the curve passes through zero (i.e., $\partial z/\partial \theta_0 = \partial z/\partial p_0 = 0$), and these points are marked with black crosses.

When internal waves are added to the ray calculation, the eigenray plot can develop complex structure because ray chaos leads to an exponential growth of eigenrays and therefore of caustics. Theoretical approaches such as weak fluctuation theory and path integrals that expand about the unperturbed ray assume that the eigenrays created by the internal-wave fluctuations remain close to the unperturbed ray. This is not always a good assumption (more in Chapter 5).

Lagrangian Manifold

Yet another way of interpreting the ray acoustic field is to plot the Lagrangian manifold, that is a depth versus momentum display (Figure 2.7). At the initial range the manifold is a horizontal line at the source depth. As the rays propagate out in range, each point on the manifold (representing different rays with different launch angles) start rotating in a closed loop; the loop is closed because the rays are periodic with double loop length R_L. Because low grazing angle rays have shorter double loop lengths than high grazing angle rays, the center of the manifold rotates more rapidly as the rays propagate out in range. This loop dispersion (i.e., grazing angle dependence of loop length) then creates the swirling pattern of the manifold. The Lagrangian manifold can become folded and stretched but it cannot break owing to phase space area preservation (Liouville's theorem). Note that in

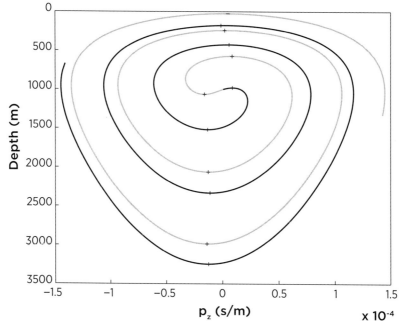

Figure 2.7. Lagrangian manifold for the rays in Munk canonical profile at a range of 400 km. Initially upward/downward rays are colored gray/black. Caustics are marked with vertical black crosses.

the Lagrangian manifold display in Figure 2.7 caustics are not located at $p_z = 0$, and this means that the cusps are slightly tilted.

The range evolution of the Lagrangian manifold is closely related to an equivalent set of canonical variables termed action-angle variables (Brown et al., 2003). For the case in which $c = c(z)$, there exists a canonical transformation from (z, p_z) to (I, θ) that yields a Hamiltonian that depends on the action coordinate I, and not the angle coordinate θ. Thus Hamilton's equations become

$$\frac{d\theta}{dx} = \frac{\partial H}{\partial I} \equiv K_L(I), \quad \frac{dI}{dx} = -\frac{\partial H}{\partial \theta} = 0. \tag{2.82}$$

The solution to these equations is simply $I = $ constant and $\theta(x) = K_L(I)x + \theta(0)$. The factor $K_L(I)$ is equal to $2\pi/R_L(I)$ and gives us the rotation rate of the points along the Lagrangian manifold as the rays propagate out in range.[8] The action variable is written

$$I = \frac{1}{2\pi} \oint p(z)dz = \frac{1}{\pi} \int_{z^-}^{z^+} p(z)dz, \tag{2.83}$$

where the integration is over one cycle of the ray's periodic trajectory. Action angle variables will be discussed in more detail in Chapter 5.

---

[8] The notation here is a departure from the literature which uses ω instead of K_L.

When internal-wave sound-speed fluctuations are added to the problem, the Lagrangian manifold, like the eigneray plot, will develop complex structure. The rays are no longer periodic, and the action is not constant. In Chapter 5 it will be shown that the degree of complexity in the Lagangian manifold and other acoustic quantities will be strongly influenced by the factor dK_L/dI, where K_L is computed from the background sound-speed profile. Physically this factor tells us how strongly the background profile shears the manifold: Strongly sheared regions of the manifold are associated with the strongest acoustic variability. This sensitivity to the background sound-speed profile was one of the major discoveries of the last 30 years of ocean acoustics research.

Ray Intensity

Lastly the ray TL computed from the stability equations (see Figure 2.8) is shown. Here it is evident that the TL becomes exceedingly low at the caustics, a result of the divergence of the ray amplitude there. The focusing effect of the waveguide is also evident. Rays with small grazing angles are gathered together near the sound channel axis and experience smaller loss. Higher grazing angle rays are not as effectively focused by the waveguide and experience greater loss.

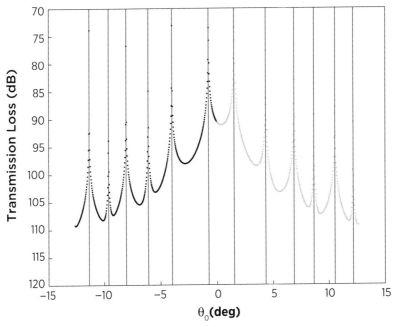

Figure 2.8. Transmission loss (TL) versus initial ray angle for the rays in the Munk canonical profile at a range of 400 km. Caustics are marked with vertical black lines.

2.5 Fresnel Zones and Ray Tubes

Ray theory tells us that sound travels along specific ray paths that connect the source and receiver, but this physical picture cannot be entirely true because it has been seen that when nearby ray paths get too close, singularities occur in the predicted wave field amplitude. Further progress with this problem can be made if one considers the possibility that nearby rays, not necessarily eigenrays, can interfere with one another. This idea will be investigated here using the notion of a Fresnel zone, that is, a region of interfering paths near a particular eigenray. These nearby paths can be physically associated with diffraction. It is important to stress that this treatment of Fresnel zones will not solve the problem of ray theory singularities, but it will allow us to think beyond the idea of a narrow ray, a concept that is critical in the Born and Rytov scattering theory (Section 2.6 and Chapter 6) as well as path integral theory (Section 2.7 and Chapter 7).

To start the analysis requires a way to quantify acoustic phase for an arbitrary path, and here ray theory is useful. At this point in the analysis it is best to examine the simple case in which the sound speed is only a function of depth. The effects of internal waves will be discussed at length in subsequent chapters. Accordingly the accumulated phase, $\Theta(\mathbf{x}_0, \mathbf{x}_R)$, along a ray path Γ starting at position $\mathbf{x}_0$ and ending at location $\mathbf{x}_R$ will be given by

$$\Theta(\mathbf{x}_0, \mathbf{x}_R) = \omega \int_\Gamma \frac{ds}{c} = q_0 \int_\Gamma \left[\frac{\dot{z}^2}{2} - V(z) \right] dx, \qquad (2.84)$$

where the last line follows from the parabolic approximation (see Eq. 2.49), with $\dot{z} = dz/dx$ and $q_0 = \omega/c_0$. Here tradition is followed and the Fresnel zone calculation is carried out in the parabolic approximation (Flatté et al., 1979; Flatté, 1983). While this approximation is not absolutely necessary, it simplifies the treatment from a pedagogical standpoint and allows a greater appreciation of the literature. In fact, to the best of our knowledge no one has carried out the calculation without the parabolic approximation, though it would be simple to do so.

An equation is sought for the phase function in the vicinity of a geometric ray path, and the focus initially is on nearby paths that are separated vertically from the geometric path. This leads to the vertical Fresnel zone. The vertical Fresnel zone will be seen later to be important because of the ocean waveguide and the small-scale vertical structure of internal waves. For a point source the vertical separation of a nearby ray is given by the stability matrix element q_{21}. In the parabolic approximation, the ray-tube equation (Eq. 2.64) gives this separation.

To find the phase of ray paths near a particular ray $z_r(x)$, one writes $z = z_r(x) + \xi(x)$, where it is required that ξ goes to zero at the source and receiver, that is, the nearby path must connect the source and receiver. Plugging this expression into

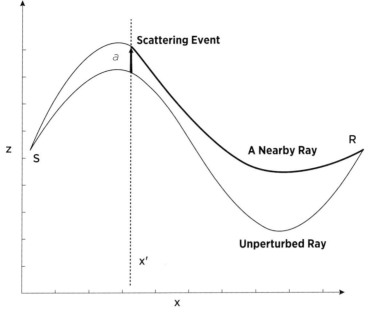

Figure 2.9. The Fresnel zone geometry for curved rays in a waveguide is shown. The broken ray at the scattering event is displaced vertically by a distance a.

Eq. 2.84 and expanding around the ray to second order gives the result

$$\Theta = \Theta_{ray} + q_0 \int_0^x \left[\frac{\dot{\xi}^2}{2} - \frac{\xi^2}{2} V''(z_r) \right] dx. \tag{2.85}$$

In Eq. 2.85 it is seen that the first-order terms disappear because the ray is an extremum of the travel time (Fermat's principle).

To compute the vertical Fresnel zone, consider a point at some intermediate range x' between the source and receiver that is displaced a vertical distance a from the unperturbed ray under consideration (see Figure 2.9). This new point $(x', z_r + a)$ helps define a new acoustic path connecting the source and receiver. This new path is actually composed of two rays obeying the ray equations: one propagating from the source to $(x', z_r + a)$ and the other joining $(x', z_r + a)$ to the receiver. At the range x' the new path has a discontinuity in slope that is sometimes interpreted as a virtual scatting event, but here it is clearly seen that the change in slope is necessary so that the path can end up at the receiver. Be that as it may the ray-tube functions ξ for these two displaced paths must be established. Because the ray-tube equation is a second-order equation the solution can be written as a linear combination of two solutions, namely

$$\xi(x) = c_1 \xi_1(x) + c_2 \xi_2(x),$$

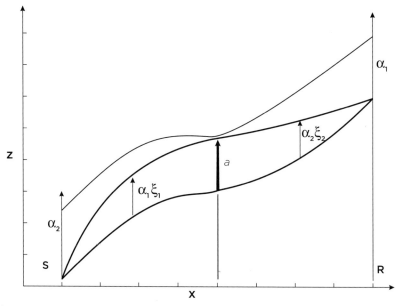

Figure 2.10. Geometry for the Fresnel zone construction from the ray tube functions ξ_1 and ξ_2.

where c_1 and c_2 are integration constants. To construct the ray tube functions it is useful to define the following boundary conditions for these solutions (see Figure 2.10),

$$\xi_1(0) = 0, \qquad \xi_1(R) = 1, \qquad \text{and} \qquad \xi_2(0) = 1, \qquad \xi_2(R) = 0.$$

Using the geometry in Figure 2.10, the displacement between the perturbed and unperturbed ray is

$$\alpha_1 \xi_1(x) \qquad 0 < x < x', \qquad \alpha_2 \xi_2(x) \qquad x' < x < R,$$

where α_1 and α_2 are now the two integration constants. Equating the solutions at x', gives

$$\alpha_1 \xi_1(x') = \alpha_2 \xi_2(x') = a. \tag{2.86}$$

Now the vertical Fresnel zone can be computed. Using Eq. 2.85 the difference in phase between the perturbed and unperturbed paths is given by

$$\Delta \Theta = q_0 \left(\alpha_1^2 \int_0^{x'} \left[\frac{\dot{\xi}_1^{\,2}}{2} - \frac{\xi_1^2}{2} V''(z_r) \right] dx + \alpha_2^2 \int_{x'}^R \left[\frac{\dot{\xi}_2^{\,2}}{2} - \frac{\xi_2^2}{2} V''(z_r) \right] dx \right). \tag{2.87}$$

Using the ray tube equation and Eq. 2.86, the integrands in Eq. 2.87 can be written as total derivatives,[9] yielding the result

$$\Delta\Theta = q_0 \frac{a^2}{2} \left[\frac{\xi_2(x') \, \partial_x \xi_1(x') - \xi_1(x') \, \partial_x \xi_2(x')}{\xi_1(x')\xi_2(x')} \right]. \tag{2.88}$$

An important case is when the phase difference is equal to π such that there is destructive interference between the unperturbed and perturbed paths. The vertical distance a from the unperturbed path that results in this destructive interference is termed the first vertical Fresnel zone and its value is given by[10]

$$R_f(x') = \left[\lambda \frac{\xi_1(x')\xi_2(x')}{\xi_2(x') \, \partial_x \xi_1(x') - \xi_1(x') \, \partial_x \xi_2(x')} \right]^{1/2}. \tag{2.89}$$

Note here that the Fresnel zone scales as one over the square root of frequency. Appendix A describes the computational aspects of obtaining the Fresnel zone by integration along the ray path, and importantly if the receiver is too near a caustic this method will fail. It should also be remembered that Eq. 2.89 was obtained in the parabolic approximation and thus describes the Fresnel zone along a PE ray, not a Helmholtz equation ray. What is seen in most of the literature, however, is a hybrid of the PE and HE theories. The actual ray paths are found using the Helmholtz Hamiltonian and the PE ray tube equations are integrated along the HE rays. Recently an internally consistent procedure has been utilized (Colosi, 2015). Here the ray tube functions are computed using the Helmholtz Hamiltonian and the ray stability equation (Eq. 2.61). The key element of that equation for the ray tube is the quantity $q_{21} = \partial z / \partial p_0$, which gives the variation of the ray depth with initial ray slowness. The wide angle ray tube function is therefore $\xi = q_{21} \cos\theta_0 / c_s$ where c_s is the sound speed at the source.

A few cases of special interest, in the PE approximation, can be easily calculated. First, for the case of a polar or constant sound-speed profile it is found that (see Appendix A)

$$R_f(x) = \left[\lambda \frac{x(R-x)}{R} \right]^{1/2}. \tag{2.90}$$

This constant sound speed Fresnel zone in fact is quite useful, not for the vertical direction but for the transverse directions around the ray where there is no waveguide. A second special case is for a parabolic sound-speed profile with the result that (again see Appendix A)

$$R_f(x) = \left[\lambda \left| \frac{\sin K_a x \, \sin K_a (R-x)}{K_a \sin K_a R} \right| \right]^{1/2}, \tag{2.91}$$

---

[9] The individual integrals are given by $\int_{x_1}^{x_2} \dot{\xi}^2 - \xi^2 V'' \, dx = \xi \dot{\xi} \big|_{x_1}^{x_2}$.

[10] The denominator in the Fresnel zone equation is called the Wronskian and can be evaluated for any value of x'.

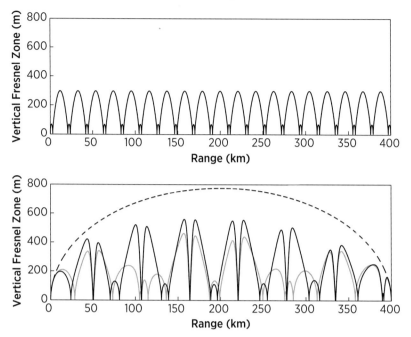

Figure 2.11. Vertical Fresnel zones in a Munk profile with the source on the axis for an axial ray (top) and near surface reflecting ray (bottom). The frequency is 250 Hz. For the axial ray the Fresnel zone is given by Eq. 2.91. For the high-angle ray, the Fresnel zone was computed using the ray tube equations in the parabolic approximation (gray) and using the variational equations (black). The constant sound speed Fresnel zone (dash) is shown for reference.

where $K_a = 2\pi/R_a = \sqrt{V''}$ is the horizontal wavenumber of the sinusoidal ray loop with double loop range R_a. This special case applies to rays traveling with small grazing angles near the deep sound-channel axis.

In general the Fresnel zone for mid-latitude or shallow-water profiles is more complicated than these special case expressions. Figure 2.11 shows the vertical Fresnel zone for 250-Hz propagation in the Munk profile for an axial ray and a near surface reflecting ray. For the axial ray a direct numerical calculation and Eq. 2.91 give identical results. For the high-angle ray a complicated variation of the Fresnel zone along the ray is evident, with zeros in the Fresnel zone where caustics exist (recall a caustic is where $\xi_1 = 0$). In the case of the high-angle ray, the Fresnel zone is computed using the PE ray tube equations and those from the variational equations. Differences in the two calculations occur because caustics are placed in different locations. Also shown in Figure 2.11 is the estimate of R_f from Eq. 2.90, and it is seen that the constant sound speed Fresnel zone roughly describes the envelope of the channeled ocean Fresnel zone.

2.6 Born and Rytov Approximations

Ray theory or geometrical acoustics provides a useful intuitive view of acoustic propagation through the stochastic internal-wave field, but as it has been seen there are important diffractive corrections that must be made especially near caustics. Later in this book, it will be found that the diffractive corrections will also be necessary due to the small-scale part of the internal-wave spectrum. In this section a perturbation method is introduced to account for the effects of diffraction in the regime in which acoustic fluctuations are small, though statistics of the wave field will not be treated (see later treatment in Chapter 6). Here it will be seen directly how the Fresnel zone enters into estimates of the acoustic pressure field.

Several inter-related methods have been applied to this situation, and are known as the Born approximation (Born and Wolf, 1999), the method of small perturbations (Tatarskii, 1971; Chernov, 1975), and the Rytov method (Rytov, 1937). Much of this work originated in the atmospheric optics context, in which line of sight propagation was considered through homogeneous isotropic turbulence. The seminal contributions in ocean acoustics were provided by Munk and Zachariasen (1976) and Flatté et al. (1979) in which the important effects of waveguide propagation and internal-wave inhomogeneity and anisotropy were correctly accounted for in the theory. In this book the Born approximation is used because it provides the simplest conceptual picture, and it provides similar accuracy to other weak fluctuation methods.

The analysis begins with the Helmholtz equation for the sound pressure field p at frequency ω, which can be written

$$\nabla^2 p + \bar{k}^2(z) \, p + 2q_0^2 \, \mu(\mathbf{r}) \, p \; = \; 0, \tag{2.92}$$

where $\bar{k} = \omega/\bar{c}(z)$, $q_0 = \omega/c_0$, $\mu = \delta c(\mathbf{r})/c_0$, and it is assumed $|\mu| \ll 1$. Using the well known Born approximation, consider the pressure field to be the sum of a series of ever-decreasing terms, where at each order n the order of smallness of the term is μ^n. This series may be convergent or divergent, depending on the magnitude of μ; however, an important physical interpretation of the series terms is that of multiple scattering. The zero-order term represents the unperturbed wave, the first-order term represents single scattering, the second double scattering, and so on. Writing the series as $p = p_0 + p_1 + p_2 + \cdots + p_n$, substituting into Eq. 2.92 and equating terms of equal order the result is

$$\nabla^2 p_0 + \bar{k}^2(z) \, p_0 = 0, \tag{2.93}$$

$$\nabla^2 p_1 + \bar{k}^2(z) \, p_1 = -2q_0^2 \, \mu \, p_0, \tag{2.94}$$

etc....

In the present treatment, in which the concern is weak fluctuations, all terms of order higher than 1 will be ignored. In this single scattering approximation, it is important to note that the equation has been transformed from one in which there is a multiplicative randomness which is a difficult problem to solve (e.g., Eq. 2.92) to one in which there is an additive randomness that is quite easily solved (e.g., Eq. 2.94). That is to say, in Eq. 2.94 the left-hand side of the equation (i.e., the deterministic component) is driven on the right-hand side of the equation by the random function $\mu\, p_0$ alone which does not involve p_1. In this case an explicit solution is available for p_1, which is simply the convolution of the Green's function with the right-hand side of the equation,

$$p_1(\mathbf{r}) = - \int_V d^3 r' G(\mathbf{r}, \mathbf{r}') \left[2 q_0^2\, \mu(\mathbf{r}')\, p_0(\mathbf{r}') \right]. \tag{2.95}$$

This equation tells us that the scattered field observed at location $\mathbf{r}$ has a contribution from scattered waves at all locations in space $\mathbf{r}'$. That is to say, the unperturbed wave propagates to location $\mathbf{r}'$, where it is scattered with strength $2 q_0^2\, \mu(\mathbf{r}')\, p_0(\mathbf{r}')$. This wave then propagates via the Green's function from $\mathbf{r}'$ to the observation location $\mathbf{r}$, where the total pressure field p_1 is the sum over all the scattered waves. In ocean acoustics where there is multipath propagation one must conceptually consider this process to occur locally near a particular ray path. Thus it should be understood that all locations in space will not contribute equally to the scattered wave; in particular it will be found that there is a distinct region around the unperturbed ray path from which the dominant contributions come. This is the Fresnel zone.

The significance of the Fresnel zone can be better appreciated by examining propagation without a waveguide. In this case the free space Green's function is $G(\mathbf{r}, \mathbf{r}') = \exp(i q_0 |\mathbf{r} - \mathbf{r}'|)/(4\pi |\mathbf{r} - \mathbf{r}'|)$, and an initial point source $p_0(\mathbf{r}) = \exp(i q_0 |\mathbf{r}|)/(4\pi |\mathbf{r}|)$ is used. For the observation point at $\mathbf{r} = (R, 0, 0)$, that is a ray along the x-axis, the Born pressure field under the small-angle scattering approximation[11] is (Colosi, 1999)

$$p(\mathbf{r}) = p_0(\mathbf{r}) \left[1 - q_0 \int_0^R dx' \int_{-\infty}^{\infty} \int_{-\infty}^{\infty} \frac{dy'\, dz'}{R_f^2(x')} \mu(\mathbf{r}') \exp\left[i\pi \frac{(y'^2 + z'^2)}{R_f^2(x')} \right] \right]. \tag{2.96}$$

Here it is seen that for $y' > R_f$ and $z' > R_f$ the exponential term will be rapidly oscillating and will not contribute to the pressure. On the other hand, for y' and z' smaller than the Fresnel radius there will be reinforcing contributions to the pressure. The Fresnel zone thus describes a region around the ray that has

---

[11] In the small-angle scattering approximation the relative distance is Taylor expanded so that $|\mathbf{r} - \mathbf{r}'| \simeq (R - x')[1 + (y'^2 + z'^2)/(2(R - x')^2)]$.

significant influence on the observed pressure. This topic and the results for various moments of the acoustical field will be discussed in Chapters 4 and 6.

2.6.1 Relation to Amplitude and Phase

Here it is shown how Eq. 2.95 describes the relative fluctuations of phase and amplitude. Writing the ratio of total pressure to the unperturbed pressure gives

$$\frac{p_T}{p_0} = 1 + \frac{p_1}{p_0} = \frac{A_0 + A_1}{A_0} e^{i(\phi_T - \phi_0)}, \tag{2.97}$$

where $p_0 = A_0 e^{i\phi_0}$, $p_1 = A_1 e^{i\phi_1}$, and $p_T = p_0 + p_1 \simeq (A_0 + A_1)e^{i\phi_T}$ because ϕ_1 is nearly equal to ϕ_0. Taking the natural log of the ratio of pressures then gives

$$\ln\left(1 + \frac{p_1}{p_0}\right) = \ln\left(1 + \frac{A_1}{A_0}\right) + i(\phi_T - \phi_0). \tag{2.98}$$

Because p_1 is a small correction to p_0, the log can be expanded in a power series and to first order the result is

$$\frac{p_1}{p_0} \simeq \frac{A_1}{A_0} + i(\phi_T - \phi_0). \tag{2.99}$$

Thus it is seen that the real part of Eq. 2.95 gives the relative amplitude fluctuation, while the imaginary part gives the relative phase fluctuation. Often the concern is the log-amplitude fluctuation, namely $\chi = \ln(A/A_0)$, but from above it is seen that $\chi \simeq A_1/A_0$.

2.6.2 Relationship between Born and Rytov Solutions

The Born approximation is often contrasted to the method of Rytov in which the pressure is expanded in an exponential series. For the Rytov method one writes $p(\mathbf{r}) = \exp(\varphi_0 + \varphi_1 + \varphi_2....) = \exp(\varphi)$, and substituting this into the Helmholtz equation gives

$$\nabla^2\varphi + (\nabla\varphi)^2 + \bar{k}^2(z) - q_0^2\mu = 0. \tag{2.100}$$

Equation 2.100 is nonlinear and is known as the Riccati equation. To zeroth and first order the Riccati equation gives

$$\nabla^2\varphi_0 + (\nabla\varphi_0)^2 + \bar{k}^2(z) = 0, \tag{2.101}$$

$$\nabla^2\varphi_1 + 2\nabla\varphi_0 \cdot \nabla\varphi_1 = 2q_0^2\mu. \tag{2.102}$$

The zeroth order equation has a solution $p_0 = \exp(\varphi_0)$, which is the solution to the unperturbed problem. To solve the first-order equation one writes $\varphi_1 = \exp(\varphi_0)u$

and substituting this into the equation gives

$$\nabla^2 u + q_0^2 u = 2q_0^2 \mu \exp(\varphi_0). \tag{2.103}$$

The solution to the first-order equation can then be written

$$\varphi_1 = \exp(\varphi_0)u = -\int_V d^3\mathbf{r}' G(\mathbf{r},\mathbf{r}')\left[2q_0^2\,\mu(\mathbf{r}')\,\frac{p_0(\mathbf{r}')}{p_0(\mathbf{r})}\right]. \tag{2.104}$$

It can be easily shown that the Born and Rytov results are related by

$$p^{Rytov}(\mathbf{r}) = p_0(\mathbf{r})e^{p_1(\mathbf{r})/p_0(\mathbf{r})} \simeq p_0\left(1 + \frac{p_1}{p_0}\right) = p_0 + p_1 = p^{Born}(\mathbf{r}), \tag{2.105}$$

where the approximation in this equation comes from a Taylor series expansion of the exponential, which is clearly valid if $p_1/p_0 \ll 1$. Therefore the Born and Rytov results are the same when $p_1/p_0 \ll 1$, which is the requirement for both perturbation expansions to be valid. While some texts claim that the Rytov result is superior to the Born result, it is clear that these claims are unfounded.

2.7 Path Integrals

The path integral has been shown to be a solution of the parabolic wave equation, and though technically sophisticated in execution (see Feynman and Hibbs, 1965), the conceptual picture of the path integral is actually quite simple. Given a source and a receiver, the acoustic field observed at that receiver will be a sum over *all* the possible acoustic paths that start at the source and end at the receiver. This approach could be considered a generalization of Huygens principle. At first glance the task of summing up this infinite set of paths seems unduly onerous, but what has been discovered is that in some cases there is a significantly smaller subset of these paths that are actually important. In fact it is those paths that lie within a few Fresnel zones of the unperturbed ray path.

   With these ideas in mind, the introduction to the path integral method is started by considering the variational approach to ray propagation, that is, a method of ray analysis often associated with "Fermat's principle" or the "principle of least time."

2.7.1 *Variational Approach to Ray Propagation*

It is well known that many of the laws of physics can be expressed in terms of variational principles, an example of which is Newton's laws of motion, which can be derived from the principle of least action (Landau and Lifshitz, 1976). In wave propagation the principle of least time, or more generally the principle that the travel time is an extremum (Gutzwiller, 1990), can be used to derive the ray trajectory. For a general 3-D path connecting the source and receiver, the travel time along that path is written

$$T(path) = \int_{path} L(z, x)\, ds, \tag{2.106}$$

where $ds = (dx^2 + dy^2 + dz^2)^{1/2} = (dx_i dx_i)^2$ is an infinitesimal path length increment, and the convention of summing over repeated indicies is used. The Lagrangian density function is given by $L = 1/c(x, z)$, thus giving the familiar ray travel time equation. To examine how the travel time changes when the path is changed, the variation of Eq. 2.106 is calculated, yielding

$$\delta T = \int_{path} (ds\, \delta L + L\, \delta ds) = \int_{path} \left(\delta x_i \frac{\partial L}{\partial x_i} + L\dot{x}_i \delta d\dot{x}_i \right) ds, \tag{2.107}$$

where an overdot represents differentiation with respect to s and the result that $\delta ds = dx_i \delta dx_i / ds = \dot{x}_i \delta d\dot{x}_i\, ds$ has been used. There is a useful physical interpretation of Eq. 2.107. The first term in the integral represents the change in travel time due to a change in the speed of sound over the path, and the second term represents the change in travel time due to the changing path length. The second term can be simplified using integration by parts, noting that the boundary terms are zero since the path perturbations must vanish at the source and receiver. The result is

$$\delta T = \int_{path} \delta x_i \left[\frac{\partial L}{\partial x_i} - \frac{d}{ds}(L\dot{x}_i) \right] ds. \tag{2.108}$$

Now if it is asserted that the variation in travel time around the "real" path taken by the sound is zero (i.e., Fermat's principle), then this condition requires the quantity in the square brackets of Eq. 2.108 to be zero. The Euler-Lagrange equations or ray equations (in vector form) result giving

$$\ddot{\mathbf{x}} = \frac{\nabla L}{L} - \dot{\mathbf{x}}\left(\dot{\mathbf{x}} \cdot \frac{\nabla L}{L} \right). \tag{2.109}$$

The vector $\dot{\mathbf{x}}$ is pointing in the direction of the ray, and the second term on the right-hand side of the equations is just the projection of the gradient of L in the direction of the ray. Thus Eq. 2.109 says that the change in direction of the ray $\ddot{\mathbf{x}}$ is affected only by the sound-speed gradients perpendicular to the direction of the ray; this is precisely the result that will be obtained from weak fluctuation theory (see Chapter 6). This equation can be interpreted more easily if 2-D propagation in the (x, z) plane is examined. In this case $ds = dx / \cos\theta$ and $\tan\theta = dz/dx$ so that the ray equations become

$$\frac{d\theta}{dx} = \tan\theta \frac{1}{c}\frac{\partial c}{\partial x} - \frac{1}{c}\frac{\partial c}{\partial z}. \tag{2.110}$$

In the case of small-angle propagation this equation reduces to

$$\frac{d\theta}{dx} + \frac{1}{c_0}\frac{\partial c}{\partial z} = 0, \tag{2.111}$$

where the factor of c in the denominator has been replaced by c_0. Using $\theta \simeq dz/dx$ the PE ray equation

$$\frac{d^2z}{dx^2} + \frac{1}{c_0}\frac{\partial c}{\partial z} = 0, \tag{2.112}$$

is recovered (see Eqs. 2.47 and 2.48).

2.7.2 Path Integrals: A Qualitative Discussion

It is now appreciated how ray theory comes out of a variational principle related to the sound energy path, but it is also known from previous discussions of the Fresnel zone that one cannot consider the ray to be infinitely thin; there is some region of paths around the ray that are important due to diffractive effects. The path integral formalism allows us to use this physical picture in a meaningful way such that the acoustical pressure at frequency ω is written

$$p(\mathbf{x}) = N \int d(path) \exp\left(i\Theta(path)\right), \tag{2.113}$$

where the integration is over all paths connecting the source and receiver ($\mathbf{x} = (R,y,z)$), Θ is the accumulated phase over the path, and the factor N is a normalization. If the case is considered in which the sound speed is only a function of depth then the primary contributions to p will be from paths within a Fresnel zone of the equilibrium or unperturbed ray. Recall that paths within a Fresnel zone arrive at the receiver with a phase less than a half cycle of the equilibrium ray phase and are thus not prone to interference or cancellation.

So, to get a better feeling for the path integral method, the form of the path integral that is useful for ocean acoustics purposes is presented, and the conceptual picture of the path integral is further developed. Application of path integrals to the specific cases with internal-wave-induced sound-speed perturbations will be deferred until Chapter 7.

2.7.3 Formulation of the Path Integral

The path integral method in ocean acoustics has been exclusively applied in cases for which the parabolic approximation can be made. In this approximation the Lagrangian density is given by (Flatté et al., 1979) (also see Eq. 2.84)

$$L\left(\frac{\partial y}{\partial x}, \frac{\partial z}{\partial x}, x, y, z\right) = \frac{1}{c_0}\left[\frac{1}{2}\left(\frac{\partial y}{\partial x}\right)^2 + \frac{1}{2}\left(\frac{\partial z}{\partial x}\right)^2 - U(z) - \mu(x,y,z)\right], \tag{2.114}$$

where it has been written $V \simeq U(z) + \mu(\mathbf{r})$ with $\bar{c}(z) = c_0(1 + U(z))$. This form is particularly pleasing because the path integral can be written as the product of two parts, one dependent on the unperturbed phase and the other due to the fluctuations. In order to simplify the notation the path integral Eq. 2.113 is written as

$$p(\mathbf{r}) = N \int d(path) \exp\left(i\Theta_0(path) - iq_0 \int_0^R \mu(x, y(x), z(x)) dx\right), \qquad (2.115)$$

where the unperturbed phase Θ_0 associated with the path is

$$\Theta_0(path) = q_0 \int_0^R \left[\frac{1}{2}\left(\frac{\partial y}{\partial x}\right)^2 + \frac{1}{2}\left(\frac{\partial z}{\partial x}\right)^2 - U(z)\right] dx. \qquad (2.116)$$

Here the factor $N^{-1} = \int d(path) e^{i\Theta_0(path)}$ is a normalization so the $p = 1$ for $\mu = 0$.

2.7.4 *Solution of the Parabolic Equation as a Path Integral*

It can be shown that the solution of the parabolic wave equation can be written as a path integral (Flatté et al., 1979), that is

$$A(\mathbf{x}) = \int d(paths) \, e^{i\Theta_0(\text{path})}. \qquad (2.117)$$

To understand this result it is useful to look at the parabolic equation, 2.11, in discretized form,

$$\frac{A(x+d,z) - A(x,z)}{d} = \left[\frac{i}{2q_0}\frac{\partial^2}{\partial z^2} - iq_0 U\right] A(x,z),$$

$$A(x+d,z) = \left(1 + i\left[\frac{1}{2q_0}\frac{\partial^2}{\partial z^2} - q_0 U\right]d\right) A(x,z),$$

$$\simeq \exp\left(\frac{id}{2q_0}\frac{\partial^2}{\partial z^2}\right)\hat{A}(x,z), \qquad (2.118)$$

where d is the range step and the waveguide, phase shifted wave function is $\hat{A}(x,z) = e^{-iq_0 U d} A(x,z)$. Next it is recognized that Eq. 2.118 can be solved by transforming to the Fourier domain thus,

$$A(x+d, k_z) = \exp\left(-i\frac{k_z^2 d}{2q_0}\right)\hat{A}(x, k_z). \qquad (2.119)$$

To complete the solution, $A(x+d, z)$ is then obtained by inverse Fourier transform. Here the exponential factor in Eq. 2.119 is termed the free propagator, and the exponential term in $\hat{A}$ is called the phase screen. Physically the transformation of the wave function from x to $x + d$ can be understood as being accomplished in two steps:

1. The first step is the phase screen in which the accumulated phase advance across the step d by the sound-speed structure is accounted for.
2. The second step is the propagation of the phase shifted wave function across the horizontal distance d using the free space propagator.

The relation to the path integral can now be clearly seen by writing out explicitly the FFT relations,

$$A(x+d,z) = \frac{1}{\sqrt{2\pi}} \int_{-\infty}^{\infty} dk_z e^{ik_z z} \, e^{-ik_z^2 d/(2q_0)} \int dz' \, e^{-ik_z z'} \hat{A}(x,z'),$$

$$= \sqrt{\frac{q_0}{id}} \int dz' \hat{A}(x,z') \exp(iq_0(z-z')^2/(2d)),$$

$$= \sqrt{\frac{q_0}{id}} \int dz' \hat{A}(x,z') \exp(i\Theta_0(x,z',x+d,z)), \qquad (2.120)$$

$$\Theta_0(x,z',x+d,z) = q_0(d^2 + (z-z')^2)^{1/2} - q_0 d \simeq q_0 \Delta z^2/(2d), \qquad (2.121)$$

where the k_z integral was done by completing the square in the exponent. Note that for the unperturbed phase one must account for the $e^{iq_0 d}$ term used to derive the PE. Importantly, Eq. 2.120 shows that the value of the wave function at $(x+d,z)$ is the integral of all the linear paths (free space) that connect the phase shifted wave function $\hat{A}(x,z')$ to the final point $A(x+d,z)$. For finite d the operations explained here correspond to solution by numerical methods, and for infinitesimal d, one obtains the solution of the parabolic wave equation. In either case the operations described here demonstrate the method of *Path Integrals*.

As in the case for rays and the Born method in this chapter, here is where the matter will be left for now. In Chapter 7 the issue of internal waves and path integrals will be taken up in detail.

2.8 Normal Modes

Using the method of rays one is able to understand geometrically the structure of wave fronts in the ocean sound channel. Given an impulsive source, one expects a series of delta-like arrivals of ever-increasing amplitude (a crescendo) until a final cut off. However, the arrival pattern is necessarily more complex than this due to both interference effects between arrivals and diffraction. To address the issue of interference and diffraction, the Born approximation and the method of path integrals have been described. Continuing in this spirit, the method of acoustic normal modes is introduced.

The normal mode approach to ocean acoustic propagation has proven to be a powerful tool for describing both deep-water and shallow-water problems, and its application to transport theory for sound propagation through the stochastic internal-wave field has provided ocean acousticians with perhaps the most accurate model to date. While ray and path integral methods have an elegant physical interpretation, mode methods are more algebraic and abstract due to the fact that only a sum over modes provides real physical information. In this section the deterministic theory of normal modes applied to ocean acoustics will be developed, and the interested reader who wants to delve deeper into the subject is referred to the ample reference material on this topic (see Frisk, 1994; Jensen et al., 1994; Munk et al., 1995).

2.8.1 Coupled Mode Equations

The coupled mode equations are easily derived from the wave equation when 2-D propagation in the depth-range plane is considered. The treatment of 3-D effects is somewhat more cumbersome, and a brief account is given at the end of this section (Penland, 1985; Voronovich and Ostashev, 2006). The starting place is close to Eq. 2.92, but because there is interest in applying the normal mode approach in both deep and shallow-water environments the background density profile denoted as $\rho_0(z)$ is retained. In cylindrical coordinates the Helmholtz equation becomes

$$\frac{1}{r}\frac{\partial}{\partial r}r\frac{\partial p}{\partial r} + \rho_0(z)\frac{\partial}{\partial z}\left(\frac{1}{\rho_0(z)}\frac{\partial p}{\partial z}\right) + \bar{k}^2(z)p = 2q_0^2\mu p. \qquad (2.122)$$

For the fluctuation problem it is expedient to expand the pressure in terms of the modes of the unperturbed problem so that

$$p(r,z;\omega) = \sum_n \frac{a_n(r)\,\phi_n(z)}{\sqrt{k_n r}}, \qquad (2.123)$$

where

$$\rho_0(z)\frac{\partial}{\partial z}\left(\frac{1}{\rho_0(z)}\frac{\partial\phi_n}{\partial z}\right) + \left(\bar{k}^2(z) - k_n^2\right)\phi_n = 0, \qquad (2.124)$$

is the unperturbed mode equation for the eigenmodes ϕ_n and eigenwavenumbers k_n. The boundary conditions for the mode equation are $\phi_n(0) = 0$, and continuity of pressure and normal velocity at the water/seafloor interface (depth D) or other layers in the bottom. The modes obey the othonormality relation

$$\int_0^\infty \frac{\phi_n(z)\phi_m(z)}{\rho_0(z)}dz = \delta_{mn}. \qquad (2.125)$$

It is important to note that in the treatment of mode scattering by internal waves the primarily concern will be with the behavior of *trapped* modes, that is the modes with discrete mode numbers that are confined by the ocean waveguide and sea surface. In principle the *continuum* modes that are also a solution of Eq. 2.124 could be used, but because these modes are not confined by the ocean waveguide they tend to attenuate quickly and become irrelevant.

Note here that due to the use of the unperturbed mode function basis all the variability due to sound-speed fluctuations is captured in the mode amplitude term $a_n(r)$. This formulation will prove quite useful when addressing acoustic field statistics.

In the mode approach the second-order partial differential equation of Eq. 2.122 can be reduced to a set of coupled ordinary differential equations. This objective is accomplished by substituting Eq. 2.123 into Eq. 2.122, multiplying the resulting equation by ϕ_m/ρ_0 and integrating over depth. The final result (switching the m and n indices) is

$$\frac{d^2 a_n}{dr^2} + k_n^2 a_n = 2k_n \sum_m a_m \rho_{mn}(r), \tag{2.126}$$

where

$$\rho_{mn}(r) = \frac{q_0^2}{\sqrt{k_n k_m}} \int_0^\infty \frac{\phi_n(z)\phi_m(z)}{\rho_0(z)} \mu(r,z)\, dz, \tag{2.127}$$

is the symmetric coupling matrix. When $\mu = 0$ then $\rho_{mn} = 0$ and solution of Eq. 2.126 gives the expected plane wave result $a_n(r) = a_n(0)\exp(\pm ik_n r)$. A few general statements can be made about the coupling matrix. First, this matrix tells us how strongly mode n is coupled to mode m. Because $|\mu| << 1$ in the ocean there is small-angle scattering. Since coupling from mode n to m physically represents a change in the vertical angle, then small-angle scattering implies that the coupling matrix will be strongly peaked along the diagonal, that is to say, near-neighbor coupling will be dominant. In the next section on the adiabatic approximation, it will be shown that the diagonal elements of the coupling matrix physically represent local perturbations to the modal wavenumber. Lastly, it is seen that Eq. 2.127 scales approximately as acoustic frequency; thus coupling is expected to be strongest for high frequency.

It is seen that Eq. 2.126 is a second-order equation and thus admits both forward and backscattered waves. Since internal-wave-induced sound-speed perturbations cause small-angle forward scattering it is useful to convert to a one-way equation. It is also useful to remove the rapid oscillations in the mode amplitude function. To accomplish both these objectives one writes $\hat{a}_n = a_n e^{-ik_n r}$ and this is substituted into Eq. 2.126. To an excellent approximation the phase variation of the mode amplitude is much more rapid than the amplitude variation; thus the second-order range derivative term can be ignored yielding (Dozier and Tappert, 1978a)

$$\frac{d\hat{a}_n}{dr} = -i \sum_m \hat{a}_m \, e^{ik_{mn}r} \rho_{mn}(r), \tag{2.128}$$

where $k_{mn} = k_m - k_n$. The re-modulated equation is also of use so it is written down here for completeness.

$$\frac{da_n}{dr} - ik_n a_n = -i \sum_m a_m \rho_{mn}(r). \tag{2.129}$$

The effects of attenuation on coupled mode behavior have been addressed by a few investigators (see Dozier, 1983; Creamer, 1996; Colosi, 2008). In this case the modal wavenumber has a small complex component that is defined as $l_n = k_n + i\alpha_n$, where α_n is generally computed using perturbation theory,

$$\alpha_n = \frac{\omega}{k_n} \int_0^\infty \frac{\alpha(z)}{\bar{c}(z)} \frac{\phi_n^2(z)}{\rho_0(z)} dz. \tag{2.130}$$

In Eq. 2.130 the quantity $\alpha(z)$ is the complex part of the total wavenumber $k(z) = \omega/\bar{c}(z) + i\alpha(z)$. Modal attenuation is critical for shallow-water problems due to the acoustic interaction with the lossy seafloor. In deep-water cases attenuation can be a factor if the acoustic frequency is sufficiently high, or if some bottom interaction exists (e.g., bottom limited propagation). In shallow-water situations, modal attenuation is generally higher for low frequencies and higher mode numbers since in both these cases the modes tend to penetrate further into the lossy seabed. For weak attenuation the coupled mode equations become to first order (Dozier, 1983; Creamer, 1996)

$$\frac{d\hat{a}_n}{dr} = -i \sum_{m=1}^N \rho_{mn}(r) e^{il_{mn}r} \hat{a}_m(r). \tag{2.131}$$

Here the coupling matrix ρ_{mn} is unchanged to first order, and the difference wavenumber is now, $l_{mn} = k_{mn} + i\alpha_{mn}$, with $\alpha_{mn} = \alpha_m - \alpha_n$. In the absence of attenuation Eq. 2.128 conserves energy (Dozier and Tappert, 1978a), but with the complex wavenumber it is found that

$$\frac{d}{dr} \sum_n |\hat{a}_n|^2 = \sum_n \hat{a}_n \frac{d\hat{a}_n^*}{dr} + \hat{a}_n^* \frac{d\hat{a}_n}{dr} \simeq -\sum_n 2\alpha_n |\hat{a}_n|^2. \tag{2.132}$$

The analysis just presented can also be carried out for 3-D propagation primarily in the r direction (i.e., small horizontal angle scattering) in which the modes can couple in both the range and cross range directions. The result for $\hat{a}_n(r, y)$, which is simply quoted here gives

$$\frac{\partial \hat{a}_n}{\partial r} + \frac{1}{2ik_n}\frac{\partial^2 \hat{a}_n}{\partial y^2} = -i \sum_m \rho_{mn}(r,y)\, e^{il_{mn}r}\, \hat{a}_m, \qquad (2.133)$$

where

$$\rho_{mn}(r,y) = \frac{q_0^2}{(k_m k_n)^{1/2}} \int_0^D \frac{\phi_n(z)\phi_m(z)}{\rho_0(z)}\mu(r,y,z)\,dz. \qquad (2.134)$$

The solution of this equation poses a much more difficult task. Because the horizontal gradients of internal waves are relatively weak, the cross range coupling is generally assumed to be significant only at extremely long range and high frequency (Penland, 1985); however this issue remains an open scientific question (Voronovich and Ostashev, 2006).

The mode equations represent an initial value problem, dictating the evolution of the mode amplitude from the initial conditions. The mode amplitude initial condition can be derived from an initial acoustic pressure distribution $p(r = 0, z)$, such that

$$a_n(0) = \int_0^\infty p(r=0,z)\frac{\phi_n(z)}{\rho_0(z)}dz. \qquad (2.135)$$

For a point source at depth z_s with $p(r = 0, z) = \delta(z - z_s)$, the initial mode amplitude is $a_n(0) = \phi_n(z_s)/\rho_0(z_s)$.

2.8.2 Adiabatic Theory

A surprisingly useful solution to the coupled mode equations can be obtained using the adiabatic approximation. In the adiabatic approximation the coupling matrix is assumed to be so strongly peaked along the diagonal that mode coupling is insignificant and the equations are essentially uncoupled. Several conditions have been derived to denote the regime of validity of the adiabatic approximation (Milder, 1969), none of which have proved terribly useful in the study of sound transmission through ocean internal waves. Suffice it to say the approximation is generally better for lower acoustic frequencies, but some caution is warranted when using the approximation, and testing is recommended. The adiabatic mode amplitude is given by

$$a_n(r) = a_n(0)e^{il_n r}\exp\left(-i\int_0^r \rho_{nn}(r')dr'\right), \qquad (2.136)$$

where $a_n(0)$ is the initial mode amplitude at the source. Here the diagonal of the coupling matrix is seen to physically represent a local wavenumber perturbation, that is, the total local wavenumber is $k_n - \rho_{nn}(r)$. This effect can also be interpreted as a local speeding up/slowing down of the mode. These adiabatic modal phase fluctuations can be contrasted to phase fluctuations induced by mode coupling. In many cases, as will be discussed later, the adiabatic phase fluctuations constitute an equal or dominate contribution to modal phase randomization, especially in shallow-water situations.

2.8.3 Vertical Modes, Horizontal Rays

The subject of vertical modes and horizontal rays was introduced into the ocean acoustics literature by Weinburg and Burridge (1974). The physical picture is that adiabatic mode theory applies in the vertical, but the modes can refract horizontally. This means that vertical changes in angle are assumed small but horizontal angle changes are not. Using Eq. 2.133 with the adiabatic approximation yields the horizontal wave equation for mode n (Wolfson and Tappert, 2000)

$$\frac{i}{k_n}\frac{\partial \hat{a}_n}{\partial r} = -\frac{1}{2k_n^2}\frac{\partial^2 \hat{a}_n}{\partial y^2} + \frac{\rho_{nn}(r,y)}{k_n}\hat{a}_n. \qquad (2.137)$$

Note that this equation has the same form as the parabolic equation, and the mode rays are seen to propagate through the random wavenumber perturbation field $\rho_{nn}(r,y)$. Introducing the mode phase velocity $c_p(n) = \omega/k_n$, a useful ray Hamiltonian function akin to Eq. 2.46, is given by

$$H = \frac{c_p(n)p_y^2}{2} + \frac{\mu_p(n,r,y)}{c_p(n)}, \qquad (2.138)$$

where $\rho_{nn}(r,y)/k_n \simeq \delta c_p(n,r,y)/c_p(n) = \mu_p(n,r,y)$ and the canonical variables are $(y, p_y = \tan\vartheta/c_p(n))$ with ϑ being the horizontal angle. From this Hamiltonian function all information concerning the ray structure of $\hat{a}_n$ can be derived (see Section 2.4). It is important to point out that the physical interpretation of the mode rays is different from that of the "classical" rays and their associated wave fronts. Here the rays do not travel at the local group speed, but roughly at the local phase speed. It is not exactly the phase speed because the small-angle approximation has been made. In order to compute wave front quantities in the ray-mode approach, the fields must be computed for a range of frequencies and subsequently Fourier synthesized to create a wave front.

The vertical mode, horizontal ray approach has been used to study acoustic scattering by ocean mesoscale structure (Wolfson and Tappert, 2000). While the technique has not been applied in the study of stochastic internal-wave scattering, it could be a useful tool to understand 3-D scattering effects. The technique in slightly different form than presented here has found wide application in global-scale acoustics (Heaney et al., 1991; Dushaw, 2008), shallow-water acoustic propagation through nonlinear internal-wave packets (Katznelson and

Pereselkov, 2000; Lynch et al., 2010; Katznelson et al., 2012), and in basin-scale internal tide propagation studies (Rainville and Pinkel, 2006).

2.8.4 Modes in a Range-independent Ocean

The modal acoustic field in the case of a range-independent sound-speed profile provides important information from which to judge the effects of internal waves. The structure of the wave front, shadow zones, and interference patterns will all be altered by the scattering effects of internal waves, and so the subject of a range-independent ocean is briefly examined.

When sound speed is only a function of depth, the Helmholtz equation is separable, allowing us to write (Jensen et al., 1994)

$$p(r,z;\omega) = \frac{i}{4} \sum_n a_n(0) H_0^1(l_n r) \phi_n(z), \qquad (2.139)$$

where H_0^1 is the Hankel function. For most cases the asymptotic form of the Hankel function is adequate, and one can write

$$H_0^1(l_n r) \simeq \sqrt{\frac{2}{\pi k_n r}} e^{i(l_n r - \pi/4)}, \qquad (2.140)$$

where the requirement is $k_n r \gg 1$ and $\alpha_n \ll k_n$. For a point source, to acquire TL one simply normalizes by the pressure at $r = 1$ m, using the spherical wave $p_0(r) = e^{ik_s r}/(4\pi r)$ where $k_s = \omega/c_s$. For single-frequency the result is $TL = -20 \log_{10}(|p(r,z;\omega)|/|p_0(r=1;\omega)|)$.

For transient signals an integration over frequency is required so that

$$p(r,z,t) = \sum_n p_n(r,z,t), \qquad (2.141)$$

$$p_n(r,z,t) = \int_{-\infty}^{\infty} W(\omega) a_n(0;\omega) H_0^1(l_n r;\omega) \phi_n(z;\omega) e^{-i\omega t} d\omega, \qquad (2.142)$$

where $W(\omega)$ is the source frequency response function. The quantity $p_n(r,z,t)$ is referred to as the mode pulse, that is, the time arrival pattern for mode n. The mode pulse has garnered some attention as an observable for ocean acoustic tomography (Munk et al., 1995), and its stability in the presence of stochastic internal waves has also been investigated (Colosi and Flatté, 1996; Udovydchenkov and Brown, 2008; Udovydchenkov et al., 2012). With regard to broadband TL, a point source normalization similar to the single-frequency case is used, except here the peak pressure for the spherical wave $p_0(r,t)$ at a one-meter range is used.

Modal phase and group velocities are important quantities that can be treated simply in the range-independent case. These horizontal velocities are defined as

$$c_p(n) = \frac{\omega}{k_n}, \quad c_g(n) = \frac{d\omega}{dk_n}. \tag{2.143}$$

They are closely related to the ray loop horizontal velocities discussed in Section 2.4.6. A useful expression for the group speed is the Rayleigh formula (Munk et al., 1995)

$$c_g(n)c_p(n) = \left(\int_{-\infty}^{0} \frac{1}{c^2(z)} \frac{\phi_n^2(z)}{\rho_0(z)} \, dz \right)^{-1}. \tag{2.144}$$

Modal Structure and Kinematics

Because the coupled mode equations have been written in terms of the modes in the range-independent ocean, these eigenmodes and eigen-wavenumbers will play a critical role in the theory of acoustic fluctuations and they dictate the sensitivity of the acoustic field to the various scales of sound-speed structure in the ocean. In particular, the shape of the modes $\phi_n(z)$ relative to the vertical distribution of sound-speed fluctuation variance $\langle \mu^2(z) \rangle$ will dictate the strength of coupling (e.g., related to Eq. 2.127). Furthermore, due to a resonance condition it will be found that only ocean structure with wavelengths equal to $k_n - k_m$ will enter the problem. This being the case, it is worthwhile to delve into the acoustic mode structure for deep and shallow-water environments where the mode approach is routinely applied.

Much can be learned about the structure of the modes and the character of the eigenvalues from an examination of the mode equation (Eq. 2.124). To simplify the analysis the mode functions are scaled by the density such that $\hat{\phi}_n = \phi_n / \sqrt{\rho_0}$, so that the mode equation becomes

$$\frac{d^2\hat{\phi}_n}{dz^2} + (\bar{k}_{eff}^2(z) - k_n^2)\hat{\phi}_n = 0, \tag{2.145}$$

where the effective background wavenumber is given by

$$\bar{k}_{eff}^2(z) = \frac{\omega^2}{\bar{c}^2} + \frac{1}{2} \left[\frac{1}{\rho_0} \frac{d^2\rho_0}{dz^2} - \frac{3}{2\rho_0^2} \left(\frac{d\rho_0}{dz} \right)^2 \right]. \tag{2.146}$$

From Eq. 2.145 it is recognized that the modes have a local vertical wavenumber given by

$$k_z(z) = \sqrt{\bar{k}_{eff}^2(z) - k_n^2}. \tag{2.147}$$

In depth regions where k_z is imaginary the mode functions will be evanescent, that is exponentially decaying, and in depth regions where k_z is real the modes will be oscillating. The transition depth between oscillating and evanescent mode shape ($k_z = 0$) is called the mode turning depth denoted as $z^\pm$ where the $\pm$ refers to upper and lower turning points akin to the ray turning points. In most cases the effect of density on the vertical wavenumber is minimal, and from the turning point condition it is seen that $k_n/\omega = 1/c(z^\pm)$. For the deep-water mid-latitude problem where the modes are trapped along a deep sound-channel axis, there are two turning points. In this case the upper turning point goes away when the mode becomes strongly surface interacting, a situation similar to the polar profile where all the modes are surface interacting. For the shallow-water case in which the sound speed in the bottom is larger than that in the water column, the lower turning point will be close to the water depth (independent of density), and there may or may not be an upper turning point depending on the degree that the mode is surface interacting.

At this point application of WKB methods are quite useful, and those not familiar with the technique should refer to Appendix B. Examining the mode equation (Eq. 2.145), one looks for solutions of the form $\phi_n \propto A(z)e^{i\varphi(z)}$. The WKB envelope and phase functions are

$$A(z) \propto \frac{1}{\sqrt{k_z(z)}}, \tag{2.148}$$

$$\varphi(z) = \int_{z^-}^{z} k_z(z')\,dz', \tag{2.149}$$

where the phase function is defined only over the oscillatory modal depth region between the lower and upper turning points. The formal requirement for the validly of the WKB approximation is presented in Appendix B, but stated in another way the condition implies that over a vertical oscillation of the mode the sound-speed profile is not changing dramatically. As such, the WKB approximation generally improves as mode number increases. Note also that Eq. 2.148 tells us that away from the turning points the mode function envelope scales like $(c^2(z)/(c^2(z^\pm) - c^2(z)))^{1/4}$; thus it is expected that the mode functions will have amplitudes that slightly increase as the depth points move away from the sound-channel axis.

An important quantity is the vertical phase difference between the upper and lower turning points, termed the phase integral ($\varphi_n^\pm$),

$$\varphi_n^\pm = \int_{z_n^-}^{z_n^+} k_z(z)\,dz = \omega \int_{z_n^-}^{z_n^+} \sqrt{c^{-2}(z) - c^{-2}(z_n^\pm)}\,dz. \tag{2.150}$$

In the evanescent regions of the mode, there are in fact exponentially growing solutions that can exist. To maintain a finite solution that is normalizable then imposes the quantization conditions (Brekhovskikh and Lysanov, 1991)

$$\varphi_n^{\pm} = \pi(n - 1/2), \quad \text{two turning points,} \tag{2.151}$$

$$\varphi_n^{\pm} = \pi(n - 1/4), \quad \text{one turning point.} \tag{2.152}$$

For deep-water problems the two turning point case is more typical, except when the mode is strongly surface interacting or the profile is closer to the Arctic profile in which case only one, lower turning point exists. There are many useful applications of the quantization equations (Munk et al., 1995) that need not be discussed here. However, one practical use in the present case is in the estimate of the number of nonsurface interacting modes that fit into the waveguide (obtained by setting $z^+ = 0$ and solving for n). Here it is seen that the number of modes fitting into the waveguide scales linearly with the frequency. The number of modes will be an important numerical consideration when addressing mode-based fluctuation theories.

Some modes for the Munk canonical sound-speed profile for frequencies of 75 and 250 Hz are shown in Figure 2.12. Mode 1 is seen to be confined near the sound-channel axis, and as mode number increases the modes are seen to spread out vertically. Mode turning points are evident where the modes transition from oscillatory to exponentially decaying. The turning depths are a strong function of frequency. Clearly mode 1 will be influenced only by ocean sound-speed structure near the sound-channel axis. As mode number increases, it is found that modes are sensitive to the sound-speed structure near their turning points (more in Chapters 4 and 8).

Because the modal eigenvalues k_n are given by $\omega/c(z^{\pm})$ it is apparent that k_n decreases as n increases. That is to say as n increases the propagation can be considered in some senses to become more vertical. An effective mode grazing angle can be defined as

$$\tan \theta_g(n) = \frac{k_z(z_a)}{k_n} = c(z_n^{\pm}) \sqrt{c^{-2}(z_a) - c^{-2}(z_n^{\pm})}, \tag{2.153}$$

and in this sense one can associate mode propagation with the corresponding ray with the same grazing angle and turning points. This association, often termed the "Ray-Mode Duality," has been known for some time and has been discussed at length by many authors (see Munk and Wunsch, 1983; Brekhovskikh and Lysanov, 1991, and references therein).

Figure 2.13 shows example modes for the canonical shallow-water sound-speed profile (Eq. 2.30) displayed in Figure 2.3. Here the acoustic frequencies are 400 and 1000 Hz. Because this profile has a mid-water sound-speed minimum, it is seen that mode 1 is confined near the depth of the axis. As mode number increases, the modes first become bottom interacting and then surface interacting. As before, the turning depths of the modes are a strong function of frequency.

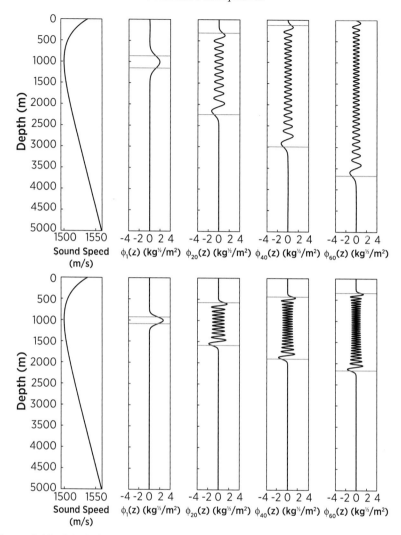

Figure 2.12. Mode functions for mode numbers 1, 20, 40, and 60 for the Munk canonical sound-speed profile shown at left. The top/bottom panels display modes for 75/250 Hz. Horizontal gray lines indicate upper and lower turning depths.

Next there is the matter of the eigenvalue spacing. It will be found in the theory of stochastic mode coupling that the strength of interaction between modes m and n will depend on the spectral strength of internal waves at the horizontal wavenumber $\kappa_{iw} = k_m - k_n = k_{mn}$. Because internal waves cause small-angle scattering, the strongest interactions will be between near neighbor modes. Thus it is useful to define the mode cycle distance

$$L_1(n) = \frac{2\pi}{k_n - k_{n-1}}, \qquad (2.154)$$

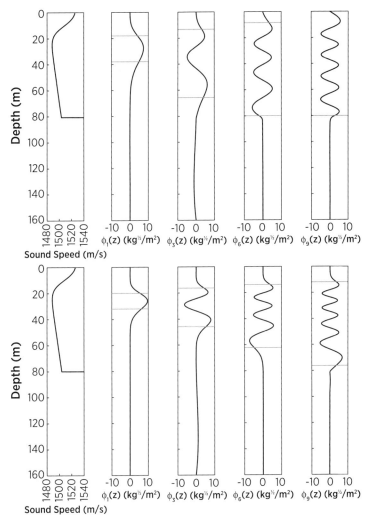

Figure 2.13. Mode functions for mode numbers 1, 3, 6, and 9 for the shallow-water canonical sound-speed profile shown at left. The bottom sound speed here is 1700 m/s and the bottom density is 1500 kg/m^3. The top/bottom panels display modes for 400/1000 Hz. Horizontal gray lines indicate upper and lower turning depths.

which physically represents the distance between nulls in the beat pattern between the neighboring modes n and $n - 1$. This distance also has a correspondence with the ray double loop length for the ray with turning points $z_n^\pm$. Modes are expected to be most strongly influenced by internal waves with this horizontal wavelength. This is a modal manifestation of the ray loop resonance condition (Cornuelle and Howe, 1987; Munk et al., 1995), which will be discussed in Chapters 4 and 5. Because mode interactions do not drop to zero after neighboring modes, another

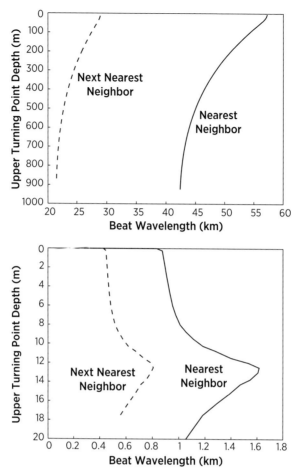

Figure 2.14. Nearest neighbor (solid) and next nearest neighbor (dash) beat wavelength ($L_1(n) = 2\pi/k_{n,n+1}$ and $L_2(n) = 2\pi/k_{n-1,n+1}$) for the Munk (top) and shallow-water canonical (bottom) sound-speed profiles.

useful quantity is the cycle distance of next nearest neighbor modes, that is

$$L_2(n) = \frac{2\pi}{k_{n+1} - k_{n-1}}. \tag{2.155}$$

This distance represents a harmonic of the ray loop distance. Figure 2.14 shows $L_1(n)$ and $L_2(n)$ plotted versus mode upper turning depth for the deep-water Munk profile and the canonical shallow-water profile previously discussed. As frequency changes specific modes move along the curves shown in Figure 2.14. The loop distances are indeed a function of mode number, so one expects the strength of acoustic scattering to also be a function of mode number. For the deep-water case the cycle distances are between 40 and 60 km for nearest neighbor interaction and they are between 20 and 30 km for next nearest neighbor interaction: These scales are ones with relatively high internal-wave spectral energy. On the other hand, for the shallow-water case the nearest neighbor cycle distance is between 500 and 1600 m while the next nearest neighbor values are between 200 and 800 m. These scales have relatively low internal-wave spectral energy, and thus acoustic coupling due to internal waves is expected to be lower in shallow water.

Mid-Latitude Deep-Water Example

Figure 2.15 shows a mode calculation of a time front at 400-km range and the range evolution of single-frequency TL for propagation though the Munk canonical sound-speed profile. The mode-derived time front shows the same double accordion pattern displayed in the ray calculation (see Figure 2.5), but here the total acoustic field TL is shown. High intensity (low loss) is apparent where the time front branches connect. These are caustic regions. At the caustics, interference patterns are evident due to the interference between the two time front branches that join together. Similar interference is seen where the time front branches cross and in the final cut-off region where many branches are close together. If the bandwidth of the calculation were to be increased, the interference would correspondingly diminish. Away from caustics and interference regions, peak TL values along the time front branches in Figure 2.15 match with the ray estimates from Figure 2.8.

The modal interpretation of the time front is aided by a calculation of the mode phase and group speeds (see Figure 2.16). The mode phase speed, which is equal to the sound speed at the mode turning depth, is seen to mirror the upper part of the sound-speed profile. The mode group speed, on the other hand, rises gently from low modes with turning points near the axis and speeds near the axial speed (1500 m/s) to high modes with turning points near the surface and speeds near 1505 (m/s). This difference in group speed means that the early part of the time front arrival pattern is composed of high modes extending over a broad depth range of the water column, and the late arriving part of the pattern is composed of lower modes that are slower and confined in a narrower region around the sound-channel axis. A simple calculation based on these group speeds and the range of propagation predicts a time spread of roughly 0.85 s, which is indeed what is seen in Figure 2.15.

The single-frequency transmission loss fields shown in Figure 2.15 demonstrate refraction generated shadow zones, focusing regions, and a complex interference pattern consistent with the ray paths shown in Figure 2.4. Along the edges of shadow zones are caustic surfaces where the intensity is high. The degree to which acoustic energy penetrates into the shadow zone (i.e., the evanescent region of the caustic) depends on the acoustic frequency with higher frequencies (smaller wavelengths) showing less penetration. The spatial pattern of the interference pattern is also frequency dependent, with higher frequencies showing smaller separation between maxima and minima in the pattern.

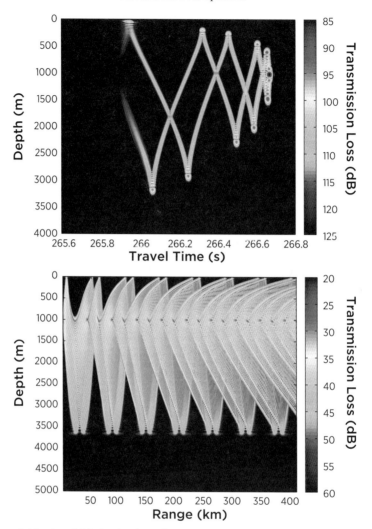

Figure 2.15. A mid-latitude time front at 400-km range (top) and the range evolution of single-frequency TL (bottom). For the calculations a Munk profile was used with a source depth of 1000-m (axial). For the time front 200 frequencies between 200 and 300 Hz were synthesized with a Hanning window to produce the front. A frequency of 250 Hz was used for the single-frequency calculation. For both the time front and single-frequency cases 200 normal modes were used. Full TL is shown for the time front, while TL without cylindrical spreading is displayed for the single-frequency.

When internal-wave-induced sound-speed perturbations are added to the problem, it is found that the early part of the time front, where the branches are well separated in time, remains relatively stable, with small deviations in wave front travel time and larger intensity variations along the front. However, in the late part of the arrival pattern, significant deviations are seen in the overall shape

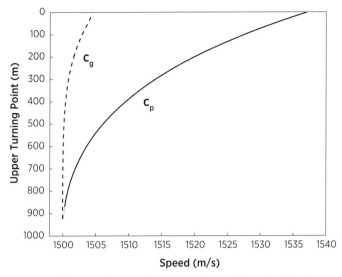

Figure 2.16. Modal phase (c_p) and group (c_g) speed for the Munk sound-speed profile as a function of mode upper turning depth.

of the front. Here separate branches of the front are overlapping resulting, in a complex interference pattern somewhat like the single-frequency example. For the entire time front, internal wave scattering will extend the ensonified region of the front in depth, leading to more sound in the shadow zones. For single frequencies both shadow zones and the interference pattern will be strongly affected by internal-wave-induced sound-speed perturbations. Sound will be scattered into the shadow zones, and the coherent interference pattern seen in Figure 2.15 will be replaced by a strongly fluctuating interference pattern.

Shallow-Water Example

Figure 2.17 shows a mode calculation of a time front at 20-km range, and the range evolution of single-frequency transmission loss. Unlike the deep-water mid-latitude case, the shallow-water time front bears little resemblance to an actual front as might be predicted from ray theory. The pattern instead resembles a series of mode arrivals, with the most energetic ones arriving first. In fact near travel times of 13.5 s, clearly separated arrivals for modes 7 and 8 are seen. This simple mode interpretation is one of the reasons why the theory of normal modes and not ray theory is used so extensively in shallow-water acoustics. As was the case for the deep-water example, the modal interpretation of the shallow-water time front is aided by a calculation of the horizontal phase and group speeds (see Figure 2.18). Here it is seen that the low modes with turning depths between 12 and 20 m have about the same group speeds, but as the mode number increases and the turning depths extend above 12 m, the group speed goes down precipitously.

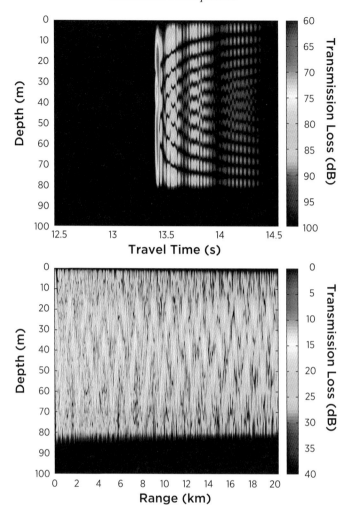

Figure 2.17. A shallow-water time front at 20-km range (top) and the range evolution of single-frequency transmission loss (bottom). For the calculations a canonical shallow-water profile (see Figure 2.3) was used with a source depth of 20 m (axial). For the time front 400 frequencies between 350 and 450 Hz were synthesized with a Hanning window to produce the front. A frequency of 400 Hz was used for the single-frequency calculation. For both the time front and single-frequency cases 18 normal modes were used. Full TL is shown for the time front, while TL without cylindrical spreading is displayed for the single-frequency. The bottom sound speed and density in this calculation are 1700 m/s and 1500 kg/m³, and the bottom attenuation constant is 0.2 dB/λ.

Thus in this case the low modes get down range fastest, and the higher modes trail behind. The low modes are also the most energetic because they are trapped in the gentle sound-speed minimum and therefore do not interact strongly with the lossy seabed. It is not generally the case that the low modes have such small

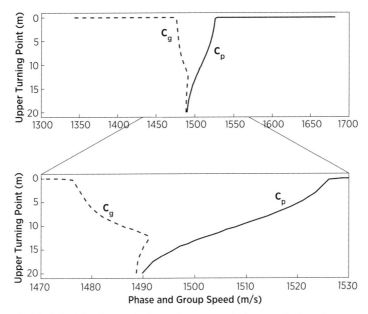

Figure 2.18. Modal phase (c_p) and group (c_g) speed for the canonical shallow-water sound-speed profile (see Figure 2.3) as a function of mode upper turning depth.

bottom interaction. A useful interpretation of this arrival pattern is in terms of the mode-pulses described in Eq. 2.142.

The range evolution of the single-frequency transmission loss field in Figure 2.17 shows significantly less structure than the corresponding deep-water case. In the shallow water, a complicated interference pattern is present over the entire spatial region, and there are no clear shadow zones, caustic surfaces, or focusing regions. As range increases, the interference pattern is seen to simplify significantly. This effect is due to the attenuation of the higher modes from bottom interaction, often termed "mode stripping," and thus at longer range the interference pattern is primarily due to the low modes.

When internal waves are included, shallow-water acoustic fields are seen to change significantly due to mode coupling and cross mode decorrelation or loss of coherence. Intensity patterns can shift dramatically as coupling transfers acoustic energy back and forth between low and high modes, and modal phase randomization occurs. The result of these two effects is that the coherent mode arrival pattern seen in the shallow-water time front is washed out, leaving a complex interference pattern, not unlike the single-frequency example. In the single-frequency case mode coupling and phase randomization lead to a highly fluctuating multimode interference pattern.

Appendix A Green's Functions and the Fresnel Zone

Consider the second-order ray tube Green's function equation that follows the ray $z_r(x)$,

$$\left(\frac{d^2}{dx^2} + V''(x)\right)g(x,x') = \delta(x - x').$$ (A.1)

For fixed end points ($x = 0$ and $x = R$) one can write the two solutions to this second-order equation as

$$\xi_1(x) \rightarrow \quad 0 < x < x', \quad \xi_1(0) = 0, \quad g^<(x,x') = A\xi_1(x),$$ (A.2)

$$\xi_2(x) \rightarrow \quad x' < x < R, \quad \xi_2(R) = 0, \quad g^>(x,x') = B\xi_2(x).$$ (A.3)

The solutions must match at $x = x'$, giving the result

$$A\xi_1(x') = B\xi_2(x').$$ (A.4)

Next one integrates Eq. A.1 around the delta function position x', giving

$$\int_{x'-\epsilon}^{x'+\epsilon} \left(\frac{d^2}{dx^2} + V''(x)\right)g(x,x')\,dx = 1.$$ (A.5)

Here the second term in the integral is of order ϵ thus it can be ignored, leading to the result,

$$\frac{d}{dx}g(x,x')\Big|_{x'-\epsilon}^{x'+\epsilon} = \left[\frac{d}{dx}g^>(x,x') - \frac{d}{dx}g^<(x,x')\right] = 1.$$ (A.6)

So the Green's function slope discontinuity at $x = x'$ is unity. Combining Eq. A.4 and A.6 the expressions for the constants A and B are

$$A = \frac{\xi_2(x')}{W(x')}, \qquad B = \frac{\xi_1(x')}{W(x')},$$ (A.7)

where

$$W(x') = \text{Wronskian} = \xi_1\frac{d\xi_2}{dx} - \xi_2\frac{d\xi_1}{dx}.$$ (A.8)

It can also be shown that $W = $ constant. Both ξ_1 and ξ_2 satisfy the homogeneous equations

$$\left(\frac{d^2}{dx^2} + V''(x)\right)\xi_1 = 0, \qquad \left(\frac{d^2}{dx^2} + V''(x)\right)\xi_2 = 0.$$ (A.9)

Thus multiplying the first equation by ξ_2 and the second equation by ξ_1 and subtracting one gets

$$\xi_2 \frac{d^2}{dx^2} \xi_1 - \xi_1 \frac{d^2}{dx^2} \xi_2 = \frac{dW}{dx} = 0, \tag{A.10}$$

hence demonstrating that the Wronskian is constant. Thus the final expression for the Green's function can be given:

$$g^<(x,x') = \frac{\xi_1(x)\xi_2(x')}{W}, \qquad g^>(x,x') = \frac{\xi_1(x')\xi_2(x)}{W}. \tag{A.11}$$

Note that the Fresnel zone given by Eq. 2.89 can be expressed in terms of the Green's function such that $R_f^2(x) = \lambda g(x,x)$.

The calculation of the Fresnel zone requires ray tube functions ξ_1 and ξ_2 that satisfy the boundary conditions at the source and the receiver. In practice these ray tube functions are constructed from two other solutions ξ_{IV1} and ξ_{IV2} that satisfy a specific initial value problem. For the initial and boundary value problems consider the following:

$$\xi_{IV1}(0) = 0 \;\; \xi_{IV1}{}'(0) = 1, \xi_{IV2}(0) = 1 \;\; \xi_{IV2}{}'(0) = 0, \tag{A.12}$$

$$\xi_1(0) = 0 \;\; \xi_1(R) = 1, \xi_2(0) = 1 \;\; \xi_2(R) = 0, \tag{A.13}$$

and so combining these solutions gives

$$\xi_1 = \frac{\xi_{IV1}(x)}{\xi_{IV1}(R)}, \xi_2 = \xi_{IV2}(x) - \xi_{IV1}(x)\frac{\xi_{IV2}(R)}{\xi_{IV1}(R)}. \tag{A.14}$$

Here it is chosen to evaluate the Wronskian W at $x = 0$ which gives us the result $W(x = 0) = -1/\xi_{IV1}(R)$. Thus the Fresnel zone is given by

$$R_f^2(x) = \lambda \xi_1(x)\xi_2(x)\xi_{IV1}(R). \tag{A.15}$$

It is important to note that there are problems if the receiver is near a caustic, that is $\xi_{IV1}(R) = 0$.

Examples

Consider the case of a polar sound-speed profile given by $c = c_0(1 + \gamma_a z)$, then $V \simeq \gamma_a z$ and $V'' \simeq 0$. This case also applies to the constant sound-speed profile. The ray tube equation therefore is the form $d^2\xi/dx^2 = 0$, which has solutions $\xi = Ax + B$. The ray tube functions are then given by

$$\xi_{IV1}(x) = x, \qquad \xi_1(x) = \frac{x}{R}, \qquad \xi_2(x) = \frac{(R - x)}{R}, \tag{A.16}$$

thus giving the Fresnel zone

$$R_f^2(x) = \lambda \xi_1(x) \xi_2(x) \xi_{IV1}(R) = \lambda \frac{x(R-x)}{R}. \tag{A.17}$$

This is the point source Fresnel zone for a constant linear profile, and for a constant background sound speed.

Next consider a quadratic sound-speed profile such as one might encounter for a mid-latitude sound-speed profile near the channel axis. Here V'' is roughly a constant, and the ray tube functions are

$$\xi_{IV1}(x) = \frac{\sin K_a x}{K_a}, \qquad \xi_1(x) = \frac{\sin K_a x}{\sin K_a R}, \qquad \xi_2(x) = \sin K_a (R-x), \tag{A.18}$$

where $K_a = 2\pi/R_a = \sqrt{V''}$ is the horizontal wavenumber of the sinusoidal ray loop with double loop range R_a. The Fresnel zone is thus

$$R_f^2(x) = \lambda \left| \frac{\sin K_a x \, \sin K_a (R-x)}{K_a \sin K_a R} \right|. \tag{A.19}$$

For the Munk canonical profile at the axis $K_a = \sqrt{2\gamma_a/B}$.

Appendix B WKB Modes

WKB analysis of the vertical mode equations for both acoustic and internal waves is essential to the clear exposition of this book. The acoustic and internal wave mode equations (Eqs. 2.124 and 3.26) can be cast in a general form given by

$$\frac{d^2 B}{dz^2} + k_z^2(z) B = 0, \tag{B.1}$$

where k_z is considered to be real. For the deep-water acoustic problem, water-borne modes obeying Eq. 2.145 clearly fit this general form. When acoustic bottom interactions are involved, Eq. 2.124 can be mapped to the general form by the substitution $\hat{\phi}_n = \phi_n/\sqrt{\rho_0}$.

The WKB method consists of examining solutions to the mode equation that are of the form $B(z) \propto A(z) e^{i\varphi(z)}$. Plugging this form into the mode equation yields two equations from the real and imaginary parts, written

$$\frac{d^2 A}{dz^2} - A\left(\frac{d\varphi}{dz}\right)^2 + k_z^2(z) A = 0, \tag{B.2}$$

$$A\frac{d^2\varphi}{dz^2} + 2\frac{dA}{dz}\frac{d\varphi}{dz} = 0. \tag{B.3}$$

The solution to the second equation is given by

$$\frac{d\varphi}{dz} = \frac{C}{A^2},\tag{B.4}$$

where C is an integration constant. Substituting this equation into the first yields

$$\frac{1}{A}\frac{d^2A}{dz^2} - \frac{C^2}{A^4} + k_z^2(z) = 0.\tag{B.5}$$

From a local plane wave approximation clearly $k_z(z) = d\varphi/dz$. Thus it is apparent that the last two terms in Eq. B.5 are proportional to k_z^2, while the first term is roughly independent of k_z. If the mode phase φ is changing more rapidly than the envelope function $A(z)$, then the first term can be ignored. Formally this is the condition $k_z^2(z) \gg (1/A)d^2A/dz^2$. This case applies mores readily to higher order modes. The WKB envelope and phase functions are therefore

$$A(z) \propto \frac{1}{\sqrt{k_z(z)}},\tag{B.6}$$

$$\varphi(z) = \int_{z^-}^{z} k_z(z')\,dz',\tag{B.7}$$

where the mode phase has been written relative to a lower turning point z^- with $k_z(z^-) = 0$. Clearly in depth regions where k_z is complex, the modes will be evanescent and this analysis will not apply. Also near turning points where $k_z(z) = 0$ special treatment is required (i.e., Airy Solutions (see Garrett and Munk, 1972)).

3

Stochastic Ocean Internal Waves

3.1 Introduction

While ocean internal waves are critical to acoustic propagation, the topic is also of intense oceanographic interest due to the fact that internal waves provide an important pathway for the transfer of energy and momentum from large scales (of order tens to hundreds of kilometers) to small scales (of order centimeters to meters). In providing this pathway, internal waves form an important link between the largely two-dimensional motions at large scale and the three-dimensional flows known to exist at smaller scale. Indeed, internal-wave breaking and subsequent mixing has been implicated in shaping the smoothly varying ocean thermocline that is observed throughout the world's oceans (Munk, 1966; Munk and Wunsch, 1998). Internal waves at tidal frequencies (internal tides) are generated by barotropic tidal flows over rough or abrupt topography, and are a form of tidal dissipation that leads to perturbations in the earth–moon system (Munk, 1998). Additionally, internal waves radiate momentum and thus exert a stress on large-scale ocean motions, which is fundamentally different from other processes that diffuse momentum (Müller, 1976). Internal waves are also known to disperse and advect chemical and biological tracers, processes that can have large ecological and public health implications (Garrett, 1979; Young et al., 1982; Duda and Farmer, 1999).

Generally speaking, ocean internal waves fit into four broad categories: inertial waves, internal tides, internal solitary waves, and broadband stochastic or random internal waves. Figures 3.1 and 3.2 show typical spectra of displacement and horizontal velocity obtained in deep ocean and continental shelf regions. Inertial waves are the energetic waves evident in the horizontal current spectra near the local inertial frequency f. These waves are largely wind-driven and therefore propagate downward from the sea surface (D'Asaro, 1984). Curiously, however, due to unknown physical processes, observations show that there is nearly as much upward propagating energy as downward propagating energy. Kinematics dictate

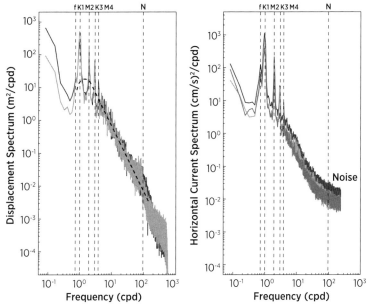

Figure 3.1. Spectra of vertical displacement (left panel) and horizontal velocity (right panel) observed in the Philippine Sea. The displacement spectra are averaged over the depth regions 130–285 m (blue) and 300–455 m (green), and a GM spectral form is shown for reference (dashed). The horizontal current spectra shown are for the east–west component (green), north–south component (red), and the sum of these two (blue). The current spectra are an average over observations taken in the depth range 170–320 m. Important frequencies are shown at the tops of the graphs with associated vertical dashed lines: f, the Coriolis frequency; N, a typical buoyancy frequency; K1, a diurnal tide frequency; M2, a semidiurnal tide frequency; K3, the third harmonic of K1; and M4, a second harmonic of M2.
Source: From Colosi et al. (2013b).

that inertial waves do not have a strong vertical displacement (see displacement spectra in Figures 3.1 and 3.2), and thus they are considered of little importance to ocean acoustics except in the case where reciprocal transmissions are made so as to isolate horizontal current effects.

Internal tides can be regularly observed near continental shelves and rough-abrupt topography and they can propagate large distances across and into the deep ocean basins (Garrett and Kunze, 2007). For latitudes less than 30°, both diurnal and semidiurnal internal tides can exist. Above 30° only the semidiurnal internal tides exist as a propagating wave, and higher than 75° the semidiurnal internal tides are largely absent due to a cut-off in the important M2 tidal species. Figure 3.1 shows spectra from the Philippine Sea (latitude $\simeq$ 20° N) that reveal strong diurnal and semidiurnal internal tide spectral lines, as well as several internal tide harmonics. This region of the world has some of the most energetic internal tides

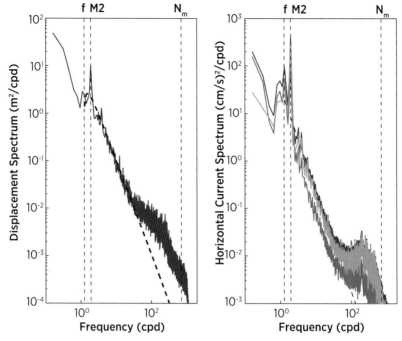

Figure 3.2. Spectra of vertical displacement (left panel) and horizontal velocity (right panel) observed on the New Jersey continental shelf. The displacement spectrum is averaged over the depth region 20–60 m, and a GM spectral form is shown for reference (dashed). The horizontal current spectra shown are for the across-shore component (green), along-shore component (red), and the sum of these two (blue). The current spectra are an average over observations taken in the depth range 20–70 m. Important frequencies are shown at the top of the graphs with associated vertical dashed lines: f, the Coriolis frequency; N, a typical buoyancy frequency; and M2, a semidiurnal tide frequency.
Source: Colosi et al. (2012).

that have been measured. By contrast, the Mid-Atlantic Bight continental shelf spectra (Figure 3.2, latitude $\simeq 39°$N) only shows a semidiurnal spectral peak and much fewer tidal harmonics. Internal tides can have large vertical displacements, which in shallow water can lead to complicated nonlinear behavior and the formation of internal solitary waves. In deep water, however, because of their relation to astronomical tides, internal tides tend to be more deterministic (Dushaw et al., 2011). Due to their relatively large-scale structure and stable time behavior their impacts in terms of randomizing acoustical fields are not generally strong, though there are some suggestions that internal tides may advect smaller-scale ocean sound-speed structure, thereby causing rapid acoustic fluctuations. In any case due to their simple impact on acoustic travel times, acoustic inversion methods are routinely applied to estimate internal-tide parameters (Dushaw et al., 1995; Dushaw and Worcester, 1998).

Internal solitary waves are nonlinear internal waves that are primarily generated on continental shelves and sills from bottom interacting tidal flows (Duda and Farmer, 1999; Apel et al., 2007; see Figure 1.8). There are rare examples of deep-water internal solitary waves (Pinkel, 2000; Chu and Hsieh, 2007), though the South China Sea is an exception to this rule (Ramp et al., 2004; Alford et al., 2010). These waves generally have a vertical mode 1 structure, and have small horizontal scales, of order tens to hundreds of meters, and their propagation speeds can be quite large, commonly exceeding a knot or more. In shallow-water spectra (Figure 3.2) these waves are apparent from the elevated energy levels between 1 cph and the buoyancy cut-off frequency. In shallow water, internal solitary waves, and the nonlinear internal tides from which they often derive, are of critical importance to acoustic propagation. These waves have been studied extensively and are the subject of a recent shallow-water acoustics textbook (Katznelson et al., 2012).

Of importance to the present work are the stochastic internal waves that occupy the frequency bands away from tidal and inertial frequencies, and are not associated with internal solitary waves. Quite remarkably, these stochastic waves are observed everywhere there is stratification. No one has ever reported an internal calm, even in the Arctic. The waves are deemed "stochastic" because they possess little coherence across frequency, wave, and mode number and behave essentially like Gaussian random noise (Müller et al., 1986). The stochastic internal waves are most evident in spectra along the so-called powerlaw or continuum region extending from the semidiurnal frequency up to the buoyancy cut-off (Figure 3.1). This region is also evident in the shallow-water spectra (Figure 3.2), but the power law shape is broken at high frequencies because of internal solitary waves. Based on observations and kinematic considerations, Garrett and Munk (1972) proposed a universal spectral form for these waves, and although there have been some small refinements over the years (Garrett and Munk, 1975; Munk, 1981; Levine, 2002) there is little doubt that this model provides a first-order description of the waves (Briscoe, 1975; Müller et al., 1978; Wunsch and Webb, 1979). Regarding a dynamical description of the stochastic internal-wave field, the state of affairs is rather disheartening. Nonlinear internal-wave wave–wave interaction theories, much like those that describe the saturation spectrum of surface gravity waves (Phillips, 1977), have proved incomplete (Müller et al., 1986; Polzin and Lvov, 2011). Recently it has been proposed that the spectral shape is the result of Doppler smearing of the inertial and tidal lines by background currents (Pinkel, 2008). The present state of understanding of the sources and sinks of internal waves is equally unsatisfactory, not from a lack of possible mechanisms but from an overabundance of them that are not easily disentangled (Thorpe, 1975; Wunsch, 1976).

Excellent reviews covering the random internal-wave field have been given by Munk (1981), Müller et al. (1986), and most recently by Polzin and Lvov (2011). Any reader familiar with the subject of ocean internal waves and their spectral characterization will find little new or of interest in this chapter. The purpose here is to provide important background material to those acousticians who are unfamiliar with the topic and to introduce material that will be used in the acoustical analysis of subsequent chapters.

3.2 Fundamental Equations of Hydrodynamics

The linear hydrodynamics equations for which internal-wave motions are derived typically involve several assumptions. Under the "Traditional Approximation" (Eckart, 1960), consider a fluid otherwise at rest and assuming a flat bottom, in the f-plane, with the Boussinesq approximation the following linear equations of motion are obtained (Phillips, 1977):

$$\frac{\partial u}{\partial t} - fv + \frac{1}{\rho_0}\frac{\partial p}{\partial x} = 0, \tag{3.1}$$

$$\frac{\partial v}{\partial t} + fu + \frac{1}{\rho_0}\frac{\partial p}{\partial y} = 0, \tag{3.2}$$

$$\frac{\partial w}{\partial t} + \frac{1}{\rho_0}\frac{\partial p}{\partial z} + b = 0, \tag{3.3}$$

$$\frac{\partial b}{\partial t} - wN^2(z) = 0, \tag{3.4}$$

$$\frac{\partial u}{\partial x} + \frac{\partial v}{\partial y} + \frac{\partial w}{\partial z} = 0, \tag{3.5}$$

where $b = \rho g/\rho_0$, $N(z)$ is the buoyancy frequency profile described in Chapter 2 (Eq. 2.22), and $f = 2\Omega \sin(\text{latitude})$ is the Coriolis parameter. Here the first three equations come from Newton's law in a rotating reference frame, the fourth equation is from mass conservation, and the last equation is the incompressibility relation. These linear equations clearly do not describe internal solitary wave motions. Some manipulation gives a single wave equation for the vertical displacement, ζ, namely

$$\left[\left(\frac{\partial^2}{\partial t^2} + N^2\right)\left(\frac{\partial^2}{\partial x^2} + \frac{\partial^2}{\partial y^2}\right) + \left(\frac{\partial^2}{\partial t^2} + f^2\right)\frac{\partial^2}{\partial z^2}\right]\zeta(\mathbf{r}, t) = 0. \tag{3.6}$$

Here some interesting quantities derived from the linear equations are

$$\text{vertical velocity: } w = \frac{\partial \zeta}{\partial t}, \tag{3.7}$$

$$\text{horizontal velocity divergence: } \frac{\partial u}{\partial x} + \frac{\partial v}{\partial y} = -\frac{\partial^2 \zeta}{\partial t \partial z}, \tag{3.8}$$

vertical component of vorticity: $\dfrac{\partial v}{\partial x} - \dfrac{\partial u}{\partial y} = f\dfrac{\partial \zeta}{\partial z}$, (3.9)

pressure fluctuation: $\dfrac{1}{\rho_0}\nabla^2 p = f^2\dfrac{\partial \zeta}{\partial z} - \dfrac{\partial(N^2\zeta)}{\partial z}$, (3.10)

pressure fluctuation hydrostatic: $\dfrac{\partial p}{\partial z} = -N^2\rho_0\zeta$, (3.11)

density fluctuation: $\rho = \dfrac{\rho_0}{g}N^2\zeta$. (3.12)

Equation 3.9 has implications for the potential vorticity of internal waves, which is an important conserved quantity in geophysical flows (Pedlosky, 1987). Potential vorticity pv is defined as (Müller et al., 1986)

$$f + pv = (f\hat{\mathbf{z}} + \nabla \times \mathbf{u}) \cdot \nabla(z - \zeta),$$ (3.13)

where in the linear limit the result is

$$pv = \hat{\mathbf{z}} \cdot \nabla \times \mathbf{u} - f\frac{\partial \zeta}{\partial z}.$$ (3.14)

The first term is the vertical component of the fluid vorticity and the second term is due to vortex stretching as described by Kelvin's theorem and the vorticity equation (Pedlosky, 1987). The potential vorticity is zero for linear internal waves, but it will be found that there is one special internal-wave solution with zero frequency, termed the vortical mode, which will carry nonzero pv (see Section 3.5.2). The absence of potential vorticity for internal waves is markedly different from all other large-scale motions such as Rossby, Kelvin, and Poincaré waves and eddies that carry significant potential vorticity (Pedlosky, 1987).

3.2.1 Ray Theory: Local Plane Waves

The prognostic variables in the internal-wave equations are the horizontal velocity components (u, v) and the vertical displacement (ζ), because the vertical velocity and the pressure and density fluctuations can be derived from the displacement using Eqs. 3.7, 3.10, and 3.12. If each internal wave variable has the plane wave dependence $e^{i(\kappa \cdot \mathbf{r} - \sigma t)}$, then the five hydrodynamic equations (Eqs. 3.1–3.5) reduce to five coupled linear algebraic equations. Solving these equations requires a zero determinant, but since there are only three prognostic variables the resulting determinant is a cubic equation for the frequency σ. Two roots of the equation give the well-known internal-wave dispersion relation, written

$$\sigma^2(\kappa) = \frac{N^2(k^2 + l^2) + f^2 m^2}{k^2 + l^2 + m^2} = N^2\cos^2\theta + f^2\sin^2\theta,$$ (3.15)

where the wavenumber is $\kappa = (k, l, m)$ and the angle $\theta = \tan^{-1}(m/\sqrt{k^2 + l^2})$ is the inclination of the internal wave with respect to the horizontal. When nonlinear effects are considered, some deviation from this relation can occur. The third root is $\sigma = 0$, describing what is termed the "vortical mode" because this mode, unlike classical internal-wave motions, carries potential vorticity (Müller et al., 1986). The vortical mode will be discussed in Section 3.5. Returning to internal waves, for the case of interest in the ocean, where $f < N$, there are solutions for $f < \sigma < N$. This bounding of the wave-like solutions of internal waves means that waves of given frequency σ will be trapped vertically between turning depths where $\sigma = N(z_{turn})$, and they will be trapped meridionally at turning latitudes where $\sigma = f(\text{lat}_{turn})$. Importantly the dispersion relation tells us that for constant N, internal waves at a fixed frequency will propagate at a fixed angle $\pm\theta$. When N is a function of depth, in the ray approximation the wave angle will also be a function of depth consistent with the dispersion relation. Some internal-wave observables for the case of constant N at frequency σ and wavenumber κ are given by

$$\zeta(\mathbf{r}, t) = \zeta_0 \cos(\kappa \cdot \mathbf{r} - \sigma t + \varphi), \tag{3.16}$$

$$w(\mathbf{r}, t) = \zeta_0 \sigma \sin(\kappa \cdot \mathbf{r} - \sigma t + \varphi), \tag{3.17}$$

$$u(\mathbf{r}, t) = \zeta_0 m \frac{(\sigma^2 k^2 + f^2 l^2)^{1/2}}{k^2 + l^2} \cos(\kappa \cdot \mathbf{r} - \sigma t + \varphi + \theta_u), \tag{3.18}$$

$$v(\mathbf{r}, t) = \zeta_0 m \frac{(\sigma^2 l^2 + f^2 k^2)^{1/2}}{k^2 + l^2} \cos(\kappa \cdot \mathbf{r} - \sigma t + \varphi + \theta_v). \tag{3.19}$$

Here ζ_0 is the displacement amplitude, φ is an arbitrary phase angle, and the u and v phase angles are given by

$$\theta_u = \tan^{-1}\left(\frac{\sigma k}{-fl}\right), \qquad \theta_v = \tan^{-1}\left(\frac{\sigma l}{fk}\right). \tag{3.20}$$

In wavenumber space, surfaces of constant σ are cones because the frequency depends only upon the inclination of the wave and not on its azimuth (Figure 3.3). This can be contrasted with the acoustic dispersion relation, $\omega = c(k_x^2 + k_y^2 + k_z^2)^{1/2}$, where the surfaces of constant frequency are spherical. For the limit $m \to 0$ or $\theta \to 0$ one obtains $\sigma \to N$, that is, these are buoyancy waves. On the other extreme in the limit $k, l \to 0$ or $\theta \to \pi/2$, the result is $\sigma \to f$, that is, inertial waves are obtained.

The motion of water parcels for internal waves is easy to derive but hard to visualize. Velocity vectors for a near-inertial wave and a near-buoyancy wave are shown in Figure 3.4. The fluid velocity vector is seen to be perpendicular to the wave vector ($\mathbf{u} \cdot \kappa = 0$). In the limit of inertial waves, water parcels move primarily in horizontal circles with little vertical velocity or vertical displacement. Inertial waves therefore provide little sound-speed fluctuation (Eq. 2.32). Near-buoyancy waves, by contrast, move water parcels primarily in the vertical direction with

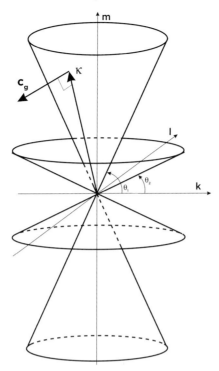

Figure 3.3. Dispersion relation surfaces for two different internal-wave frequencies. For a fixed frequency all the allowed internal-wave wavenumbers lie on a cone. The group velocity, $\mathbf{c_g}$, is perpendicular to the cone and to the wavenumber $\boldsymbol{\kappa} = (k, l, m)$.

little horizontal motion. Inertial waves are often termed circularly polarized, while buoyancy waves are termed linearly polarized. For frequencies away from the extremes, the fluid parcels move in tilted ellipses. Note that the ellipticity of the horizontal currents is a function of frequency. Aligning the wave propagation direction with the x-axis by setting $l = 0$, it is seen that the ratio $u/v \propto \sigma/f$.

Wave Packet Propagation

Internal waves in the ocean occupy a continuum of frequencies between f and N and thus wave packet or group propagation is possible. If the background ocean density structure is not changing too rapidly, the internal-wave wave packets can be described using ray theory, that is, Eqs. 2.38 and 2.39 with the dispersion relation Eq. 3.15. Internal-wave group velocity is given by

$$\mathbf{c_g} = \frac{m^2}{|\boldsymbol{\kappa}|^2} \frac{N^2 - f^2}{\sigma} \left(k, l, -\frac{k^2 + l^2}{m} \right). \tag{3.21}$$

Here it is seen that the group velocity is perpendicular to the wavenumber (i.e., $\mathbf{c_g} \cdot \boldsymbol{\kappa} = 0$) and thus to the direction of phase propagation. Internal-wave energy

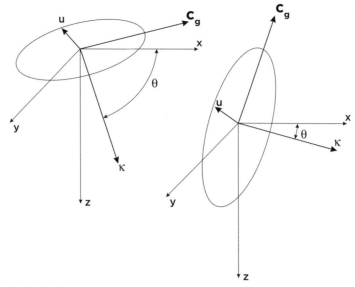

Figure 3.4. Hodographs for the particle velocity for near-inertial waves (left) and near-buoyancy waves (right). Also shown are the wavenumber κ and group velocity $\mathbf{c_g}$.

therefore oddly propagates along the crests of the waves and not perpendicular to the crests, as is obtained for sound waves. In addition, the vertical component of the group velocity has the opposite sign from that for vertical phase propagation. Thus internal-wave energy propagating away from a source, say the ocean surface due to wind forcing, will move downward, but the phase propagation will be upward. Useful limits are

$$\sigma \to f : \mathbf{c_g} \to 0, \quad \kappa \to (0,0,m), \tag{3.22}$$

$$\sigma \to N : \mathbf{c_g} \to 0, \quad \kappa \to (k,l,0). \tag{3.23}$$

Given the odd kinematic behavior of internal waves it will come as little surprise that internal waves do unexpected things when they interact with sloping boundaries. Consider waves interacting with a boundary of slope β. For acoustic waves the tangential component of the wavenumber (i.e., the component parallel to the slope) will remain unchanged, and thus the angle of the wave with respect to the horizontal will change by 2β. This is a consequence of the rule of equality of the incident and reflected angles relative to the normal to the slope. Internal waves, on the other hand, have their horizontal angle fixed by the wave frequency. Thus the angle of an internal wave ray is unchanged after interaction with a slope. However, it is also required that the net flow $\mathbf{u}$ from the incident and reflected wave must be parallel to the reflecting surface, which leads to the requirement that the component of the wavenumber parallel to the slope is conserved on reflection. These two conditions dictate that a wave interacting with a slope β can be reflected either shoreward or back. Defining $\alpha = \pi/2 - \theta$ as the angle of group propagation, it is found that the ratio of the magnitudes of the reflected, and transmitted wavenumber are given by

$$\frac{|\kappa_r|}{|\kappa_i|} = \sin(\alpha + \beta) \begin{cases} \frac{1}{\sin(\alpha-\beta)}, & \alpha > \beta \\ \frac{1}{\sin(\beta-\alpha)}, & \alpha < \beta \end{cases}, \tag{3.24}$$

where $\alpha > \beta$ gives reflection shoreward and $\alpha < \beta$ gives back reflection. When the angle of group propagation is equal to the bottom slope, the magnitude of the reflected wave wavenumber becomes infinite and the bottom slope is termed critical. This critical bottom slope, β_c, is given by

$$\tan\beta_c = \left[\frac{\sigma^2 - f^2}{N^2 - \sigma^2}\right]^{1/2}. \tag{3.25}$$

For waves propagating shoreward with a bottom slope less than β_c, internal-wave energy continues to be reflected upslope with ever-increasing energy density, which eventually leads to wave breaking and mixing (Wunsch and Hendry, 1972). For bottom slopes greater than β_c the wave abruptly cannot propagate forward, and the wave reflects back out to deep-water or in some cases does not survive the reflection process. For waves that are close to critical, the growing magnitude of the reflected wave wavenumber means that energy density and shear increase dramatically, also leading to wave breaking and mixing (Wunsch and Hendry, 1972).

3.2.2 Modal Solutions

The behaviors of internal waves that are due to the depth structure of the buoyancy frequency profile, $N(z)$, are now analyzed. The modal and WKB solutions to the internal-wave equations will prove quite useful, and they will be the building blocks for the ocean models of stochastic internal waves (e.g., the Garrett-Munk spectrum). Examining modal solutions of Eq. 3.6 with the form $\zeta(\mathbf{r}, t) \propto \psi_j(z) \exp(i(kx + ly - \sigma t))$ gives the linear internal-wave mode equation

$$\frac{d^2\psi_j}{dz^2} + \kappa_j^2 \left(\frac{N^2 - \sigma^2}{\sigma^2 - f^2}\right)\psi_j = 0. \tag{3.26}$$

This equation is written with the understanding that the internal-wave frequency has been chosen and one must solve for the horizontal wavenumber eigenvalue κ_j and the eigen-mode ψ_j, both of which will be a function of the frequency. The boundary conditions for the modes are generally taken to be zero displacement at both the seafloor and the sea surface. In some cases a free surface boundary condition given by

$$\frac{d\psi_j}{dz} - \kappa_j^2 \frac{g}{\sigma^2 - f^2} \psi_j = 0, \qquad \text{at } z = 0, \tag{3.27}$$

can be of some use.[1] Here it is seen that internal waves have a small but measurable surface signature (Ray and Mitchum, 1996). This surface expression is acoustically insignificant but oceanographically quite important. An order of magnitude calculation gives the surface elevation for mode j to be

$$\zeta(0) \simeq \zeta_{max} \frac{\Delta\rho}{\rho} \frac{(-1)^j}{\pi j}, \tag{3.28}$$

where $\Delta\rho/\rho$ is the fractional change in density over the entire water column, and ζ_{max} is the maximum internal displacement for mode j. For the internal-wave mode functions, ψ_j, the orthogonality condition for fixed frequency and horizontal wavenumber eigenvalue $\kappa_j = (k^2 + l^2)^{1/2}$ is

$$\int_0^D \psi_j(z)\psi_m(z) \frac{N^2 - \sigma^2}{\sigma^2 - f^2} \, dz = 0, \tag{3.29}$$

where $m \neq j$. Often Eq. 3.29 is written without the factor in the denominator because it is independent of depth. Several different normalization conditions can be used but here the potential energy normalization, that is, $\int_0^D N^2 \psi_j^2(z) \, dz = 1$, is generally adopted. Another normalization is discussed in Section 3.2.3. The mode equation is used to yield the modal dispersion relation, that is $\kappa_j(\sigma)$, although WKB approximations are most often used (e.g., Eq. 3.44).

Internal-wave observables at frequency σ and mode number j are given by

$$\zeta(\mathbf{r},t) = A_0 \psi_j(z) \cos(kx + ly - \sigma t + \varphi), \tag{3.30}$$

$$w(\mathbf{r},t) = A_0 \sigma \psi_j(z) \sin(kx + ly - \sigma t + \varphi), \tag{3.31}$$

$$u(\mathbf{r},t) = A_0 \frac{(\sigma^2 k^2 + f^2 l^2)^{1/2}}{k^2 + l^2} \psi_j'(z) \cos(kx + ly - \sigma t + \varphi + \theta_u), \tag{3.32}$$

$$v(\mathbf{r},t) = -A_0 \frac{(\sigma^2 l^2 + f^2 k^2)^{1/2}}{k^2 + l^2} \psi_j'(z) \cos(kx + ly - \sigma t + \varphi + \theta_v). \tag{3.33}$$

Here A_0 is the displacement amplitude with units of (rad m$^{3/2}$/s), φ is an arbitrary phase angle, and the u and v phase angles are given by

$$\theta_u = \tan^{-1}\left(\frac{fl}{-\sigma k}\right), \qquad \theta_v = \tan^{-1}\left(\frac{fk}{\sigma l}\right). \tag{3.34}$$

The internal-wave displacement mode function vertical derivative is ψ'.

---

[1] This boundary condition comes from the linear condition that $p = \rho_0 g \zeta$ at $z = 0$. Taking the time derivative and horizontal Laplacian of this equation and using Eqs. 3.5 and 3.7–3.9, the modal free surface boundary condition can be derived.

3.2.3 WKB Analysis

The Garrett-Munk internal-wave spectral model that will be used involves a number of WKB results; thus a review of this material is warranted. Appendix B in Chapter 2 has an overview of WKB modal analysis. The starting point of the WKB analysis will be the internal-wave mode equation (Eq. 3.26), where solutions of the form $\psi_j \propto A(z)e^{i\varphi(z)}$ are sought. The WKB envelope and phase functions are found to be

$$A(z) \propto \frac{1}{\sqrt{m(z)}}, \tag{3.35}$$

$$\varphi(z) = \int_{z_L}^{z} m(z')\,dz', \tag{3.36}$$

where the phase function is defined only over the oscillatory modal depth region between the lower and upper turning points, that is, where $\sigma = N(z_L) = N(z_U)$. From the internal-wave mode equation the local vertical wavenumber is given by

$$m(z) = \kappa_j \left(\frac{N^2(z) - \sigma^2}{\sigma^2 - f^2} \right)^{1/2}. \tag{3.37}$$

Clearly in depth regions where the modes are evanescent and near turning points where $m(z) = 0$ special treatment is required (e.g., Airy Solutions (Garrett and Munk, 1972)). For σ not close to f or N, Eq. 3.35 gives the important information that the displacement and vertical velocity modal envelope scales as $1/\sqrt{N}$. It can also be shown that the horizontal current envelope scales as $\sqrt{N}$. Formally the WKB validity condition is $m^2(z) \gg (1/A)d^2A/dz^2$, but stated in another way the condition implies that over a vertical oscillation of the mode the buoyancy frequency does not change dramatically. As previously mentioned, WKB results apply more readily to higher order modes.

Because κ_j scales like frequency and considering the mode function normalization, the WKB representation of the internal-wave displacement can be written (Phillips, 1977)

$$\zeta(\mathbf{r}, t) \propto \frac{1}{\sigma} \left(\frac{\sigma^2 - f^2}{N^2(z) - \sigma^2} \right)^{1/4} e^{i(\boldsymbol{\kappa} \cdot \mathbf{r} - \sigma t)}, \tag{3.38}$$

and other quantities such as the horizontal currents can be easily derived from the displacement using the vorticity and horizontal divergence equations (Eqs. 3.8 and 3.9). These WKB results provide important scaling relationships for horizontal kinetic energy and gravitational potential energy that will be used to scale different quantities in the GM spectrum. Here the results are simply quoted (Phillips, 1977):

$$\frac{\rho}{2}(\bar{u^2} + \bar{v^2}) \propto \left(\frac{N^2(z) - \sigma^2}{\sigma^2 - f^2}\right)^{1/2} \frac{\sigma^2 + f^2}{\sigma^2}, \tag{3.39}$$

$$\frac{\rho}{2}N^2(z)\bar{\zeta^2} \propto \frac{N^2(z)}{\sigma^2}\left(\frac{\sigma^2 - f^2}{N^2(z) - \sigma^2}\right)^{1/2}. \tag{3.40}$$

No Turning Point Approximation

To better understand the WKB approximations utilized in the GM spectral model, and to describe Monte Carlo numerical simulation techniques common to the ocean acoustics community (Appendix B, Colosi and Brown, 1998), some further development is required. This development involves a treatment of internal waves in which little attention is paid to the influences of the oceanic boundaries and to modal turning points. The creators of the GM model make no secret of this treatment in which they refer to the model assumptions of a topless and bottomless ocean as "The Tijuana boundary condition." Inspired by the WKB envelope and phase functions, Eq. 3.26 is scaled using the following equations (Bell, 1974; Colosi and Brown, 1998):

$$\hat{\psi}_j(z) = \left(\frac{N(z)}{N_0}\right)^{1/2}\psi_j(z), \qquad \hat{z}(z) = \frac{1}{N_0 B}\int_D^z N(z')\,dz', \tag{3.41}$$

where $N_0 B = \int_D^0 N(z')\,dz'$ so that $0 \le \hat{z} \le 1$. Here D is the water depth, and this transformation is understood to approximately map the mode functions ψ_j onto constant amplitude sinusoids. With the WKB-like approximation that $2\pi j \gg N_0/N$, the mode equation becomes (Colosi and Brown, 1998)

$$\frac{d^2\hat{\psi}_j}{d\hat{z}^2} + \kappa_j^2\frac{(N_0 B)^2}{\sigma^2 - f^2}\hat{\psi}_j = 0, \tag{3.42}$$

subject to the boundary conditions $\hat{\psi}_j(0) = \hat{\psi}_j(1) = 0$, thus giving modal solutions

$$\psi_j(z) = \left(\frac{2}{N_0 B N(z)}\right)^{1/2}\sin(\pi j\hat{z}(z)), \tag{3.43}$$

with the requirement that

$$\sigma^2 = f^2 + \kappa_j^2\left(\frac{N_0 B}{\pi j}\right)^2. \tag{3.44}$$

The WKB condition $2\pi j \gg N_0/N$ is seen to be hardest to satisfy for low modes and in the deep ocean where N_0/N is the largest. In any case, the deep ocean is not significant from an acoustics standpoint because the potential sound-speed gradient is extremely small there. Equation 3.43 gives the WKB modes that are seen in this approximation to be independent of internal-wave frequency. Consistent with the GM approach, there is no attempt to treat mode turing points (see the next section). These modes are normalized such that $\int_D^0 \psi_j^2 N^2 \, dz = 1$. The WKB dispersion relation is given by Eq. 3.44, and from this relation it is also evident that the internal-wave vertical wavenumber is given by

$$m(z) = \frac{\pi j N(z)}{N_0 B}. \tag{3.45}$$

Equations 3.43–3.45 have been used extensively in the ocean acoustics literature and will be used throughout this book.

Treatment of Turning Points

The previous development can be easily modified to accommodate mode turning points, and a reasonable increase in accuracy is attained. Following the development of Levine (2002) but using the notation of Colosi and Brown (1998), the mode equation is transformed using the new mapping:

$$\hat{\psi}_j(z) = \left(\frac{N(z)}{N_0}\right)^{1/2} \psi_j(z), \qquad \hat{z}(z) = \frac{1}{N_0 B(\sigma)} \int_{z_L}^{z} N(z') \, dz', \tag{3.46}$$

where $N_0 B(\sigma) = \int_{z_L}^{z_U} N(z') \, dz'$. In this approximation the mode functions are defined only between the turning points and setting the value of the mode functions at the turning points to zero, the modal solution is obtained:

$$\psi_j(z, \sigma) = \left(\frac{2}{N_0 B(\sigma) N(z)}\right)^{1/2} \sin(\pi j \hat{z}(z)), \quad \text{for} \quad z_L \le z \le z_U, \tag{3.47}$$

where it is now evident that the modes are an explicit function of frequency. These modes are normalized such that $\int_{z_L}^{z_U} \psi_j^2 N^2 dz = 1$. This treatment of the turning points also influences the WKB dispersion relation that is now

$$\sigma^2 = f^2 + \kappa_j^2 \left(\frac{\pi j}{N_0 B(\sigma)}\right)^2, \tag{3.48}$$

and similarly the vertical wavenumber becomes

$$m(z, \sigma) = \frac{\pi j N(z)}{N_0 B(\sigma)}. \tag{3.49}$$

Figure 3.5 shows comparisons of internal-wave mode shapes computed for a deep-water environment using the mode equation, and the WKB approach with and without turning point corrections. The WKB methods handle higher modes better than lower modes, and WKB with turning point corrections improves the fit at higher frequencies. In the upper ocean, which is most critical for acoustics, all

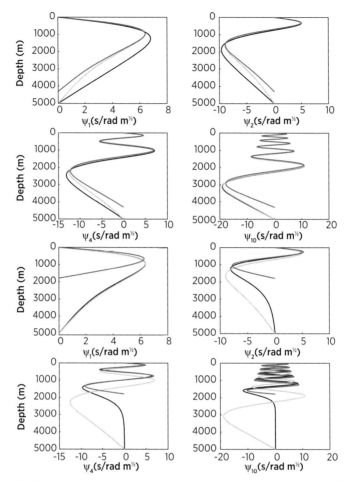

Figure 3.5. Modes functions for a canonical buoyancy frequency profile $N(z) = N_0 \exp(z/1000)$ with $N_0 = 6$ cph and a latitude of $30°$. Results are shown derived from the mode equation (black), and WKB approximations with (dark gray) and without (light gray) turning point corrections. The upper four panels show results for $\sigma = 1/12.42$ (cph), the M2 frequency, and the lower four panels are for $\sigma = 1$ (cph).

three methods give reasonable results. It should also be mentioned here that this particular treatment of turning points leads to discontinuities in horizontal velocity since this velocity depends on the vertical derivative of the mode. On the matter of the dispersion relation, Figure 3.6 shows dispersion curves computed from the same three approaches. Here the WKB with turning point corrections does a much better job at the higher frequencies.

3.3 Garrett-Munk Internal-Wave Model

The Garrett-Munk internal-wave model describes a subset of the internal waves in the ocean for which their behavior appears to be temporally and spatially highly

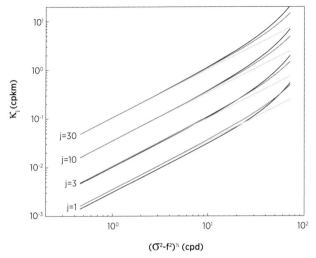

Figure 3.6. Dispersion curves derived from the mode equation (black), and WKB approximations with (dark gray) and without (light gray) turning point corrections.

random but statistically quite stable. The model derives from a large number of spatial and temporal measurements of horizontal velocity, and temperature and salinity that, after proper WKB scaling, yield spectral forms with a remarkable similarity in shape and energy level. This unexpected stationarity and homogeneity of the ocean internal-wave spectrum suggest that the waves are in some equilibrium with their sources and sinks of energy, although a dynamical description of the spectrum is still lacking. Locations where significant deviations from the universal form are seen include the Arctic, the Equator, near submarine canyons and seamounts, and the upper ocean near the mixed layer (Wunsch, 1976; Wunsch and Webb, 1979). Near abrupt topography the spectral form is seen to rapidly relax back to its equilibrium form at short distances from the topographic features. Notable situations where significant deviations from the GM model are absent include continental shelf regions, below and near boundary currents and storms, and in the vicinity of intense eddies.

Through the GM internal-wave model it is seen that the ocean displacements are a linear superposition of plane waves with random amplitudes such that

$$\zeta(\mathbf{r}, t) = \int \int \int \hat{\zeta}(\boldsymbol{\kappa}) e^{i(\boldsymbol{\kappa} \cdot \mathbf{r} - \sigma(\kappa) t)} d\boldsymbol{\kappa}, \tag{3.50}$$

where the internal-wave amplitudes $\hat{\zeta}(\boldsymbol{\kappa})$ obey the homogeneous statistics:

$$\langle \hat{\zeta}(\boldsymbol{\kappa}) \rangle = 0, \tag{3.51}$$

$$\langle \hat{\zeta}(\boldsymbol{\kappa}) \hat{\zeta}^*(\boldsymbol{\kappa}') \rangle = F_\zeta(\boldsymbol{\kappa}) \delta(\boldsymbol{\kappa} - \boldsymbol{\kappa}'), \tag{3.52}$$

$$\langle \hat{\zeta}(\boldsymbol{\kappa}) \hat{\zeta}(\boldsymbol{\kappa}') \rangle = \langle \hat{\zeta}^*(\boldsymbol{\kappa}) \hat{\zeta}^*(\boldsymbol{\kappa}') \rangle = 0. \tag{3.53}$$

Here $F_\zeta(\kappa)$ is the displacement wavenumber spectrum with the mean square displacement given by

$$\langle \zeta^2 \rangle = \int \int \int F_\zeta(\kappa)\, d\kappa. \qquad (3.54)$$

For acoustical purposes it serves the objectives best to express the GM spectrum foremost as a displacement spectrum.[2] For simplicity the Garrett-Munk spectrum is written in terms of frequency and mode number yielding a factorized spectrum of the form

$$F_\zeta(\sigma, j) = \zeta_0^2 \frac{N_0}{N(z)} B(\sigma) H(j), \qquad (3.55)$$

$$B(\sigma) = \frac{4}{\pi} \frac{f}{\sigma^3} \frac{\sqrt{\sigma^2 - f^2}}{\sigma^3}, \qquad (3.56)$$

$$H(j) = \frac{1}{N_j} \frac{1}{j^2 + j_*^2}, \qquad (3.57)$$

where the normalizations are

$$\int_f^N B(\sigma)\, d\sigma = \frac{2}{\pi} \left[\frac{\sqrt{(N/f)^2 - 1}}{(N/f)^2} + \tan^{-1}\left(\frac{1}{\sqrt{(N/f)^2 - 1}} \right) \right] \simeq 1, \qquad (3.58)$$

$$N_j = \sum_{j=1}^J \frac{1}{j^2 + j_*^2}. \qquad (3.59)$$

Here $\zeta_0 = 7.3$ m is a reference displacement, $N_0 = 3$ cph is a reference buoyancy frequency, j_* is a modal bandwidth parameter (close to 3 in deep water and close to 1 on continental shelves), and J is a maximum mode number. In many calculations that are not sensitive to the higher modes, J is taken to be infinite. When a finite cut-off is required (and this is often the case), there are reasonable choices that can be made as discussed in Section 3.3.2. In some cases in the literature and in this book the normalization of the mode spectrum, N_j, is taken to approximately be $\pi/(2j_*)$ (Flatté et al., 1979). The fact that the spectrum scales linearly with f unfortunately leads to problems near the equator (discussed in detail in Section 3.3.4). This model spectrum is seen to have a power law slope that goes as frequency and mode number to the minus two power.

The horizontal current spectrum is obtained from the displacement spectrum using WKB scalings of mean square horizontal current and displacement (Eqs. 3.39 and 3.40), the ratio of which is $N^2(\sigma^2 + f^2)/(\sigma^2 - f^2)$. The total

---

[2] It should be noted that the GM spectrum was originally based on an exponential buoyancy frequency profile $N(z) = N_0 \exp(-z/B)$. In the following presentation N_0 will be used as a reference buoyancy frequency value and B will not be used. The WKB quantity $N_0 B$ is used extensively. The notation is not ideal.

horizontal current spectrum is thus

$$F_{u_h}(\sigma, j) = F_u(\sigma, j) + F_v(\sigma, j) = \zeta_0^2 N_0 N(z) \frac{4}{\pi} \frac{f}{\sigma^3} \frac{(\sigma^2 + f^2)}{\sqrt{\sigma^2 - f^2}} H(j). \qquad (3.60)$$

Mean square quantities derived from WKB analysis are given by

$$\langle \zeta^2 \rangle = \zeta_0^2 \frac{N_0}{N(z)}, \qquad (3.61)$$

$$\langle u^2 \rangle + \langle v^2 \rangle = u_0^2 \frac{N(z)}{N_0}, \qquad (3.62)$$

where $u_0^2 = 3\zeta_0^2 N_0^2 \simeq 44$ (cm/s)^2. Figures 3.1 and 3.2 show comparisons between observed and GM displacement frequency spectra in both shallow and deep-water environments, and the fits are seen to be rather good. Geographic and temporal variations seen in observed spectra generally show internal-wave variances to be within a factor of 2 or 3 of the GM value, and spectral slopes vary on the order of 10–20% (Polzin and Lvov, 2011). More experimental results are shown later in the chapter.

3.3.1 Other Useful Forms of the GM Spectrum

Translation of the GM spectrum into wavenumber space is typically done using the WKB dispersion relation (Eq. 3.44) and the WKB vertical wavenumber (Eq. 3.45). Defining $\hat{\kappa}_j = \pi j f / N_0 B$ and writing $\kappa_h = \kappa_j$ as the continuous horizontal wavenumber, the WKB dispersion relation becomes

$$\sigma^2 = f^2 \left(1 + \frac{\kappa_h^2}{\hat{\kappa}_j^2} \right). \qquad (3.63)$$

Useful forms of the GM displacement spectrum for acoustical purposes are

$$F_\zeta(\kappa_h, j) = \zeta_0^2 \frac{N_0}{N(z)} H(j) \frac{4}{\pi} \frac{\hat{\kappa}_j \kappa_h^2}{(\kappa_h^2 + \hat{\kappa}_j^2)^2}, \qquad (3.64)$$

$$F_\zeta(k, l, j) = \zeta_0^2 \frac{N_0}{N(z)} H(j) \frac{2}{\pi^2} \frac{\hat{\kappa}_j \sqrt{k^2 + l^2}}{(k^2 + l^2 + \hat{\kappa}_j^2)^2}, \qquad (3.65)$$

$$F_\zeta(k, j) = \int_{-\infty}^{\infty} F_\zeta(k, l, j; z) \, dl,$$

$$= \frac{\zeta_0^2 N_0}{N(z)} \frac{H(j)}{\pi^2} \left[\frac{2\hat{\kappa}_j}{k^2 + \hat{\kappa}_j^2} + \frac{k^2}{\sqrt{(k^2 + \hat{\kappa}_j^2)^3}} \ln\left(\frac{\sqrt{k^2 + \hat{\kappa}_j^2} + \hat{\kappa}_j}{\sqrt{k^2 + \hat{\kappa}_j^2} - \hat{\kappa}_j} \right) \right],$$

(3.66)

$$F_\zeta(k,l,m) = \zeta_0^2 \frac{N_0}{N(z)} \frac{4}{\pi^3} \frac{m_*}{m^2 + m_*^2} \frac{mf}{N} \frac{(k^2 + l^2)^{1/2}}{(k^2 + l^2 + (mf/N)^2)^2},$$

(3.67)

where $m_* = \pi j_* N / N_0 B$. Note in this last spectral form with the vertical wavenumber m, the spectrum is normalized such that $\int_0^\infty dm \int_{-\infty}^\infty dk dl F_\zeta = \langle \zeta^2 \rangle$. Importantly it is seen that the scaled wavenumber, $\hat{\kappa}_j$, has an important physical meaning, that is, its inverse tells us roughly the horizontal correlation length of the internal-wave field for mode j (see Appendix A).

It is also useful to express the GM spectrum with a variable power law such that

$$F_\zeta(\sigma, j) = \zeta_0^2 \frac{N_0}{N(z)} H(j) N_\sigma \frac{f^{p-2} \sqrt{\sigma^2 - f^2}}{\sigma^p},$$

(3.68)

where N_σ is a normalization factor and the frequency spectral slope is $-p + 1$. Translating into a wavenumber spectrum using the dispersion relation the result is

$$F_\zeta(\kappa_h, j) = \zeta_0^2 \frac{N_0}{N(z)} H(j) N_\sigma \frac{\hat{\kappa}_j^{p-2} \kappa_h^2}{(\kappa_h^2 + \hat{\kappa}_j^2)^{(p+1)/2}}.$$

(3.69)

Here the spectral slope at large wavenumbers goes as κ_h^{-p+1}. The standard GM results are obtained for $p = 3$.

Lastly, results are presented relevant to the horizontal velocity. In acoustical calculations one is generally not concerned with the spectrum of both components of the current (e.g., Eq. 3.60) but with only that component in the direction of the acoustical path. Equations 3.18 and 3.19 give the scaling relationship between displacement and horizontal flow. Using the WKB dispersion relation (Eq. 3.44), the horizontal wavenumber, mode spectra of the currents are found to be

$$F_u(k,l,j) = F_\zeta(k,l,j) \frac{m^2(j)}{k^2 + l^2} f^2 (1 + k^2/\hat{\kappa}_j^2),$$

(3.70)

$$F_v(k,l,j) = F_\zeta(k,l,j) \frac{m^2(j)}{k^2 + l^2} f^2 (1 + l^2/\hat{\kappa}_j^2).$$

(3.71)

Note that these spectra are not isotropic because they do not depend strictly on $k^2 + l^2$.

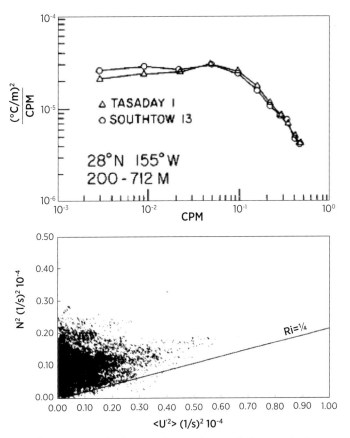

Figure 3.7. The top panel shows observations of the vertical wavenumber spectrum of the vertical temperature gradient. The bottom panel shows a scatter plot of squared buoyancy frequency versus mean square shear. The sloping line shows the Richardson number equal to 1/4.
Sources: Gregg (1977) and Eriksen (1978).

3.3.2 Maximum Internal-Wave Mode Number

For several acoustic observables, like intensity variance, and vertical and horizontal coherence, a maximum mode number, J, is required to obtain nonsingular results. Observations do indeed reveal a cut-off in the spectrum such as shown in Figure 3.7 (also see Duda and Cox (1989)). In this example the cut-off wavenumber is roughly $m_{max} = 2\pi \times 0.2$ rad/m (Gregg, 1977). Therefore a useful empirical relation for J using WKB results is

$$J = m_{max}N_0B/(\pi\bar{N}),\qquad(3.72)$$

where $\bar{N}$ is an average buoyancy frequency value in the main thermocline. The physical reasoning behind the mode cut-off is shear instability as discussed by Munk (1981) and Gregg (1977).

One can be a little more precise on this matter by examining the inverse Richardson number, that is, the mean square shear divided by N^2, or

$$Ri^{-1} = \frac{\langle (du/dz)^2 \rangle + \langle (dv/dz)^2 \rangle}{N^2(z)}. \tag{3.73}$$

Here a value greater than 4 indicates wave breaking due to shear instability Munk (1981). Using the Garrett-Munk spectrum the inverse Richardson number can be computed, giving

$$Ri^{-1} = \frac{3\pi^2 \zeta_0^2 N_0 N(z)}{(N_0 B)^2} \sum_{j=1}^{J} j^2 H(j) \simeq \frac{6\pi \zeta_0^2 N_0 N(z) J j_*}{(N_0 B)^2}. \tag{3.74}$$

Figure 3.7 shows observations in which the Richardson number rarely drops below the critical value of 1/4. Taking into account the statistical nature of the waves so that cases in which $Ri^{-1} > 4$ are extremely rare, the mode cut-off J can be computed by equating the above Garrett-Munk relation to 1 (Munk, 1981; Duda and Cox, 1989). The result is then

$$J = \frac{(N_0 B)^2}{6\pi \zeta_0^2 N_0 \bar{N} j_*}. \tag{3.75}$$

where again as in the case of Eq. 3.72, $N(z)$ has been replaced by $\bar{N}$, a typical main thermocline value.[3] This value of $\bar{N}$ is chosen because the instability is most likely to occur in regions of large buoyancy frequency as shown by Eq. 3.74. Using standard GM parameters ($\zeta_0 = 7.3$ m and $j_* = 3$), and values of $N_0 B = 8.7$ rad-m/s and $\bar{N} = 3$ cph (typical of mid-latitudes), a value of $J = 909$ is obtained. In the main thermocline where $N = 3$ cph this corresponds to a WKB vertical wavenumber of 0.25 cpm or a 4-m wavelength. Thus internal waves in the ocean do indeed go down to small vertical scales

In Eq. 3.75 it is seen that the cut-off depends on the internal-wave displacement variance, ζ_0^2, as well as the modal bandwidth j_*. Observations by Duda and Cox (1989) show indeed that the vertical cut-off depends inversely on the internal-wave energy and the buoyancy frequency. In most acoustic calculations there is a logarithmic singularity associated with $J \to \infty$ so the acoustic results are generally not highly sensitive to J computed from either Eq. 3.72 or 3.75.

3.3.3 Internal-Wave Correlation Scales

Sound transmission through the stochastic internal-wave field is strongly influenced by the correlation scales of the internal-wave displacement field and in

---

[3] See a similar calculation carried out by Desaubies (1978).

particular by the anisotropy of the fluctuations. The horizontal correlation function is written as the inverse cosine transform of the spectrum:

$$C(x) = \sum_{j=1}^{\infty} \int_f^N \frac{1}{2\pi} \int_{-\pi}^{\pi} B(\sigma)H(j)\cos(\kappa_j(\sigma)x\cos\alpha)\,d\sigma d\alpha,$$

$$= \sum_{j=1}^{\infty} \int_f^N B(\sigma)H(j)J_0(\kappa_j(\sigma)x)\,d\sigma \simeq 1 - \frac{8j_*}{\pi}\frac{f|x|}{N_0B}\left[\log\left(\frac{N}{f}\right) - \frac{1}{2}\right].$$

$$(3.76)$$

Similarly, the vertical and temporal correlation functions are written as

$$C(z) = \sum_{j=1}^{\infty} H(j)\cos(m(j)z) \simeq 1 - (\pi j_* - 1)\frac{N|z|}{N_0B}, \qquad (3.77)$$

$$C(t) = \int_f^N B(\sigma)\cos(\sigma t)\,d\sigma \simeq 1 - \frac{2Nf|t|}{N-f}. \qquad (3.78)$$

These correlation functions (Munk and Zachariasen, 1976) suggest the following order of magnitude correlation lengths and times,

$$L_h = \frac{\pi(N_0B/f)}{8j_*[\log(N(z)/f) - 1/2]}, \quad L_z = \frac{N_0B}{N(z)(\pi j_* - 1)}, \quad T_{iw} = \frac{N-f}{2Nf}. \qquad (3.79)$$

The horizontal correlation length can also be estimated using an approximate form of the 1-D spectrum (see Appendix A). In this case the horizontal correlation function is exponential for each mode with e-folding length $1/(\sqrt{2}\hat{\kappa}_j)$. Averaging over the modes the result is (Munk and Zachariasen, 1976)

$$L_h = \sum_{j=1}^{\infty} \frac{1}{\sqrt{2}\hat{\kappa}_j}H(j) \simeq \frac{N_0B}{\sqrt{2}\pi f}\frac{\log(4j_*^2 + 1)}{\pi j_* - 1}, \qquad (3.80)$$

which is independent of depth. The loss of depth dependence in Eq. 3.80 occurs because of the use of the WKB dispersion relation that for large wavenumbers allows waves with frequencies greater than N (see Appendix A).

Typical values for the correlation lengths and times are shown in Table 3.1. Internal-wave fluctuations are seen to be strongly anisotropic, with horizontal scales exceeding vertical scales by one to two orders of magnitude. This anisotropy has important acoustical consequences, as the interaction with these features will depend critically on orientation. This means that scattering will be a strong function of ray angle or mode number.

The correlation times of internal waves are seen to be on the order of hours. One might be tempted to conclude that acoustic fields are stable over periods comparable to an hour. This is not so, as will be demonstrated in subsequent chapters.

Table 3.1. *Correlation scales for Garrett-Munk internal waves computed from Eq. 3.79*
for different latitudes and buoyancy frequency values.

Latitude	$N = 6$ cph	$N = 3$ cph	$N = 1$ cph	$N = 0.5$ cph
10	(4.9, 0.059, 5.5)	(5.6, 0.12, 5.5)	(7.3, 0.35, 5.4)	(8.9, 0.71, 5.3)
30	(2.1, 0.059, 1.9)	(2.5, 0.12, 1.9)	(3.5, 0.35, 1.8)	(4.7, 0.71, 1.7)
50	(1.5, 0.059, 1.2)	(1.8, 0.12, 1.2)	(2.7, 0.35, 1.2)	(3.9, 0.71, 1.1)
70	(1.3, 0.059, 1.0)	(1.6, 0.12, 1.0)	(2.5, 0.35, 0.9)	(3.7, 0.71, 0.9)

Here a deep-water value of $N_0 B = 5.24$ is used. Results are displayed as (L_h, L_z, T_{iw}) in units of km, km, and hours respectively.

3.3.4 Modifications to GM

Over the years the GM model has evolved consistent with Garrett and Munk's philosophy of "planned obsolescence" (Garrett and Munk, 1972). Most of the changes occurred over the first decade of the model's existence. However, in 2002 a significant change to the model was proposed by Levine in an effort to rectify two issues, namely the scaling of the spectrum with latitude and treatment of mode turning points in depth such that the model would be more readily applicable in shallow water and other noncanonical environments. The problem of scaling with latitude is immediately obvious from Eq. 3.56, where the frequency spectrum scales linearly with f thereby going to zero at the equator. There is also the issue that the variance is fixed over the frequency band f to N which is not strictly observed. To rectify these problems Levine (2002) suggested a new spectral form given by

$$\hat{B}(\sigma) = \frac{\pi}{4} \frac{B(\sigma)}{C} \begin{cases} 1, & \sigma_{S2} \leq \sigma \leq N \\ \left(1 + \frac{f}{\sigma_{S2}}\right) \frac{(\sigma/\sigma_{S2})^3}{(\sigma/\sigma_{S2})^{2.5} + (f/\sigma_{S2})}, & f \leq \sigma \leq \sigma_{S2} \end{cases}, \qquad (3.81)$$

where σ_{S2} is the semidiurnal frequency equal to 2 cpd. Importantly the normalization constant C is determined from the high-frequency part of the spectrum such that $\int_{\sigma_{S2}}^{N} \hat{B}(\sigma) d\sigma = 1$, giving

$$C = \frac{1}{2} \left[\tan^{-1}\left(\frac{1}{\sqrt{(\sigma_{S2}/f)^2 - 1}}\right) - \tan^{-1}\left(\frac{1}{\sqrt{(N/f)^2 - 1}}\right) + \right.$$
$$\left. \frac{\sqrt{(\sigma_{S2}/f)^2 - 1}}{(\sigma_{S2}/f)^2} - \frac{\sqrt{(N/f)^2 - 1}}{(N/f)^2} \right]. \qquad (3.82)$$

Because the normalization is for the high frequency or continuum part of the spectrum, the integral over the entire frequency range from f to N will be larger than 1 and will vary with latitude. This change of internal-wave variance with latitude appears to be more consistent with the observations (Levine, 2002). It

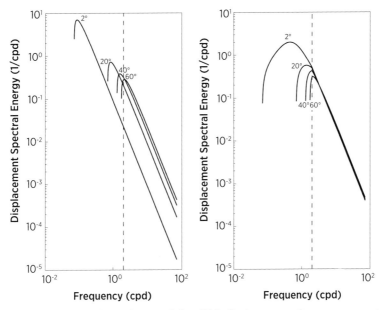

Figure 3.8. Latitude dependence of the GM displacement frequency spectrum (left panel) and the modified spectrum of Levine (2002) (right panel). Spectra are shown for latitudes of 2°, 20°, 40°, and 60°, and are unit normalized. Dashed line indicates the S2 frequency (two cycles per day).

should also be noted that the new model has a break in spectral slope at σ_{S2} such that the high frequencies decay as σ^{-2} while the lower frequencies show a whiter spectral fall-off. This behavior too is based on observations, but it is evident only at latitudes much less than 30°.

Figure 3.8 shows frequency spectra $B(\sigma)$ and $\hat{B}(\sigma)$ for a variety of latitudes. The whitening of the spectra for frequencies below σ_{S2} is evident at latitudes below 30°. At high latitudes the two models are virtually identical.

A more careful treatment of mode turning points has many oceanographic implications, but from the standpoint of acoustical applications the key changes have to do with the mode shapes, the dispersion relation, and the vertical wavenumber. As such Eqs. 3.43–3.45 can be replaced with Eqs. 3.47–3.49. This change generally helps to treat more accurately the high-frequency waves and there is little improvement with regard to the low-frequency waves. Importantly, the turning point corrected WKB dispersion relation and vertical wavenumber can be used to translate the frequency and mode number spectra into wavenumber spectra of various forms, but this will not be done here.

3.4 Observations

Over the decades technological innovations such as Acoustic Doppler Current Profilers (ADCPs) and sensitive moored and profiling Conductivity, Temperature,

Depth (CTD) sensors have improved the ability to observe ocean internal waves, but given the wide range of space and times scales of these waves it is safe to say that they are still greatly under sampled. Acoustic remote sensing methodologies applied to internal-wave spectra (internal-wave tomography) have yet to make a contribution to the field, but could prove insightful since the sampling properties are quite different from traditional ones already mentioned (Munk et al., 1981; Flatté, 1983). In any case, because of the observational limitations there remain many unanswered questions regarding the stochastic ocean internal-wave field. What follows is largely an overview of the measurements of internal-wave spectra taken in various ocean environments with an eye to giving the reader a sense of the observational variability. An excellent historical account has been given by Munk (1981), and more modern treatments are given by Polzin and Lvov (2011) and Pinkel (2008). It is also fair to say that the topic of the ocean internal-wave spectrum has been somewhat neglected in the last few decades.

Be that as it may, the scientific consensus remains that the Garrett-Munk internal-wave spectrum does indeed provide a first-order description of the distribution of internal waves, but there are anomalies and deviations. For the purposes of ocean acoustics it is important to point out these anomalies because various acoustic observables can be sensitive or insensitive to these deviations. Because different issues exist for different locations in the ocean, the discussion is separated into five distinct ocean environments: (1) deep ocean mid-latitude; (2) slopes and seamounts; (3) continental shelves and shallow water; (4) the Arctic; and (5) the Equator.

3.4.1 Deep Ocean: Mid-Latitude

Data from the deep-ocean provided descriptions of internal-wave spectra that led to the first Garrett-Munk model in 1972. Shortly thereafter an ambitious field effort in the North Atlantic Ocean, simply called the Internal Wave Experiment (IWEX), was carried out to test some of the assumptions of the GM model (Briscoe, 1975; Müller et al., 1978). Using a three-legged mooring of current meters and thermistors, IWEX was designed to measure deep-ocean, three-dimensional current and temperature fields over a wide range of horizontal and vertical separations. In the frequency range between the semidiurnal tides and 1 cph (i.e., the continuum region of the spectrum), IWEX largely confirmed many of the GM assumptions, that is: (1) horizontal velocity and vertical displacement were consistent with a random field of waves; (2) the waves were horizontally isotropic and vertically symmetric; (3) the spectral form was roughly consistent with GM; and (4) WKB scaling was valid. Important deviations uncovered by the IWEX were: (1) the vertical bandwidth factor, equivalent to j_*, varied somewhat with frequency, which means that the frequency mode number spectrum

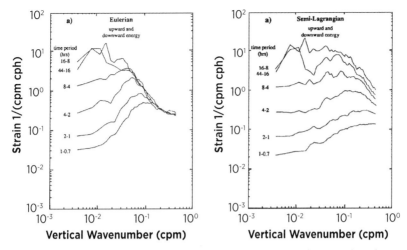

Figure 3.9. Strain vertical wavenumber spectra at different frequencies observed during the PATCHEX experiment. Eulerian/semi-Lagrangian estimates are respectively shown on the left- and right-hand panels. Spectral estimates have been logarithmically smoothed in vertical wavenumber.
Source: Sherman and Pinkel (1991).

is not strictly separable; and (2) measurements of displacement derived from temperature were contaminated by temperature fine structure, later attributed to density-compensated temperature and salinity structure or spice (Section 3.5).

With the advent of the ADCP and high-speed CTD profiling from the stable ocean platform R/P FLIP, a series of experiments carried out between 1983 and 2002 revealed a more detailed view of the internal-wave field (Sherman and Pinkel, 1991; Pinkel, 2014). These measurements, which were obtained in the North Pacific Ocean, again raised the issue of the nonseparability of the spectrum in terms of frequency and mode number. Figure 3.9 shows vertical wavenumber spectra as a function of frequency for strain ($d\zeta/dz$). The spectra are seen to be a strong function of frequency. The GM strain spectral form is $m^2/(m^2 + m_*^2)$. However, there are fundamental disagreements as to the source of the observed nonseparability. One explanation attributes the deviation to small-scale vortical motions (Section 3.5) that have nothing to do with internal waves (Lien and Müller, 1992). A recent re-analysis of several of these experiments has isolated the vortical motions but the nonseparability appears to be intrinsic to the internal waves (Pinkel, 2014). Yet another suggestion is that vertical advection of small-scale internal-wave motions by larger scales (nonlinearity) makes it impossible to interpret the Eulerian measurements in terms of linear dynamics (largely the foundation of the GM model). When the analysis is done in an isopycnal following coordinate system, the spectra appear much more separable (Figure 3.9) (Sherman and Pinkel, 1991). Analysis of a large moored array of CTD sensors in the Philippine Sea (Colosi et al., 2013b) also reveals nonseparability of the

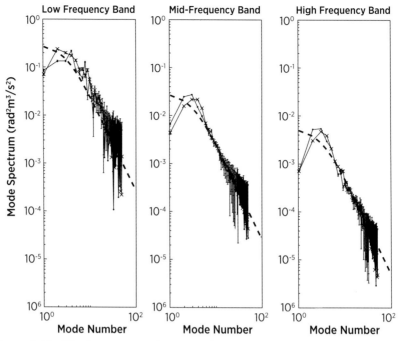

Figure 3.10. Displacement mode spectra in the Philippine Sea observed during the spring of 2009. The three different frequency bands are given by $f < \sigma < 0.1$ cph, $0.4 < \sigma < 0.6$ cph, and $0.9 < \sigma < 1.1$ cph. The two data curves with dots and crosses are for two different moorings separated by 200 km. The dashed curve shows the GM spectral form, $1/(j^2 + j_*^2)$, with $j_* = 3$.
Source: Colosi et al. (2013b).

frequency/mode spectrum, and anomalously low mode 1 energy (Figure 3.10). In these measurements the displacement mode spectrum over the continuum region is relatively independent of frequency, but differs greatly from the near-inertial band where the modal bandwidth is much greater.

Measurements over the last several decades have also shown the rich tidal harmonic structure seen in Figure 3.1 over the continuum region of the spectrum (Pinkel, 1983, 1984), suggesting that energy appearing in the continuum region may be due to nonlinear interactions associated with the primary tidal and inertial peaks.

The interested reader should consult Polzin and Lvov (2011), who have catalogued dozens of deep-water internal-wave spectra and their variability. In this effort they find that (1) total internal-wave energies vary by factors of 2–3 geographically and over seasonal time scales; (2) frequency-vertical wavenumber spectra show nonseparability properties with near-inertial waves showing greater modal bandwidth (although nonseparability is less pronounced for more energetic sites); (3) distinct geographic variations in frequency and vertical wavenumber power law slopes are evident and co-vary (whiter frequency spectra accompany redder vertical wavenumber spectra and vice versa); and (4) the modal bandwidth parameter, j_*, for the vertical wavenumber spectrum varies widely from 1 to over 20.

3.4.2 Seamounts, Slopes, and Canyons

Because internal-wave propagation angles are set by the wave frequency and there can be critical slope effects (Section 3.2.1), significant changes to the internal-wave spectrum near topographic features such as continental slopes, seamounts, and canyons may be expected. These regions are further complicated by highly variable stratification and strong background currents. Measurements by Eriksen (1998) in the vicinity of Fieberling Seamount demonstrated how the spectrum can be distorted by topographic interactions (Figure 3.11). For observations near the slope (F321, 1435m), a significant amount of energy accumulates near the critical slope frequency. There is no doubt that wave instability, breaking, and mixing are occurring in these regions. Interestingly, not far from the slope the spectra have "relaxed" back to the GM form (F301, 95m), suggesting that whatever physical process is shaping the spectrum acts rather quickly to equilibrate the waves. Measurements on continental slopes (Wunsch and Hendry, 1972) as well as in canyons (Hotchkiss and Wunsch, 1982; Kunze et al., 2002) also show significant distortions of the low-frequency part of the spectrum near the critical slope frequency due to topographic effects. Observations of the distortion of the vertical wavenumber or modal spectrum near topographic features have not been carried out.

3.4.3 Continental Shelves

Continental shelf regions, like those near seamounts, slopes, and canyons, can involve complicated topography as well as highly variable stratification and background currents, but the critical factor on the continental shelf is the close proximity of the surface and bottom. This means that the waves constantly interact with the ocean boundaries, where important forcing and dissipation processes occur. Given the large differences between the deep ocean and the shelf, one might think that a Garrett-Munk-like model would be of little use, but this is not the case.

Figure 3.2 shows frequency spectra from the Mid-Atlantic Bight. A clear GM spectral shape is seen between the inertial frequency and 1 cph. At higher frequencies the spectrum is dominated by nonlinear internal solitary waves. Other shelf measurements also show frequency spectra with the GM form (Apel et al., 1997; Pringle, 1999; Levine, 2002), although there is some indication that the continuum spectrum at the highest frequencies is elevated relative to the GM prediction even in the absence of internal solitary waves. However, on the shelf

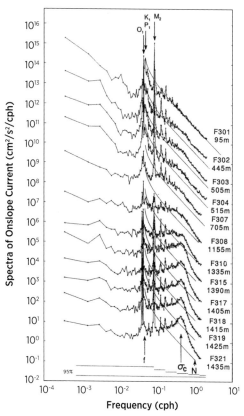

Figure 3.11. Autospectra of onslope (NorthEast) current. The ordinate labels are for the spectrum of the deepest instrument (F321, 20 m off the bottom); successive spectra are offset by one decade. Thin smooth curves between the inertial and buoyancy frequencies at each depth indicate the GM prediction. Diurnal and semidiurnal tidal frequencies are marked O1, P1, K1, and M2. Spectra within a few hundred meters of the bottom are enhanced near the local critical frequency σ_c. Horizontal bars indicate the 95% confidence intervals.
Source: Eriksen (1998).

the stochastic internal-wave field is composed of a complicated mix of locally generated waves and waves propagating in from deep water. Because of the deep-water component, the spectral levels of along-shore current are generally lower than those in the across-shore direction, meaning that the spectrum is not horizontally isotropic (Figure 3.2, right-hand panel; also see Pringle (1999)).

There are few reports of mode or vertical wavenumber spectra on the shelf (Pringle, 1999; Colosi et al., 2012). Measurements from the mid-Atlantic Bight made in the summer of 2006 (Colosi et al., 2012) show a mode spectrum close to the GM form but with a modal bandwidth closer to $j_* = 1$ (Figure 3.12). Here it is suspected that issues related to spectral separability do indeed exist but this has not been investigated.

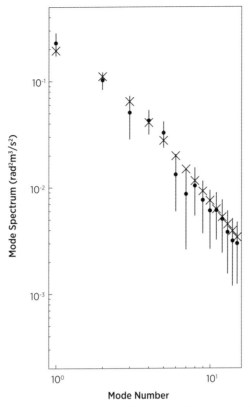

Figure 3.12. Displacement mode spectra from the Mid-Atlantic Bight observed in the summer of 2006 for stochastic internal waves in the frequency band $f < \sigma < 1$ cph, where the mode shapes are independent of frequency . The crosses show the GM spectral form, $1/(j^2 + j_*^2)$, with $j_* = 1$.
Source: Colosi et al. (2012).

3.4.4 Arctic and High Latitude

Internal-wave spectra from the Arctic show some of the most dramatic differences from the GM model. In the 1985 Arctic Internal Wave Experiment (AIWEX) in which instruments were suspended below a 3-m thick ice pack in the Beaufort Sea, internal-wave spectra were observed to be strongly nonstationary and importantly total internal-wave energy levels were seen to be lower than GM by a stunning factor of 0.03–0.07 (Levine et al., 1987). Figure 3.13 shows some examples of frequency (Levine et al., 1987) and vertical wavenumber spectra (D'Asaro and Morehead, 1991) observed during AIWEX. The spectral slope in frequency is observed to be −1 instead of −2 as predicted by the GM model, and the observation of localized coherent internal-wave structures means that the vertical modal bandwidth (i.e., j_*) is larger than the canonical value ($j_* = 3$) by roughly a factor of 10 (Levine, 1990; D'Asaro and Morehead, 1991). On the other hand, the

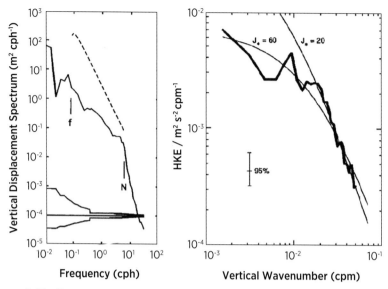

Figure 3.13. Frequency and vertical wavenumber spectra from AIWEX. In the left-hand panel the dashed curve is the GM spectrum, and in the right-hand panel the GM spectral form $1/(j^2 + j_*^2)$ is given for two values of j_*.
Sources: Levine et al. (1987); D'Asaro and Morehead (1991).

waves in the spectral continuum behave like a random wave field with horizontal isotropy, consistent with the GM model (Levine, 1990).

Observations by Plueddemann (1992) using a drifting buoy locked into the pack ice that drifted from the Nansen Basin, over the Yermak Plateau, and then south to the Greenland Sea revealed large spatial variations of the internal-wave spectrum. In the deep water of the Nansen Basin, frequency spectral levels and shape were similar to those from AIWEX, but near topographic features, like the Yermak Plateau, the spectral levels and shape were more GM-like. In the Greenland Sea the spectra were intermediate between the AIWEX result and the GM model. Measurements analyzed by Wijesekera et al. (1993) in a similar area near the Yermak Plateau also showed GM-like spectral shapes associated with the topographic features and enhanced modal bandwidths.

Other anomalous high latitude spectra have been reported by Eckert and Foster (1990) in the Fram Strait, and by Lynch et al. (1996) in the Barents Sea Polar Front region.

3.4.5 Equator

With the vanishing of the Coriolis parameter at the equator, one might expect that low-frequency internal waves will behave quite differently from the GM model. Indeed, below tidal frequencies the spectral slope is less steep (Figure 3.1; Levine (2002), and Eriksen (1980)). Figure 3.14 shows frequency spectra of horizontal

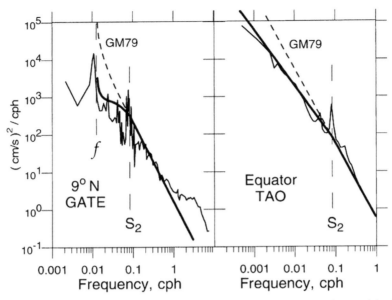

Figure 3.14. Moored spectra of horizontal velocity (thin curves) observed in the Equitorial Pacific Ocean. The GM spectrum scaled with f at a latitude of $30°$ is dashed, while the modified GM spectrum due to Levine (2002) (Eq. 3.81) is plotted with a bold line.
Source: Levine (2002).

current in the Pacific Ocean at North $9°$ and at the equator. Here the modified GM spectrum of Levine (2002) is seen to fit the observations much better than the GM model. At frequencies above the semidiurnal frequency (i.e., the continuum region), the power law shape of the spectrum is close to the GM model (Eriksen, 1985; Levine, 2002). The consistency of the spectral form in the continuum region and the flattening of the spectra at low frequency has led oceanographers to suspect a connection between the internal tides and high-frequency internal waves (Munk and Wunsch, 1998).

Other equatorial spectral anomalies in the continuum region have been reported by Boyd et al. (1993), including (1) horizontal anisotropy with eastward energy fluxes four times larger than westward fluxes; (2) mode spectra that show non-GM behavior for the low modes similar to that seen in Figure 3.10; (3) high vertical coherence of modes suggesting a nonrandom wave field; and (4) unusual ratios of kinetic to potential energy and of zonal to meridional kinetic energy.

3.5 Other Sources of Stochastic Sound-Speed Structure

While internal waves have been directly linked to acoustical statistics, there are other processes that generate small-scale random sound-speed fluctuations. Two important process that have garnered some attention are intrusive fine structure or spice and vortical internal-wave motions. Interestingly both of these processes have essentially no intrinsic time evolution (e.g., zero frequency) and thus temporal behavior is dictated by advection. The acoustic impacts of these processes have not been studied to the degree that internal-wave scattering has been investigated because the space–time scales are not well known.

3.5.1 Spice

In the linear approximation, the equation of state of seawater allows one to write the variation of sound speed and density in terms of the variations in temperature and salinity such that

$$\frac{\Delta C}{C} = a\Delta T + b\Delta S, \qquad \frac{\Delta \rho}{\rho} = -\alpha\Delta T + \beta\Delta S, \tag{3.83}$$

where typical values of the constants (which depend on pressure, T, and S) are (Munk, 1981)

$$a = 2.0 \times 10^{-3}\,^\circ\mathrm{C}^{-1}, \quad b = 0.74 \times 10^{-3}\mathrm{PSU}^{-1},$$
$$\alpha = 0.25 \times 10^{-3}\,^\circ\mathrm{C}^{-1}, \quad \beta = 0.75 \times 10^{-3}\mathrm{PSU}^{-1}.$$

Thus compensating anomalies of temperature and salinity along an isopycnal (e.g., a surface of constant density) are reinforcing in sound speed. These anomalies along isopycnals were termed spiciness to reflect the variation from cold and fresh (weak spice) to hot and salty (strong spice) (Munk, 1981). These spicy sound-speed anomalies are created where there is a co-mingling of different oceanic water masses, and there is adequate dynamical forcing to drive mixing. The anomalies are in fact remnants of the mixing process. Regions where strong spice is expected include boundary current regions, marginal seas, continental shelves and slopes, and near the air/sea interface. Because spice is density-neutral it is dynamically irrelevant and will be advected around the ocean as a passive tracer.

Spice variations are easily seen in CTD data when potential temperature is plotted versus salinity (i.e., a T/S diagram; Figure 3.15). Regions in the T/S diagram where data points are scattered along contours of constant potential density indicate large spice variation, for example in Figure 3.15 in the upper ocean along the density contours 1024-1026 kg/m^3. Some spice is present in the deeper layers near the salinity minimum; little spice is present in the main thermocline and in the deep ocean. Temporal variations of spice (and the associated sound-speed fluctuations) at a fixed point in the ocean tend to be quite intermittent (Figure 3.15, lower panels). This behavior is due to the complex spatial structure of the spicy features and the advection of these features by internal-wave and other currents. Work by Ferarri and Rudnick (2000) and Dzieciuch et al. (2004) suggests that

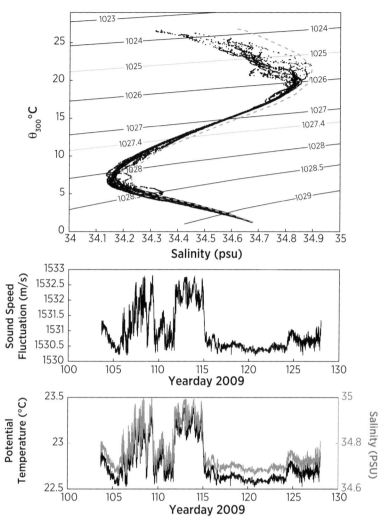

Figure 3.15. (Upper) Potential temperature versus salinity for 16 CTDs taken in the Philippine Sea in the spring of 2009 with contours of potential density. The potential temperature is referenced to a pressure of 300 decibars. The gray-dashed curve is from climatology. Spread of data points along contours of constant potential density reveals spice variability. In this region spice is largest in the upper 200 m of the ocean. (Lower) Timeseries of spicy sound speed (middle) and T/S variability (lower) along isopycnal 1025.1 kg/m^3.
Source: Colosi et al. (2013b).

the spatial structure of spice is front-like, that is T-S variations are seen to be step-like. Thus as a random medium, the spice sound-speed field is quite unlike the internal-wave field.

Frequency spectra of spicy sound-speed variations have been observed in both deep and shallow-water environments (Figure 3.16). In these examples from the Philippine Sea and the Mid-Atlantic Bight, the shallow-water spectral levels are

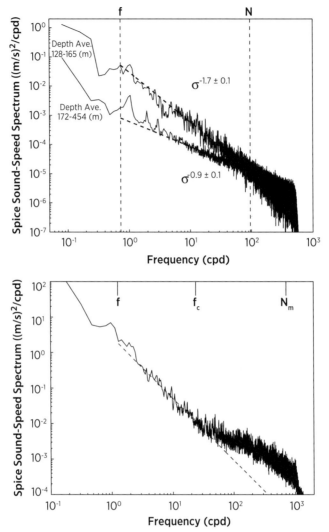

Figure 3.16. Frequency spectra of spicy sound-speed fluctuations measured in the Philippine Sea (top), and in the Mid-Atlantic Bight (bottom). The slopes of the continental shelf spectrum and the upper ocean Philippine Sea spectrum are both −1.7. The deep spectrum from the Philippine sea with a slope of −0.9 is in a region of weak spice.
Sources: From Colosi et al. (2012, 2013b).

seen to be much higher. In fact for the Mid-Atlantic Bight case spice was the largest source of observed high frequency sound-speed variations (Colosi et al., 2012). The spectral shapes and slopes of these two cases are remarkably similar with a power law slope of −1.7, which is close to what one would expect from flont-like steps being advected past the observing array. (The magnitude square of the Fourier Transform of a step has a slope of −2.)

3.5.2 Vortical Motions

As was seen in the beginning of this chapter, there exists a solution to the internal-wave hydrodynamic equations for which the frequency is identically zero (Section 3.2.1). These solutions are called the vortical mode with vertical displacement and horizontal velocity given by (Müller et al., 1986)

$$\zeta(\mathbf{r}) = A_\kappa \cos(\boldsymbol{\kappa} \cdot \mathbf{r}), \tag{3.84}$$

$$u(\mathbf{r}) = A_\kappa \frac{N^2}{f} \frac{l}{m} \cos(\boldsymbol{\kappa} \cdot \mathbf{r}), \tag{3.85}$$

$$v(\mathbf{r}) = -A_\kappa \frac{N^2}{f} \frac{k}{m} \cos(\boldsymbol{\kappa} \cdot \mathbf{r}), \tag{3.86}$$

where A_κ is the displacement amplitude for the wave with wavenumber $\boldsymbol{\kappa}$. These motions are in geostrophic and hydrostatic balance, have zero horizontal divergence, and satisfy the thermal wind equations:

$$\frac{\partial u}{\partial z} = \frac{N^2}{f} \frac{\partial \zeta}{\partial y} \quad \text{and} \quad \frac{\partial v}{\partial z} = -\frac{N^2}{f} \frac{\partial \zeta}{\partial x}. \tag{3.87}$$

It is seen that the vortical mode has few constraints on its vertical structure, so the terminology of a mode is somewhat misleading.

Figure 3.17 shows a decomposition of strain into internal-wave and vortical mode components for the N. Pacific PATCHEX experiment. The strain variations for the two processes are seen to be comparable, although the internal-wave record shows much more high-frequency variability. The time evolution of the vortical structure is due to Doppler spreading or advection of the vortical structure. The internal-wave strain variance is seen to grow with increasing depth (e.g., decreasing N) while the vortical strain variance shows some tendency to decrease with increasing depth (Pinkel, 2014). The space/time scales of the vortical motions are also of great interest for acoustical purposes. Figure 3.17 shows the vertical wavenumber and frequency spectra of the strain field and the vortical motions. The internal-wave strain spectra show shapes similar to GM, although the frequency spectra has a rather flatter roll-off in frequency close to $\sigma^{-1.5}$. The vortical motion strain energy is concentrated at high wavenumber and low frequency. The effects of the vortical motion on acoustics are, at this point, unknown.

Appendix A An Internal-Wave Model with an Exponential Correlation Function

A modified version of the GM spectrum is useful for some theoretical calculations because it has nice correlation function properties. To this end, consider the

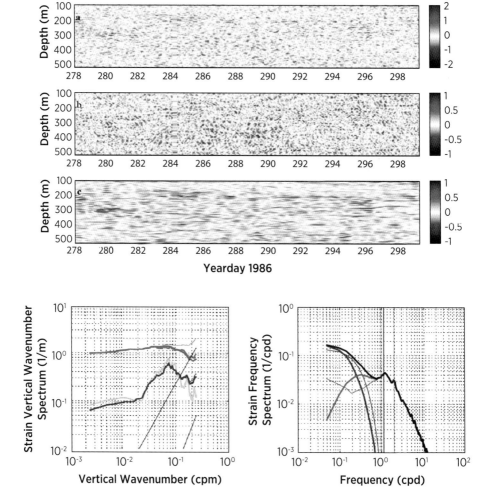

Figure 3.17. Upper panel: Observed water column strain from PATCHEX (top) and a subjective separation of internal waves (middle) and vortical motions (bottom). Lower panel: Vertical wavenumber and frequency spectra of internal wave and vortical motion strain from the PATCHEX experiment. The red/blue curves denote the internal wave/vortical mode spectra. In the left panel the spectral noise estimate is denoted by the black curve. In the right panel the black curve gives the total strain spectrum, and reference lines for the inertial (black), diurnal (magenta), and semidiurnal (magenta) frequencies are indicated.
Source: Pinkel (2014).

following "pure" power law form of the frequency spectrum:

$$F_\zeta(\sigma, j) = \zeta_0^2 \frac{N_0}{N(z)} H(j) \frac{f}{\sigma^2}. \tag{A.1}$$

This spectrum clearly violates the kinematic condition that $\zeta \to 0$ when $\sigma \to f$, but the model is intended to aid in the testing of acoustic propagation theories, not to compare to observations. A slightly modified form of the dispersion relation is also used, that is,

$$\sigma^2 = f^2 \left(1 + \frac{\kappa_h^2}{\tilde{\kappa}_j^2}\right), \tag{A.2}$$

where $\tilde{\kappa}_j = \sqrt{2}\pi j f / N_0 B$. The modified dispersion relation is used so that the wavenumber spectra of the present model, which will be presented next, more closely matches the wavenumber spectra of the standard GM model.

Translating to a wavenumber spectrum the result is

$$F_\zeta(\kappa_h, j) = \zeta_0^2 \frac{N_0}{N(z)} H(j) \frac{\kappa_h \tilde{\kappa}_j}{(\kappa_h^2 + \tilde{\kappa}_j^2)^{3/2}}, \tag{A.3}$$

which can be compared to the GM spectrum:

$$F_\zeta(\kappa_h, j) = \zeta_0^2 \frac{N_0}{N(z)} H(j) \frac{4}{\pi} \frac{\kappa_h^2 \hat{\kappa}_j}{(\kappa_h^2 + \hat{\kappa}_j^2)^2}, \tag{A.4}$$

where $\hat{\kappa}_j = \pi j f / N_0 B$. Figure A.1 shows a comparison between results from Eqs. A.3 and A.4, and the differences are over all rather small except most for low wavenumbers $(\kappa_h \ll \hat{\kappa}_j, \tilde{\kappa}_j)$. In Cartesian coordinates the pure power law spectrum is

$$F_\zeta(k, l, j) = \zeta_0^2 \frac{N_0}{N(z)} H(j) \frac{1}{2\pi} \frac{\tilde{\kappa}_j}{(k^2 + l^2 + \tilde{\kappa}_j^2)^{3/2}}. \tag{A.5}$$

Integrating out the l contributions the spectrum along an ocean slice is

$$F_\zeta(k, j) = \int_{-\infty}^{\infty} F_\zeta(k, l, j) dl = \zeta_0^2 \frac{N_0}{N(z)} H(j) \frac{1}{\pi} \frac{\tilde{\kappa}_j}{k^2 + \tilde{\kappa}_j^2}. \tag{A.6}$$

The wavenumber part of Eq. A.6 is a Lorentzian which has the nice property that its inverse Fourier transform is an exponential. Thus the horizontal correlation function of the jth mode is written

$$C_\zeta(\Delta x, j) = \int_{-\infty}^{\infty} F_\zeta(k, j) \cos(k\Delta x) dk = \zeta_0^2 \frac{N_0}{N(z)} H(j) \exp(-\tilde{\kappa}_j |\Delta x|). \tag{A.7}$$

Appendix B Monte Carlo Simulation

Monte Carlo simulation of random fields of internal-wave displacements has proven useful for the interpretation of acoustical observations via Monte Carlo parabolic equation simulations (Colosi et al., 1994; Wage et al., 2005; Van Uffelen

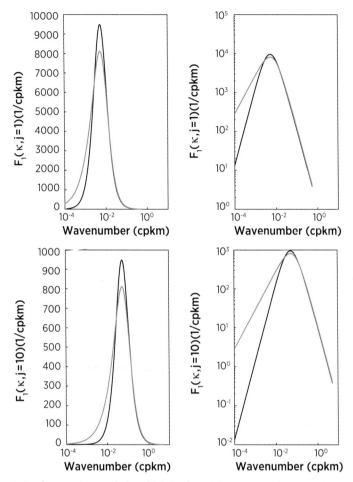

Figure A.1. Comparison of the GM horizontal wavenumber spectrum (black, Eq. A.4) and the pure power law approximation (gray, Eq. A.3) for internal-wave mode number 1 (top) and 10 (bottom). The spectra are normalized so that the area under the curves is 1. Left and right panels show semi-log and log-log displays of the same spectra.

et al., 2009) and for validating theory (Flatté and Tappert, 1975; Flatté and Colosi, 2008; Colosi and Morozov, 2009). Utilizing the WKB eigenmode expansion described in Section 3.2.3, a computationally efficient numerical treatment is possible (Colosi and Brown, 1998). Since WKB analysis is built into the GM spectral model, any shortcomings in this approach are deemed forgivable. This method has been utilized quite extensively in the ocean acoustics literature.

The linear modal expansion for the displacement is given by

$$\zeta(x,y,z,t) = \mathrm{Re}\left[\int dk \int dl \sum_{j=1}^{J} g(k,l,j)\, \psi_j(z,\sigma_j(\kappa))\, e^{i(kx+ly-\sigma_j(\kappa)t)} \right], \qquad (\mathrm{B.1})$$

where the amplitudes $g(k,l,j)$ are zero mean Gaussian random variables with variance related to Eq. 3.65. Utilizing the nonturning point–corrected WKB modes that are independent of frequency, a simple expression amenable to Fourier analysis is obtained. Here is an expression in a form that is convenient for numerical analysis which is given by

$$\zeta(x,y,z,t) = \text{Re}\left[\sum_{j=1}^{J} \sin(\pi j \hat{z}(z)) \sum_k \sum_l g(k,l,j)\, e^{i(kx+ly-\sigma_j(\kappa)t)}\right], \qquad \text{(B.2)}$$

where the amplitudes are drawn from a complex Gaussian random number generator such that $\langle |g(k,l,j)|^2 \rangle = 4F(k,l,j)\Delta k \Delta l$. Here $\hat{z}(z)$ is the normalized WKB depth coordinate, and $\sigma_j(\kappa)$ is given by Eq. 3.44. The factor of 4 in the random number variance comes from the fact that the average of the square of the sine function is 1/2, and only the real part of the sum is taken. The wavenumber sums are efficiently evaluated using the fast Fourier transform. The choice of the sampling in wavenumber is of course critical here. The first consideration is that the spectrum $F(k,l,j)$ is peaked where $\kappa = \sqrt{k^2 + l^2} = \hat{\kappa}_j$; thus adequate sampling around the spectral peak is required. The second consideration is that the WKB dispersion relation does not have an explicit cut-off at the buoyancy frequency. Thus care must be taken not to allow large wavenumbers such that $\sigma > N$; this condition is given mathematically by $\kappa_{max} = \hat{\kappa}_j(N_{max}/f)$ (Colosi et al., 2013a). Lastly J is chosen as described in Section 3.3.

Often it is the case that realizations of internal-wave displacements are only required in a vertical slice with no time evolution. In this case the computation is greatly simplified, yielding

$$\zeta(x,z) = \text{Re}\left[\sum_{j=1}^{J} \sin(\pi j \hat{z}(z)) \sum_k g(k,j)\, e^{ikx}\right], \qquad \text{(B.3)}$$

where the amplitudes are drawn from a complex Gaussian random number generator such that $\langle |g(k,j)|^2 \rangle = 4F(k,j)\Delta k$, with the spectrum given by Eq. 3.66.

4

Introduction to Acoustic Fluctuations

The purpose of this chapter is to outline and illustrate some of the basic conceptual models of sound transmission through a variable sound-speed field, with a particular emphasis on internal-wave-induced fluctuations. The objective is not to focus on rigorous derivations but instead to bring forward physical concepts. This approach involves simplifications and idealizations that will be mentioned but discussed in more detail only in subsequent chapters. Here the first objective is to describe the origin of phase and amplitude fluctuations using the important ray-like process of wave front distortion and the subsequent generation of microrays and interference. From the mode perspective, it is seen how mode coupling generates a form of modal multipathing and interference, the result of which is acoustic phase and amplitude fluctuations. In the limit of strong ray or modal multipathing, the notion of saturation is described where the acoustic field behaves like Gaussian random noise.

These conceptual models bring forward immediately the notion of acoustical resonances, such that the acoustical field is sensitive to select internal waves with specific space and time scales. These resonances will be a function of ray propagation angle and location in the water column, and in the mode formalism they will be a function of acoustic mode number and frequency. From the ray perspective, fully developed ray chaos modifies the simple resonance conditions. With a conceptual understanding of the propagation physics, some of the statistical characteristics of the signals that have been observed in experiments can be described. These signal statistics are described in terms of propagation regimes denoted by the terms unsaturated, partially saturated, and fully saturated.

4.1 Origin of Phase and Amplitude Fluctuations

Here building upon a wave front picture of acoustic propagation, the origin of phase and amplitude fluctuations are described. This wave front picture fits naturally into the problems of deep-water, long-range propagation, and high-frequency applications. In one way or another acoustic fluctuations result from interference effects, and in the ray picture one can envision this interference occurring due to extra ray paths that are generated by the sound-speed fluctuations. These rays are termed *microrays*. In the mode picture, which is strongly relevant to the shallow-water environment, interference occurs from mode coupling that creates new paths for acoustic energy to travel down range in mode space. The simplest case of weak fluctuations is addressed first.

4.1.1 Weak Fluctuations

In this section the initial development of acoustic fluctuations is examined, and as such these fluctuations in phase and amplitude will be considered small. A useful conceptual model is to consider a plane wave impinging on a region of weakly variable sound speed (Figure 4.1). Different parts of the plane wave encounter regions of faster and slower sound speed that advance or delay the front. In addition, if the frequency is low enough, the front may be distorted by diffraction through the sound-speed heterogeneities. In either case, the front will no longer be planar but will become corrugated . Since the focus is on weak fluctuations,

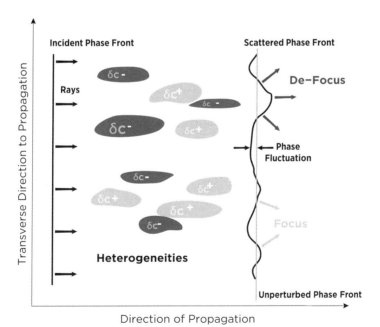

Figure 4.1. Using the Born and Rytov theory, phase fluctuations appear to first order and give rise to focusing and de-focusing (e.g., intensity variability). Focusing can be understood in terms of the bending of the phase front. The physical picture is of multiple weak multiple forward scattering.

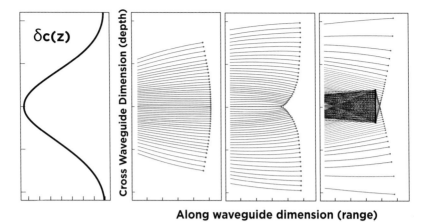

Figure 4.2. From left to right are shown three stages of wave front folding caused by the low sound-speed zone depicted at far left. Lines mark distinct ray paths, and circles at the ends of the rays show the location of the wave front. In the first frame the initial circular arc of the wave front is straightened out nearly into a plane wave by the refraction effects of the low-speed zone. By the second frame the low-speed zone produces a significant distortion of the front with a focusing zone near the low-speed axis. By frame three rays on either side of the focusing zone have crossed over one another and created a triplication.

the corrugations of the front will be much less than an acoustic wavelength. In the approximation that internal-wave currents are not important, the acoustic energy flux is normal to the front (Munk et al., 1995) and thus regions that are advanced, having a positive wave front curvature, will be defocusing, while the regions that are delayed, having a negative wave front curvature, will be focusing. This weak focusing and defocusing due to wave front curvature is the cause of amplitude variability along the wave front. Now a wave front is a surface of constant phase; however, the advance or delay of the front can be considered a phase fluctuation relative to the unperturbed planar front. Thus geometric distortion of a wave front leads to both amplitude and phase fluctuations, and this is the primary mechanism whereby ocean acoustic variability is initiated (Clifford, 1978). As the refractive and diffractive distortion of the wave front becomes more severe, new mechanisms become important, namely wave front folding and interference.

4.1.2 Strong Fluctuations: Wave Front Folding and Interference

Eventually, as a wave front is strongly distorted by variations in sound speed, a focusing region will lead to a folding or triplication of the wave front (Figure 4.2) (Duda et al., 1992; Duda and Bowlin, 1994). A triplication occurs when opposite sides of a focusing region cross one another or fold over, forming three segments of the wave front where there was once only one. The triplication process is familiar in other areas of wave propagation, particularly seismology (Chapman, 2004). At low frequencies, diffraction effects can enhance or diminish the triplication process. When a triplication does form, two permanent focusing regions or caustics are generated that lead to locally increased intensity. The mechanism of ray triplication is the same one that gives rise to the folded wave front described in Chapter 2, only in this case the triplication is generated locally by the random sound-speed structure and is superimposed on the larger wave front pattern. It is important to realize that in the triplication the three segments of the wave front may in fact be extremely close to one another and thus, depending on the signal bandwidth, there can be interference effects. In practice these triplications are extremely thin in the horizontal direction, extending perhaps a few meters, and interference effects are therefore possible. While caustics are regions of enhanced acoustic intensity, interference effects can either enhance or diminish acoustic intensity. In this regime with caustics and interference, acoustic intensity fluctuations can be large. In addition, acoustic phase fluctuations can also become large, both from the increasing wave front distortion effect described in the previous section and also from interference and caustic formation effects. After each caustic there is an added phase shift of $\pi/2$, as described in Chapter 2.

The wave front triplication process that is caused by randomly occurring local regions of low sound speed can, of course, take place repeatedly as the sound travels though the ocean. Thus triplication is an important mechanism that can create many closely spaced ray paths where there was once only one ray path; the original ray is seen to fracture into many rays. These new rays are termed microrays, and the resulting interference pattern is termed micro-multipath interference. Some important properties of microrays are discussed in the following sections.

4.1.3 Ray Micro-multipath

As was developed in Chapter 2, the Fresnel zone defines the region around a ray in which meaningful interference can occur, and so the extent to which microrays fall within or outside of this zone has a strong impact on the acoustic variability. To better understand this notion, consider the situation of a point source in a region of constant background sound speed with sound-speed fluctuations from internal waves superimposed (Figure 4.3). At some location x consider that there is a thin slab of sound-speed perturbation in the (y,z) plane that will distort the wave front which is observed at the point $(\mathbf{x} = R, 0, 0)$. In the absence of sound-speed fluctuations and considering acoustic propagation primarily in the x-direction, it was shown in Chapter 2 that the travel time of an acoustic pulse moving along the

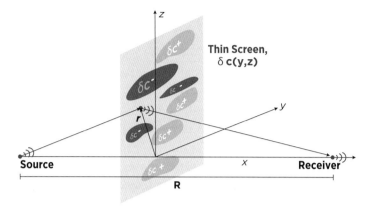

Figure 4.3. Sound propagation through a thin screen of random sound-speed fluctuations. The vector **r** is in the plane of the thin screen.

path going through the point (y, z) is to first order

$$T(y,z) = T_0 + \frac{|\mathbf{r}|^2}{R_f^2(x)}\left(\frac{T_p}{2}\right), \tag{4.1}$$

where $|\mathbf{r}| = (y^2 + z^2)^{1/2}$, $T_0 = R/c_0$ is the travel time of the path going along the x-axis, R_f is the Fresnel zone (Eq. 2.90), and $T_p = 2\pi/\omega$ is the acoustic wave period. The travel-time function $T(y,z)$ is seen to be a parabolic surface, and as such the $|\mathbf{r}| = 0$ path is the "least time" or geometric ray path (Fermat's principle). Now the sound-speed fluctuations are turned on in the thin slab, delaying or advancing the acoustic pulse as it passes through each point (y, z). This travel time change is taken to be the change that would occur over the entire triangular path (Γ) passing through (y, z) (often called the phase screen method), that is,

$$\delta T(y,z) = -\int_\Gamma \frac{\delta c(x,y,z)}{c_0^2} ds. \tag{4.2}$$

Thus the travel time along the paths is

$$T(y,z) = T_0 + \frac{|\mathbf{r}|^2}{R_f^2(x)}\left(\frac{T_p}{2}\right) + \delta T(y,z). \tag{4.3}$$

The function $\delta T(y,z)$ is a random function that depends on the strength and correlation properties of the sound-speed fluctuations, but it can be said in general that this contribution can lead to new extrema along the surface $T(y,z)$ and thus the generation of new Fermat paths or microrays. Because of the parabolic shape of the background travel-time function, microrays are most likely to occur where $|\mathbf{r}|$ is small; that is, the microrays will cluster close to the unperturbed ray. The steepness of the parabolic travel-time function is set by the Fresnel zone, and thus the Fresnel zone dictates how far away from the unperturbed ray the microrays can meander (to first approximation). However, rays represent lines along which acoustic energy travels only in the infinite-frequency limit. At finite frequencies, two microrays may not be separated far enough to be distinct from one another. Therefore, the interference and resulting phase and amplitude fluctuations are a function not only of the sound-speed field and geometry, but also of the acoustic frequency and bandwidth.

Regarding microrays clustering near the unperturbed ray, it has been found that as the distance between source and receiver becomes large, or the vertical wandering of microrays becomes comparable with the width of the ocean acoustic waveguide, a further effect known as ray chaos may become important (Section 2.2), and new techniques are needed to understand quantitatively the phase and amplitude fluctuations (Simmen et al., 1997; Beron-Vera et al., 2003). In particular, in this regime the parabolic surface $T(y,z)$ described in the previous section no longer exists and a much more complicated surface takes its place. Further discussion of this problem is deferred to Chapter 5.

The intensity that is observed at a particular point in space will depend on the interference pattern caused by the superposition of the microrays, which have both phase and amplitude variations. The amplitudes and phases of the microrays will follow random distributions. If the number of microrays is large, but their relative phase fluctuations are small compared with π radians, then the total intensity fluctuations will not be large. In order to have large intensity fluctuations, the relative phase fluctuations of the microrays must be comparable to or larger than π radians. Thus it can be concluded that phase fluctuations develop first with small amplitude variability, and only after phase fluctuations have grown to a sufficient level and triplications have occurred can significant amplitude variability result.

The previous discussion has been for the case of a constant background sound speed and thus straight line ray propagation. However, there is the issue of the effect of the sound channel. This will be addressed in detail in the subsequent chapters, but for the moment this straight ray picture can be considered to apply locally along the curved ocean acoustic ray paths.

Wave Front Distortion: Microray Scattering Along the Wave Front

One of the fundamental aspects of microray behavior in the ocean is that the scattered microrays primarily meander along the wave front rather than across it (Flatté and Colosi, 2008). A narrow pulse emitted by a source will evolve by and large into a wave front that geometrically is similar to the wave front one would expect in an ocean without sound-speed fluctuations, and thus a series of narrow pulses would be observed at a distant receiver. Figure 4.4 shows the distribution of internal-wave-induced microrays that have a fixed launch angle. These microrays have a large spread distributed along the wave front, with a small spread across

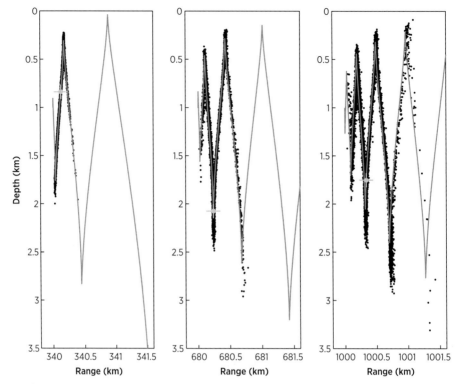

Figure 4.4. Range evolution of the ray endpoints for 12,000 realizations of ray propagation through internal-wave-induced sound-speed perturbations superimposed on a Munk canonical profile. For all realizations the initial ray angle is +7° and the initial depth is 1000 m. The gray curve shows the unperturbed wave front from a point source, and the gray plus-dot shows the location of the unperturbed +7° ray.
Source: Flatte and Colosi (2008).

the wave front. These microrays induce significant intensity variability along the wave front due to microray interference from wave front folding. In this section this phenomenon is discussed qualitatively and scaling relationships are obtained to estimate microray distributions along and across the wave front.

The dominance of scattering along the front as opposed to across the front comes from the multiple small-angle forward scatterings caused by internal waves. Thus, when a ray propagating in the x direction is scattered by a small-angle θ, the deviation of this ray from the unperturbed ray will scale as $\cos\theta$ in the x direction and $\sin\theta$ in the vertical or wave front direction. Hence in the small-angle approximation the effect will be second order across the front and first order along the wave front.

Consider the case of no waveguide. If a ray propagating in the x direction experiences an angular deviation at some location $\theta(x_1)$, then the ray will propagate to the receiver range along a straight line with a vertical deviation at the receiver of

$z(x_1) = (R - x_1)\theta(x_1)$, or $dz/d\theta = (R - x_1)$. Summing up the scattering contributions from all locations x_1, then the total vertical deviation along the unperturbed wave front will be (Flatté and Colosi, 2008)

$$Z(R) = \int_0^R \frac{dz}{d\theta} d\theta = -\int_0^R (R - x_1)\mu' dx_1, \qquad (4.4)$$

where Snell's law has been used to give the refraction angle change in terms of the sound-speed change or $d\theta \simeq -\mu' dx$ where μ' is the vertical derivative of the fractional sound speed $\mu = \delta c/c_0$. Thus it is seen that there is an important moment arm involved in the growth of Z with range R such that an angular scattering event near the source can give rise to a larger depth deviation than an equal event occurring near the receiver. It has been shown that the variance of Z is given by (Flatté and Colosi, 2008)

$$\langle Z^2 \rangle \simeq \frac{R^3 \langle \mu^2 \rangle L_H}{3L_z^2}, \qquad (4.5)$$

where L_H and L_z are the horizontal and vertical correlation lengths of the internal waves, respectively. The deviation across the front is due to the advance or delay of the microrays, which is given by $X = c_0 \delta T = -\int_0^R \mu dx$ and thus

$$\langle X^2 \rangle \simeq R \langle \mu^2 \rangle L_H. \qquad (4.6)$$

Hence the rms along wave front vertical deviation grows as $R^{3/2}$ while the across front horizontal deviation only grows as $R^{1/2}$. Using typical values for ocean internal waves, such as $L_H = 10$ km, $L_z = 0.1$ km, and $\langle \mu^2 \rangle = 10^{-8}$, and $R = 1000$ km, the result is $Z_{rms} \simeq 60$ km, and $X_{rms} \simeq 10$ m. Calculations of this sort taking into account the ocean waveguide have been done (see Chapter 5), and the results are similar with a reduction in Z_{rms} by about a factor of 10.

To understand the significance of these results, one must realize that rays in the ocean are spread out over a range of angles between $\pm15°$. A spherical wave with this angular range would have a wave front arc-length of roughly 500 km, at 1000 km range. Thus an rms along-wave-front deviation of 60 km represents a spread over roughly 10% of the wave front, which is a significant amount of spreading. This process then is extremely important to describing and quantifying the effects of microray interference that leads to amplitude and phase variability, as well as the degradation of acoustic coherence.

This microray scattering result can also be applied in the horizontal plane. In this case the microray horizontal spread is given by

$$\langle Y^2 \rangle \simeq \frac{R^3 \langle \mu^2 \rangle}{3L_H}. \qquad (4.7)$$

Using parameters typical of ocean internal waves as described above, an rms horizontal deviation at 1000 km range is predicted to be roughly 600 m. This relatively small horizontal spread indicates that one can largely ignore out of plane scattering in the deep-water problem.

Wave Front Distortion: Microray Scattering into Shadow Zones

Figure 4.4 also shows that the scattered microrays are penetrating the shadow zones past unperturbed caustic regions. This microray behavior demonstrates another fundamental aspect of ocean acoustic propagation through internal waves in both shallow and deep-water environments; that is, wave front shape near shadow zones can be significantly changed to first order (Figures 1.6 and 1.7). The internal-wave effect is not simply to introduce a fluctuation whose mean is the unperturbed acoustic field, but there can be significant biases in the intensity and phase. In shallow-water acoustic propagation, the effect is particularly striking because of the differential seafloor attenuation that exists for acoustic energy propagating at different angles with respect to the horizontal. Internal-wave-induced scattering, which alters the propagation angle, thus introduces different seafloor attenuation effects and can modify the mean properties of the signal. Experimental examples of microray scattering into shadow zones and of biases in intensity and phase will be discussed later in Chapters 5 and 8.

Microray Bundles

There is one final aspect of the microray problem that must be addressed, which is the shape of eigenray microray bundles. These eigenray bundles are formed because each of the microrays created by the fluctuations has a different ray path connecting the source and the receiver. Of critical importance to acoustic fluctuations is the vertical spread of these bundles relative to the vertical Fresnel zone, R_f, and the vertical correlation length of the sound-speed fluctuations, L_z. Figure 4.5 shows a ray simulation of a microray bundle for 1000-km propagation in a mid-latitude sound-speed profile. The bundle is seen to be spread vertically of order hundreds of meters and horizontally of order several kilometers. The important microray vertical spreading factor is denoted as R_{mr}, in analogy with the vertical Fresnel zone R_f. A precise mathematical definition will be given in Chapter 5. With regard to time, the eigenray microrays are spread of order several milliseconds and have an average travel time within a few milliseconds of the unperturbed travel time. Not shown is the unperturbed ray path, which is located roughly in the middle of the bundle and has the same number of ray turning points. The spreading of the microray bundle, just like the Fresnel zone, will be a function of the shape of the background sound-speed profile, and the grazing angle of the corresponding unperturbed ray. Unlike the Fresnel zone, R_{mr} will be a function of the strength and spectral form of the internal waves.

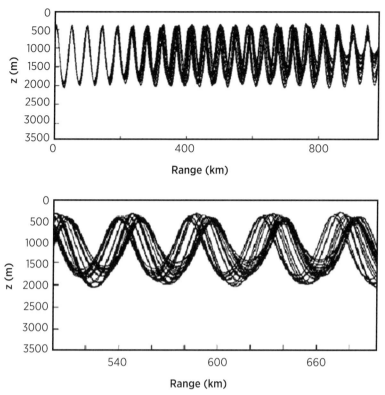

Figure 4.5. A simulation of a microray bundle for 1000-km propagation through random internal waves in a mid-latitude sound-speed profile. The source and receiver are on the sound channel axis which is 800 m depth. The microrays are spread over both depth and range.
Source: Simmen et al. (1997).

4.1.4 A Simple Model of Microray Interference

As a simple model of microray interference, consider a model in which each microray is a Gaussian pulse with center frequency ω_0 and bandwidth (standard deviation) $\Delta\omega$. Here the demodulated field at the mean travel time of the multipaths $\bar{t}$ is (Colosi and Baggeroer, 2004),

$$p(\bar{t}) = e^{-i\omega_0\bar{t}} \sum_{n=1}^{N} a_n \, e^{-\delta t_n^2 \Delta\omega^2/2} \, e^{-i\omega_0\delta t_n} = e^{-i\omega_0\bar{t}} \sum_{n=1}^{N} a_n \, e^{-\gamma^2\delta\Theta_n^2/2} \, e^{-i\delta\Theta_n}, \qquad (4.8)$$

where N is the number of microrays, $\gamma \equiv \Delta\omega/\omega_0$ is the fractional bandwidth, and $\delta\Theta_n \equiv \omega_0\delta t_n = \omega_0(\bar{t} - t_n)$ is the phase change associated with the fluctuating travel times of the pulses. To get useful analytic results a few well-justified assumptions concerning the microrays must be made. First, because significant acoustic fluctuations depend on independent microray contributions, it is assumed that there are no correlations between the N microrays; that is to say the total number of microrays may in fact be greater than N, but only the effective number of independent contributions is treated. Further, it is assumed that the phase and amplitude of the microrays are independent. This assumption is well justified because in ray theory the amplitude of a ray is controlled by sound-speed curvature, while the phase of a ray is controlled by sound speed itself and its gradients. Because these aspects of the microrays are controlled by difference scales of the sound-speed structure, the phase and amplitude are expected to be independent. So, the statistics of the microrays are such that

$$\langle a_n \delta\Theta_j \rangle = 0, \quad \langle a_n a_j \rangle = \langle a \rangle^2 (n \neq j), \quad \langle \delta\Theta_n \delta\Theta_j \rangle = \Phi^2 \delta_{n,j}. \tag{4.9}$$

With the stated assumptions, one need not specify the PDF of the amplitudes a_n, although research has shown that the log-normal distribution accurately characterizes the PDF at short and long ranges (Flatté et al., 1979; Wolfson and Tomsovic, 2001). However, the phase PDF must be specified, and a Gaussian distribution is a reasonable starting point because (1) the phase fluctuation is directly related to travel time; and (2) the random sound-speed fluctuations causing the travel-time variability are nearly Gaussian (a fundamental assumption of the GM internal-wave model).

The first important quantity to examine is the mean pressure $\langle p \rangle$, which in the narrowband limit is computed to be (Colosi and Baggeroer, 2004)

$$\langle p \rangle = \sum_{n=1}^{N} \langle a \rangle \langle e^{i\delta\Theta_n} \rangle = N\langle a \rangle \, \exp\left(-\frac{\Phi^2}{2}\right). \tag{4.10}$$

Taking into consideration the bandwidth the result is (Colosi and Baggeroer, 2004)

$$\langle p \rangle = \frac{N\langle a \rangle}{(1+\gamma^2\Phi^2)^{1/2}} \, \exp\left(-\frac{\Phi^2}{2(1+\gamma^2\Phi^2)}\right). \tag{4.11}$$

Here the critical role of the phase fluctuations is seen, which drives the mean pressure to zero exponentially. There is also an important parameter, the time-bandwidth product $\gamma\Phi$, which quantifies the amount of interference that can occur between microrays. If the bandwidth is large (i.e., large γ) then the pulses will be narrow in time, and thus two microrays need to be close to one another to interfere. Hence a large/small time bandwidth product reduces/enhances microray interference. For this particular observable, the mean pressure, a large time bandwidth product clearly slows the decay of the mean pressure to zero as the phase variance increases.

The next important quantity is the scintillation index SI which is the normalized intensity variance:

$$SI = \frac{\langle I^2 \rangle - \langle I \rangle^2}{\langle I \rangle^2}. \tag{4.12}$$

In the narrowband case and assuming large phase variance ($\Phi^2 \gg 1$), the mean intensity and mean square intensity can be written as

$$\langle I \rangle = \sum_{k=1}^{N} \sum_{j=1}^{N} \langle a_k a_j \rangle \langle e^{i(\delta\Theta_k - \delta\Theta_j)} \rangle \simeq N\langle a^2 \rangle, \tag{4.13}$$

$$\langle I^2 \rangle = \sum_{k=1}^{N} \sum_{j=1}^{N} \sum_{m=1}^{N} \sum_{n=1}^{N} \langle a_k a_j a_m a_n \rangle \langle e^{i(\delta\Theta_k - \delta\Theta_j + \delta\Theta_m - \delta\Theta_n)} \rangle \simeq N\langle a^4 \rangle + 2N(N-1)\langle a^2 \rangle^2. \tag{4.14}$$

Thus in this limit, the scintillation index is found to be (Colosi and Baggeroer, 2004)

$$SI \simeq 1 + \frac{1}{N}\left(\frac{\langle a^4 \rangle}{\langle a^2 \rangle^2} - 2\right) \qquad \text{Narrowband, } \Phi \gg 1. \tag{4.15}$$

A more complicated calculation taking into consideration the signal bandwidth gives (Colosi and Baggeroer, 2004)

$$SI \simeq 1 + \frac{1}{N}\left(\gamma\Phi\frac{\langle a^4 \rangle}{\langle a^2 \rangle^2} - 2\right) \qquad \text{Broadband, } \gamma\Phi \gg 1. \tag{4.16}$$

Here the term $\langle a^4 \rangle / \langle a^2 \rangle^2$ is a measure of the amount of amplitude fluctuation among the microrays. Large/small values for this term mean large/small microray amplitude fluctuations. Several points are noteworthy from Eqs. 4.15 and 4.16. First, as range increases it is expected that both N and $\langle a^4 \rangle / \langle a^2 \rangle^2$ will increase. The presumption is that N will grow more quickly, and thus at long range SI is expected to "saturate' at a value of 1. The notion of saturation will be central to the subsequent discussions of acoustic fluctuations. Next, Eqs. 4.15 and 4.16 show that the saturation limit can be approached from either above or below 1 depending on the strength of the microray amplitude variability. In the narrowband case, the transition from an approach from above 1 to below 1 is for $\langle a^4 \rangle / \langle a^2 \rangle^2 = 2$, while the broadband case requires $\gamma\Phi \langle a^4 \rangle / \langle a^2 \rangle^2 = 2$ with the additional requirement that $\gamma\Phi \gg 1$.

To get a feeling for the orders of magnitude, some sample PDFs for the multipath amplitudes (a) can be examined. For the idealized case of no amplitude fluctuations (i.e., where $\langle a^4 \rangle / \langle a^2 \rangle^2 = 1$), the broadband case yields $\gamma\Phi = 2$ for the transition. The narrowband case always approaches 1 from below (Dyer, 1970), and thus some variability in the amplitudes is needed to obtain saturation from above 1. As previously discussed, analysis from ray theory indicates that the PDF of a is close to log-normal (Wolfson and Tomsovic, 2001; Beron-Vera et al., 2003), and for a log-normal amplitude PDF with variance σ_χ^2 one obtains $\langle a^4 \rangle / \langle a^2 \rangle^2 = e^{4\sigma_\chi^2}$. Therefore the condition for $\langle a^4 \rangle / \langle a^2 \rangle^2 = 2$ is $\sigma_\chi = (\ln 2)^{1/2}/2 \to 1.8$ dB, which is indeed a modest value for the amplitude variation. In the ray view, therefore, an approach to saturation from above 1 is much more likely. This issue will be discussed in more detail in Section 4.3 when the important process of diffraction is introduced.

4.1.5 Modal-multipath: Mode Coupling

Multipath also appears within the mode formalism, although the physical interpretation is not quite the same. Range-dependent sound-speed structure, like that caused by internal waves, results in mode coupling, which randomizes the modes in both amplitude and phase. When a group of propagating modes interacts with a region of variable sound-speed, energy will be exchanged between the modes, and their phases will be altered due to mode–mode interaction and adiabatic effects (Dozier and Tappert, 1978a; Colosi and Flatté, 1996; Colosi et al., 2013a). Acoustic fluctuations result when the randomized modes are summed to form the acoustic pressure. Physically mode coupling results from variable sound-speed structure that significantly changes the vertical angle of acoustic propagation. Thus acoustic energy that is detected in some mode number n at the receiver will not in general be the energy that originated in mode n at the source. In fact, some proportion of the detected energy will have traveled from source to receiver via several other modes (with different phase and group velocities), and it is in this sense that mode coupling causes what could be termed "modal-multipath."

A simple example of mode coupling is obtained using the sudden approximation (Merzbacher, 1961). Assuming there are two different sound-speed profiles on either side of the origin $x = 0$ (call them region 1 and 2), then the pressure fields p_1 and p_2 are represented as a sum over the normal modes in each region, that is,

$$p_1(x,z) = \sum_{n=1}^{N} a_n(1)\phi_n(z;1)\, e^{ik_n(1)x}; \; x < 0,$$

$$p_2(x,z) = \sum_{n=1}^{N} a_n(2)\phi_n(z;2)\, e^{ik_n(2)x}; \; x > 0. \tag{4.17}$$

Continuity of pressure at $x = 0$ means that the modal amplitudes on the far side of the transition $a_n(2)$ can be expressed in terms of the incident mode amplitudes $a_n(1)$, giving

$$a_n(2) = \sum_{m=1}^{N} a_m(1)S_{mn}, \tag{4.18}$$

$$S_{mn} = \int_0^{\infty} \phi_n(z;1)\phi_m(z;2)\, dz, \tag{4.19}$$

where S_{mn} is the mode scattering matrix. In this approximation the modes are seen to change amplitude due to coupling, and the modal phases change due to the changing horizontal wavenumber (i.e., $k_n(1)$ and $k_n(2)$). The strength of the modal mulitpathing depends on the magnitude of the off-diagonal elements of the scattering matrix. For internal waves in the ocean and for both shallow and deep-water problems, the scattering matrix values are generally largest in the neighborhood of the diagonal; that is, near neighbor coupling is dominant. Because mode coupling can be considered to be a change in acoustic propagation angle, near neighbor coupling is a consequence of the generally weak, small-angle forward scattering that occurs from ocean internal waves.

In this formalism one can see that the modal multipath can become quite complex if there are even a small number of coupling events. If it is assumed that the modes primarily couple into J near neighbors, and there are M scattering events, then the total number of modal multipaths is J^M! Thus after propagation through several scattering events the modes will become randomized, and saturation will be observed. Because the pressure is a sum over all the modes (Eq. 4.17), it has a form close to the sum over random waves (Eq. 4.8) with $\gamma = 0$. Thus the approach to saturation will be of the form of Eq. 4.15 with routes from above and below 1.

4.2 Acoustic Sensitivity to Internal Waves

When an acoustic wave field interacts with the internal-wave field, the strength of the interaction will depend not only on the energy of the internal-wave field but also on the specific spatial scales and orientation of the internal waves (Munk and Zachariasen, 1976; Flatté et al., 1979). Here the term strength of interaction is used somewhat ambiguously because the interaction depends critically on the specific acoustic observable of interest. Nonetheless, the sensitivity of the acoustic field to internal waves can be discussed using the conceptual models introduced in the previous sections. From a ray perspective, it was found that internal waves whose crests are aligned with the unperturbed ray path will have much larger effects on the acoustic phase and amplitude than internal waves whose crests are perpendicular to the ray. A simple cartoon illustrates this point (Figure 4.6). When the internal-wave crest is aligned with the ray, the wave front becomes corrugated, as was described in Section 4.2 and Figure 4.1. The results of this orientation are amplitude and phase fluctuations. However, if the internal-wave crests are perpendicular to the ray path, the wave front is not corrugated but is alternately advanced and delayed as a whole; small phase fluctuations and no intensity fluctuation result. An important assumption has been made in this argument, namely that it makes sense to think about the effects of internal waves

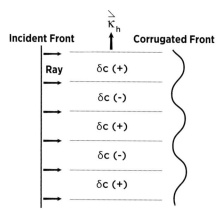

Internal Wave Crests and Troughs Parallel to the Ray

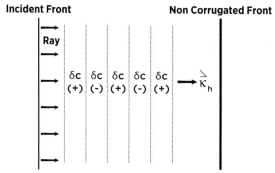

Internal Wave Crests and Troughs Perpendicular to the Ray

Figure 4.6. Plane wave sound propagation through internal waves whose crests are parallel (left) and perpendicular (right) to the rays. Plus and minus signs denote regions along the internal wave where the sound-speed perturbation is either positive or negative.

near the ray that would exist in the absence of internal waves (the unperturbed ray). In general, internal-wave-induced sound-speed fluctuations will alter the unperturbed ray path (ray chaos), and this issue, which is important in long-range acoustic propagation, will be discussed in Section 4.2.2.

Further insight for intermediate angles can be obtained by considering the acoustic observable of phase for a straight line ray path along the x-axis. An internal wave with wavenumber κ is assumed to be propagating across this ray path at some angle θ_{IW} relative to the x-axis. The fractional sound-speed perturbation along the acoustic ray path can be written as $\mu(x) = \mu_0 \cos(\kappa \cos\theta_{IW} x)$, where μ_0 is the fractional sound-speed perturbation amplitude. The phase fluctuation is then $\delta\Theta = \omega\delta t = -q_0 \int_0^R \mu(x)dx$, or explicitly

$$\delta\Theta = -q_0\mu_0 R \, \text{sinc}(\kappa\cos\theta_{IW}R) = -2\pi \, q_0\mu_0 \, \delta(\kappa\cos\theta_{IW}), \quad R \to \infty. \quad (4.20)$$

Thus in the limit $R \to \infty$ it is seen that the only nonzero contribution comes when $\theta_{IW} = 90°$; that is, the wave whose crests are aligned with the ray path or alternatively the wave whose wavenumber is perpendicular to the ray path. This "resonance" condition in which only the waves whose crests are aligned with the ray are considered will be used extensively in the theory of sound transmission through random internal-wave fields. This is discussed next in Section 4.2.1.

4.2.1 Ray/Internal Wave Resonance and Diffraction

Here results from weak fluctuation theory are presented, without derivation, with the objective of demonstrating ray/internal wave resonances and the effects of diffraction. These results will be rigorously obtained in Chapter 6, but they will be conceptually understandable based on what has been presented already related to ray paths, Fresnel zones, and the importance of internal waves that are propagating perpendicular to a ray. Assuming acoustic propagation through fields of randomized internal waves and using the weak fluctuation method (Rytov, 1937; Munk and Zachariasen, 1976) investigators have been able to obtain expressions for the variances of phase and log-amplitude expressed as integrals over the unperturbed ray path Γ_0, giving

$$\langle \phi^2 \rangle, \ \langle \chi^2 \rangle = \pi q_0^2 \int_{\Gamma_0} ds \int d\boldsymbol{\kappa}_\perp \ F_\mu(\boldsymbol{\kappa}_\perp) \left[1 \pm \cos\left(\frac{m^2(\boldsymbol{\kappa}_\perp) R_f^2(s)}{2\pi} \right) \right], \qquad (4.21)$$

where $F_\mu(\boldsymbol{\kappa}_\perp)$ is the wavenumber spectrum of the fractional sound-speed fluctuations evaluated at the wavenumbers that are locally perpendicular to the ray (i.e., crests aligned with the ray), $m(\boldsymbol{\kappa}_\perp)$ is the vertical component of the perpendicular wavenumber, q_0 is a reference acoustic wavenumber, and $R_f(s)$ is the vertical Fresnel zone. The plus sign applies to the phase, while the minus sign applies to the log-amplitude. The spectrum of μ is related to the spectrum of vertical displacement by the simple relation $F_\mu = F_\zeta (dc/dz)_p^2/c_0^2$. The variances are seen to include contributions from all positions along the acoustic path from source to receiver, and at each point along the ray all perpendicular wavenumbers for the internal waves are summed up at the strength given by the internal-wave spectrum. At each point along the ray it is seen that in addition to the resonance condition that picks out only perpendicular wavenumbers there is also the quantity in square brackets that serves as a filter function that can enhance or diminish contributions from the various perpendicular wavenumbers. This function is often termed the Fresnel filter, and it describes the effects of diffraction on the scattering.

Action of the Fresnel Filter

As has been demonstrated, the Fresnel zone is an important function that defines the region around a geometric ray path where diffraction can act (Flatté et al.,

1979). Figure 4.3 shows the Fresnel zone geometry for the case of a point source and constant background sound speed, and in Chapter 2 it was seen that when the ocean sound channel is introduced into the problem, the geometry becomes more complicated, but the conceptual picture is the same. Because of the vertical structure of the sound channel and the small vertical scale of internal waves, vertical diffractive effects are the most important.[1]

To understand the impact of the Fresnel filter on the phase and log-amplitude variance it is important to realize that the spectrum of ocean internal waves has roughly a κ^{-2} shape, and thus most of the energy is at larger scales. Therefore the first maximum of the Fresnel filter at small wavenumber $m(\kappa_\perp)$ is significant, since this is where the largest contribution to the integral comes from. For phase the first maximum occurs at $m(\kappa_\perp) = 0$ where the Fresnel filter is equal to 2. Thus phase will be sensitive to the largest vertical scales in the problem. Because the internal-wave spectrum is dominated by large vertical scales, the phase variance will almost always behave geometrically. In fact the Fresnel filter is routinely ignored (replacing it with a value of 2!) and thus to a good approximation diffraction is not important when considering phase statistics. For log-amplitude, the Fresnel filter has a zero at $m(\kappa_\perp) = 0$ and thus log-amplitude is not sensitive to large scales. The first maximum of the filter is attained when the argument of the cosine is equal to π, giving the value of m with the maximum contribution to be $m(\kappa_\perp) = \sqrt{2}\,\pi/R_f$. This result says that the most important ocean structure for log-amplitude is that with a vertical wavelength roughly equal to the vertical Fresnel zone. This result for the log-amplitude is consistent with the notion that the Fresnel zone is the scale at which diffractive paths just start to cause destructive interference, a prerequisite for amplitude variability. These results are also conceptually consistent with the formulation of ray theory in which the travel time (or phase) depends on sound-speed and its first derivative, while the ray amplitude has critical contributions from the sound-speed second derivative.

In summary, ray theory tells us that the acoustic signal is only sensitive to internal waves whose crests are aligned with the unperturbed ray, that is, only internal waves whose wavenumbers are perpendicular to the ray are important. This striking resonance condition allows a focus on a distinct subset of the ocean internal waves that are responsible for acoustic scattering. To get further traction on the issue of sensitivity to specific internal waves, one must decide which observables are of interest. Considering phase and amplitude variability, it is found from the action of the Fresnel filter that phase is sensitive to the largest scales of the ocean while amplitude is sensitive to scales near the Fresnel zone. Because the

---

[1] In Eq. 4.21 there is in fact an additional term in the argument of the cosine of the form $l^2(\kappa_\perp)R_{fy}^2(x)/(2\pi)$ where $R_{fy}(x)$ is the transverse Fresnel zone in the y-direction given by Eq. 2.90. Because $l \ll m$, this transverse term is small compared to the vertical term.

Fresnel zone changes along an ocean acoustic path, the scale sensitivity will also change along the acoustic path. Other observables can be treated. It is found that the coherence function for two receivers separated vertically will be sensitive to smaller scale internal waves, while the time coherence function, like the phase, will be sensitive to the larger-scale internal waves whose phase speeds are generally faster than the small-scale internal waves.

The Correlation Length along the Ray

In this book the resonance condition that picks out only perpendicular wavenumbers of the internal waves will often be expressed in a slightly different form than is denoted in Eq. 4.21. In particular, the integral over the perpendicular wavenumbers will be carried out analytically, resulting in a quantity $L_p(\theta, z)$ called the correlation length of the internal waves in the direction of the ray (Flatté et al., 1979; Esswein and Flatté, 1980, 1981; Flatté, 1983). Using the Garrett-Munk internal-wave spectrum, which factorizes in terms of internal-wave frequency σ and vertical mode number j, the analytical result from Eq. 4.21 is

$$\langle \phi^2 \rangle, \ \langle \chi^2 \rangle = q_0^2 \int_{\Gamma_0} ds \ \langle \mu^2(z) \rangle \ L_p(\theta, z) \ G_\pm(R_f, s), \tag{4.22}$$

$$G_\pm(R_f, s) = \frac{1}{2\langle j^{-1} \rangle} \sum_{j=1}^{\infty} \frac{H(j)}{j} \left[1 \pm \cos\left(\frac{m^2(j)R_f^2(s)}{2\pi} \right) \right], \tag{4.23}$$

where $\langle \mu^2(z) \rangle$ is the profile of fractional sound-speed variance, $L_p(\theta, z)$ is the correlation length of the internal waves in the local direction of the ray with angle θ, $H(j)$ is the vertical mode number spectrum of the internal waves, and $\langle j^{-1} \rangle = \sum_{j=1}^{\infty} H(j)/j$. The WKB vertical wavenumber is taken to be $m(j) = \pi j N(z)/N_0 B$, with $N_0 B = \int_0^D N(z)dz$. An analytic formula for L_p using the Garrett-Munk spectrum is given in Chapter 5. Figure 4.7 shows the behavior of L_p as a function of ray angle and depth for a canonical buoyancy frequency profile. L_p has its largest value at nearly zero ray angle because internal waves have a much larger horizontal correlation length than vertical correlation length (vertical–horizontal anisotropy). Now it must be remembered that L_p is computed only with the perpendicular wavenumbers and thus L_p will not be precisely equal to L_H at $\theta = 0$, and it will not be equal to L_z for $\theta = 90°$ (although the values are close). The horizontal and vertical correlation lengths in the perpendicular wavenumber condition are respectively denoted $\hat{L}_H$ and $\hat{L}_z$. The small maxima in L_p near zero angle occur because the maximum correlation length along the ray is at the angle $\theta = \tan^{-1} \hat{L}_z/\hat{L}_H$. The depth dependence of L_p is due to the depth dependence of $\hat{L}_z$ and $\hat{L}_H$.

Another way of thinking about the function L_p is as follows. Because of the wave nature of internal waves there is a direct connection between spatial

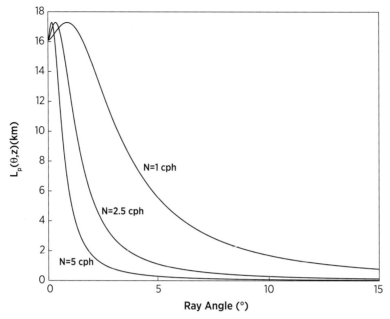

Figure 4.7. The correlation length of internal waves, $L_p(\theta, z)$, as a function of ray angle and depth, computed using the Garrett-Munk internal-wave spectrum and a canonical buoyancy frequency profile $N(z) = 5\exp(z/1000)$ cph. The latitude is taken to be $30°$. The curve furthest to the left is for a depth near the sea surface ($N = 5$ cph), the next curve is for a depth in the main thermocline ($N = 2.5$ cph), and the curve to the right is for a depth below the thermocline ($N = 1$ cph).

and temporal behaviors that is dictated by the dispersion relation. Hence the perpendicular wavenumber resonance means that only internal waves with particular frequencies will be able to interact with the acoustic field. In Chapter 5 it is shown that the perpendicular wavenumber resonance imposes a minimum internal wave frequency σ_L for interaction with the acoustic field (Munk and Zachariasen, 1976). This cut-off frequency depends on the ray angle and the buoyancy frequency and is given by

$$\sigma_L(z_{ray}) = \left(f^2 + N^2(z_{ray}) \tan^2 \theta_{ray} \right)^{1/2}. \tag{4.24}$$

An important part of this equation is the dependence on ray angle. Because of the perpendicular wavenumber resonance, rays with a large angle will be too steep to interact with the energetic low-frequency internal waves in the ocean. Recall from Chapter 3 that low-frequency internal waves have phase propagation and hence wavenumbers that are primarily in the vertical direction. If a ray is tilted at a significant angle these waves cannot be perpendicular to that ray. When the ray goes through a turning point ($\theta_{ray} \simeq 0$) then σ_L is equal to the Coriolis frequency, and thus the sound field can interact with all the energetic internal waves at low frequency. Because of this angular dependence of σ_L, L_p will be largest for near-zero angles, and decay rapidly for larger angles. Figure 4.8 shows the cut-off frequency along a steep angle ray in a mid-latitude environment. The cut-off frequency attains its largest values near the upper turning points where N is largest.

The behavior of L_p as a function of angle demonstrates the *anisotropy* of acoustic scattering by internal waves because the contributions depend strongly on ray angle. The changes in L_p and μ^2 with depth demonstrate the *inhomogeneity* of scattering by internal waves, which will be discussed in more detail in the next section.

Turning Point Contributions

Internal-wave-induced scattering involves contributions from all along the acoustic ray path, but the region of the ray near the upper turning point proves to be the most important (Flatté et al., 1979). From Eq. 4.22 the strength of scattering along the ray is determined by the function $q_0^2 \langle \mu^2 \rangle L_p$. In the previous section it was seen that L_p is largest near ray angles of zero, so there is already an indication that turning points will be important. In Chapter 2 it was shown that the sound-speed fluctuation from a vertical displacement is equal to the displacement times the potential sound-speed gradient (Eq. 2.32) and that the potential sound-speed gradient in a canonical ocean is proportional to N^2. Combining this result with the WKB depth scaling of internal-wave displacements (Eq. 3.61), a useful canonical model for the fractional sound-speed variance is

$$\langle \mu^2(z) \rangle = \mu_0^2 \frac{N^3(z)}{N_0^3}, \qquad (4.25)$$

where $\mu_0^2 \simeq 6.26 \times 10^{-8}$ and $N_0 = 3$ cph (Munk, 1974). Here the strength of the sound-speed fluctuations in the upper ocean is readily apparent. Thus ocean internal waves strongly influence the region near to the upper turning point of a ray. Figure 4.8 shows an example of this effect in which $q_0^2 \langle \mu^2 \rangle L_p$ is plotted along a curving ray path. The important scattering region around the upper turning point is quite narrow indeed. Subsequent results in Chapters 5 through 7 will show that because of ray curvature effects the important scattering region around the ray upper turning depth is somewhat larger than indicated here, but the basic importance of the upper turning point is fundamental (Colosi et al., 1999; Flatté and Rovner, 2000). Figure 4.8 also shows how different angle rays are sensitive to the internal-wave field. Steep grazing angle rays that travel near the sea surface have a narrow region near the upper turning point where L_p has significant values, and $\langle \mu^2 \rangle$ is large. If a steep grazing angle ray reflects off the sea surface, L_p is small because the ray never actually has a zero angle. Small grazing angle rays on the other hand can have significant values of L_p over their entire trajectory, but travel

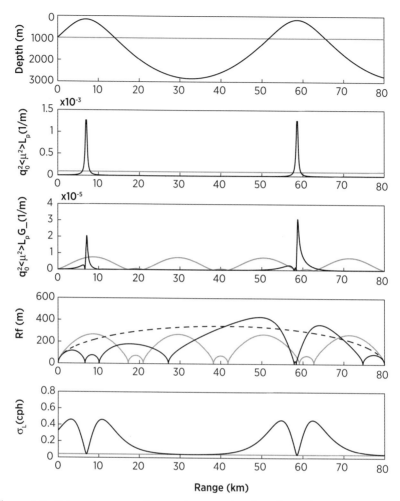

Figure 4.8. Internal wave and acoustic scattering functions along two geometric ray paths (upper panel) with launch angles of $0°$ (gray) and $10°$ (black). The calculations are done using a Munk canonical sound-speed profile and an exponential buoyancy frequency profile of the form $N(z) = 5\exp(z/1000)$ cph. The acoustic frequency is 250 Hz, and the latitude is taken to be $30°$. The second and third panels show the phase and log-amplitude variance weighting functions, $q_0^2\langle\mu^2\rangle L_p$ and $q_0^2\langle\mu^2\rangle L_p G_-$ respectively. The rms phases Φ for the $0°$ and $10°$ rays respectively are 4.3 and 2.4 radians, while the rms intensities respectively are 4.5 and 2.3 dB. The fourth panel shows the vertical Fresnel zone along the rays and the constant background sound-speed Fresnel zone displayed with a dashed line. The last panel shows the lower cut-off frequency along the ray σ_L.

in regions of smaller $\langle\mu^2\rangle$ since they are trapped near the sound-channel axis. The question as to which rays, large or small grazing angle, have the most scattering effect will depend on which acoustic observable is being considered. For phase variance, the Fresnel filter can be approximated by a constant, 2, and it is seen that the entire effect comes from $q_0^2\langle\mu^2\rangle L_p$. Because of the trade-offs just mentioned for small and large grazing angle rays the phase variance does not vary dramatically as a function of ray path, except for the case in which the ray reflects off the ocean surface. For the example shown in Figure 4.8, the computed rms phase fluctuation for the axial ray and a $10°$ grazing angle ray are 4.3 and 2.4 radian respectively. For the log-amplitude variance, the Fresnel filter cannot be easily approximated, and the entire function $q_0^2 \langle \mu^2 \rangle L_p G_-$ must be considered. Here the axial and steeper rays have rms intensities 4.5 and 2.3 dB, respectively (Figure 4.8).

4.2.2 Ray Chaos

The previous discussion concerning the effects of internal waves on acoustic wave fields made use of the important linear approximation that the sound-speed fluctuations do not significantly affect the unperturbed ray path; this approximation is often termed expansion about the deterministic or unperturbed ray. Hence Eqs. 4.21 and 4.22 involved integrating the scattering functions along the unperturbed ray path Γ_0. However, it is now well established that internal-wave-induced range-dependent sound-speed structure leads to an instability of the unperturbed path (Tappert and Tang, 1996; Simmen et al., 1997; Beron-Vera et al., 2003; Brown et al., 2003). This instability fundamentally changes the sensitivity of the acoustic field to the spatial and temporal structure of the internal-wave field; that is, the perpendicular wavenumber resonance will break down at some range and the acoustic field will be sensitive to many more wavenumbers of the ocean internal-wave field. This instability is termed ray chaos. As discussed in Chapter 2, it is directly related in many ways to dynamical systems chaos (Tabor, 1989).

As previously described, ray chaos means that ray trajectories in the presence of internal-wave-induced sound-speed fluctuations will exhibit exponential sensitivity to initial conditions. Thus two rays that start out infinitesimally close to one another will separate exponentially as they propagate out in range. A measure of how rapidly the rays diverge in range is given by the Lyapunov exponent ν (see Chapters 2 and 5 for details). For example, the depth separation between two rays infinitesimally close together at the source will increase as $\exp(\nu r)$, and therefore the ray amplitude will decrease as $\exp(-\nu r)$. Importantly, because of ray chaos, theories that expand around the unperturbed ray will only be valid at ranges before the chaos has strongly developed. A crude estimate of that breakdown range is ν^{-1}.

For sound transmission through random fields of internal waves, ν will be a probabilistic function that depends on the ray launch angle. Work by Wolfson and Tomsovic (2001) shows that the PDF of ν is nearly Gaussian. Numerical simulation of ray propagation through random realizations of internal waves confirms this Gaussian shape with some distortion in the tails. Typical values of ν for deep-water propagation are 1/100 to 1/500 km^{-1} with some indications that low grazing angle rays near the sound channel axis are more unstable than high grazing angle rays (Colosi, 2001; Beron-Vera et al., 2003).

Another important consequence of ray chaos for acoustic phase and amplitude fluctuations is that the number of microrays or eigen rays also grows exponentially with range (Tappert and Tang, 1996); in this case the growth rate is again ν. However, not all of the chaotic microrays are significant. Since the vertical separation of nearby rays grows exponentially, the amplitude of a chaotic microrays will correspondingly decrease exponentially. The most unstable rays carry the least energy. Microray generation by ray chaos follows the same wave front triplication rules described in Section 4.1.2, so there will also be an exponential increase in the number of caustics along the wave front. The implications of the chaos-induced microray processes for the approach to saturation are a fundamental research problem.

In ray chaos, rays are also exponentially sensitive to variations in the sound-speed structure; thus as the internal-wave field evolves temporally the individual chaotic microray paths will move around dramatically in response to these sound-speed changes (Flatté and Vera, 2003). However, while the chaotic microrays can move around dramatically, they do in fact stay close to the unperturbed wave front, and closely follow the along and across wave front scattering rules described in Section 4.1.3. More discussion and details of ray chaos will be given in Chapter 5.

Effects of the Background Sound Speed: α

From studies of ray chaos, it has been found that the background sound-speed profile can have a tremendous effect on magnifying or diminishing the effects of ocean internal waves on acoustic propagation (Simmen et al., 1997; Beron-Vera et al., 2003; Virovlyansky, 2003; Beron-Vera and Brown, 2004). In particular, the following discussion applies to the situation at long ranges where chaotic ray dynamics is fully developed. At shorter ranges expansion about the deterministic ray works well, and the dependence on background sound-speed profile comes in through the deterministic ray geometry over which the integrals are carried out.

Simply stated, in the chaotic ray regime, if the dispersion of unperturbed ray trajectories for a group of ray launch angles is large then these rays will experience much larger variability than another group of ray launch angles for which the dispersion is weaker. Dispersion can be interpreted in many ways but in this case it is useful to think about dispersion in terms of the change in ray double loop length as a function of ray grazing angle, ray upper/lower turning point, or ray action (see Chapter 2, Section 2.4.5). The definition of dispersion in this context then means the rate of change of ray loop length with respect to changes in any one of these equivalent parameters. It has been traditional to use action-angle variables and define the ray loop spatial frequency $K_R = 2\pi/R_L(I)$, where $R_L(I)$ is the ray double loop length as a function of the ray action I. A measure of dispersion then is the nondimensional parameter α given by

$$\alpha(I) = \frac{I}{K_R} \frac{dK_R}{dI}. \tag{4.26}$$

This α parameter should not be confused with the attenuation parameter bearing the same name (see Chapter 2, Section 2.3.3). Importantly $\alpha(I)$ also quantifies the degree of Lagrangian manifold shearing that is imposed by the background sound-speed profile (see Figure 2.7). The notion of shear leading to enhanced instability is one that is familiar to hydrodynamics (Pedlosky, 1987). When internal-wave-induced scattering is considered, a simple conceptual model is to assume that at each upper turning point the ray instantaneously changes its action to be $I+\delta I$, and thus enters a new ray orbit with loop length $R_L(I+\delta I)$ with ray loop time $T(I+\delta I)$. If α is large, then this new orbit will have significantly different loop length and time and thus can bring the scattered ray trajectory a good distance away from the unperturbed trajectory. It is in this sense that the background sound-speed profile can magnify or diminish internal-wave-induced scattering.

Examples of acoustic observables that have been shown to be controlled by α are: (1) wave front travel-time spread (Beron-Vera and Brown, 2004; Brown et al., 2005); (2) microray scattering along the wave front (Beron-Vera and Brown, 2004, 2009); (3) Lyapunov exponents (Beron-Vera and Brown, 2003); (4) microray amplitudes (Beron-Vera et al., 2003); (5) spatial and temporal spreading of narrow beams (Beron-Vera and Brown, 2009); and (6) travel-time bias and spreading of normal modes (Udovydchenkov and Brown, 2008). More discussion of α will be given in Chapter 5.

4.2.3 Mode Coupling

The previous discussion of acoustic sensitivity to internal waves centered on the ray picture of wave propagation, but the mode formalism, which considers single frequencies, has its own resonance conditions. In particular, an important inverse length scale in the mode formalism is the beat wavenumber $k_{mn} = k_m - k_n$ where k_n are the modal eigen-wavenumbers. To illustrate this point, in the absence of a nice geometrical argument like that available for the rays, a small derivation is required. A perturbation solution to the coupled mode equations (Eq. 2.128) is sought, and here the Dyson series (Dyson, 1949; Sakurai, 1985) is useful. The small parameter to be used is the order of magnitude of the coupling matrix ρ_{mn}. A heterogenous region of sound-speed centered on the origin is considered and the coupled mode equations are integrated across this region. To first order the Dyson series solution gives (Colosi, 2008)

$$\hat{a}_n(R) = \hat{a}_n(0) - i \sum_{m=1}^{N} \hat{a}_m(0) S_{mn}^{(1)} + ..., \qquad (4.27)$$

where the first-order scattering matrix between modes m and n is written as

$$S_{mn}^{(1)} = \int_{-R/2}^{R/2} e^{il_{mn}r} \rho_{mn}(r) dr = \int_{-\infty}^{\infty} \hat{\rho}_{mn}(k_r) dk_r \int_{-R/2}^{R/2} e^{i(l_{mn}-k_r)r} dr. \qquad (4.28)$$

In the last step the Fourier transform of $\rho_{mn}(r)$ along the propagation path is introduced, namely $\hat{\rho}_{mn}(k_r)$ where k_r is the horizontal wavenumber. Considering the integral over r as a filter function, $B_{mn}^{(1)}(k_r)$, on the Fourier transformed coupling matrix it is found that this filter is given by

$$B_{mn}^{(1)}(k_r) = R\,\text{sinc}[(l_{mn}-k_r)R/2],$$

$$\simeq 2\pi\delta(k_{mn}-k_r), \quad \text{for} \quad \alpha = 0, R \to \infty. \qquad (4.29)$$

In the limit of weak loss and long range there is a resonance condition $S_{mn}^{(1)} = 2\pi\hat{\rho}_{mn}(k_r = k_{mn})$ and the only scale in the coupling matrix that matters is $k_r = k_{mn}$, the beat wavenumber. For finite R and with attenuation, the resonance is broader, but for all practical purposes the asymptotic no loss result is quite accurate (Creamer, 1996; Colosi, 2008). This is yet another example of how the background sound-speed profile, through k_n, has a significant influence on the strength of interaction with the stochastic internal-wave field.

The physical interpretation of the beat wavenumber resonance is quite different from the perpendicular wavenumber resonance in weak fluctuation theory. Recall from Chapter 2 that the beat wavenumber between neighboring modes is closely related to the ray double loop spatial frequency $K_R = 2\pi/R_L$. Thus the beat wavenumber resonance can be viewed as a form of loop harmonic resonance first discussed by Cornuelle and Howe (1987). The loop harmonic resonance simply says that a ray will be strongly scattered by a sound-speed perturbation that is reinforcing each time the ray goes through an upper turning point. Thus perturbations with horizontal wavelengths equal to the ray double loop length and harmonics thereof will be the strongest scatterers.

To further illustrate the effects of the beat wavenumber and get a physical understanding of the meaning of $\hat{\rho}_{mn}(k_r = k_{mn})$, the transport equation for mode energy $|a_n(r)|^2$ (Dozier and Tappert, 1978a; Creamer, 1996) is presented:

$$\frac{\partial |a_n(r)|^2}{\partial r} = -2\alpha_n |a_n(r)|^2 + 2\pi \sum_{m=1}^{N} \langle |\hat{\rho}_{mn}(k_{mn})|^2 \rangle \left[|a_m|^2 - |a_n|^2 \right]. \qquad (4.30)$$

This equation, which will be derived in Chapter 8, tells us the range evolution of the mode energy due to attenuation (first term on the right-hand side) and coupling from stochastic internal waves (second term on the right-hand side). Some insight can be gleaned from this equation in the approximation that only nearest neighbor coupling is occurring. This is a reasonable starting place because in the ocean $|\mu| \ll 1$, and thus only small-angle scattering is seen. In this case the transport equation becomes

$$\frac{\partial |a_n(r)|^2}{\partial r} = -2\alpha_n |a_n(r)|^2 + D_n \left[|a_{n+1}(r)|^2 - 2|a_n(r)|^2 + |a_{n-1}(r)|^2 \right], \qquad (4.31)$$

where $D_n = 2\pi \langle |\hat{\rho}_{n+1,n}(k_{n+1,n})|^2 \rangle$, and it has been assumed $\langle |\hat{\rho}_{n+1,n}(k_{n+1,n})|^2 \rangle \simeq \langle |\hat{\rho}_{n-1,n}(k_{n-1,n})|^2 \rangle$. The quantity in the square brackets on the right-hand side of the equation is recognized as the discrete form of the second derivative, and so Eq. 4.31 is in the form of a diffusion equation with D_n being the diffusion constant in mode space. Thus if a narrow initial distribution of mode energy as a function of mode number n (say a Gaussian distribution) is considered, it is expected that the width of the distribution will diffusively widen (on average) as the modes propagate out in range. In fact, in the absence of attenuation and if D_n is not a strong function of n, the standard deviation of that distribution grows in range as $\sqrt{2D_n r}$ in accordance with the rules of random walks and diffusion (Landau and Lifshitz, 1980). But the width of the mode distribution cannot widen indefinitely. Ignoring attenuation and considering the ocean waveguide to only support a finite number of modes N, the asymptotic distribution of mode energy will be one of *equipartion* of energy (i.e., all modes have the same energy on average) (Dozier and Tappert, 1978a).

The solutions of this and other mode transport equations will be discussed in detail in Chapter 8, but for now one can be content with the physical picture of the internal waves at the beat wavenumber causing a diffusion of energy in mode space, and a phase fluctuation from mode coupling.

4.3 Propagation Regimes and Signal Behavior

The objective of this section is to provide a physical understanding of signal behavior for various acoustic observables from the situation of weak fluctuations out to the regime of saturation. One way to quantify acoustic signal variability is to use the scintillation index (Eq. 4.12). The behavior of *SI* in various propagation regimes is surprisingly universal and is schematically depicted in Figure 4.9. The horizontal axis in Figure 4.9 is purposefully not labeled because the ranges for which the different propagation regimes will be reached will vary as a function of background profiles, ray path, source/reciever location, and frequency. At long ranges or high frequency, *SI* saturates to 1 as expected from strong microray interference (Section 4.1.4). However, the approach to saturation can be either from above or below 1 depending on such things as phase and amplitude statistics

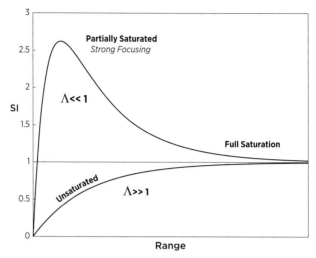

Figure 4.9. Cartoon of scintillation index *SI* variation with propagation range showing the unsaturated, saturated, and partially saturated (strong focusing) regimes for large and small diffraction parameter, Λ.

of the interferers and the signal bandwidth. The approach from above 1 shows a region called the strong focusing region or partially saturated region because the scintillation can be extremely strong. The reader familiar with the striking bands of focused light on the bottom of a swimming pool caused by sunlight passing through the irregular air/water interface has experienced the partially saturated regime. But diffraction works against strong focusing (Born and Wolf, 1999), and thus the partially saturated regime can conceptually be considered a small wavelength or high-frequency regime. This small wavelength regime is denoted with the symbol $\Lambda \ll 1$. The diffraction parameter Λ will be defined in the following section. Correspondingly the approach to saturation from below 1, in which strong focusing does not occur, must be considered a diffractive or large wavelength/low-frequency regime. This large wavelength regime is denoted with the symbol $\Lambda \gg 1$. Of course diffraction depends on the ratio of acoustical length scales to medium length scales and hence the diffraction parameter will depend not only on the acoustic frequency but also on the spectrum of internal waves. Lastly, in the small *SI* region the unsaturated regime is found that has both geometric (small wavelength) and diffractive (large wavelength) regions.

4.3.1 Definition of Φ and Λ: Ray Propagation

The acoustic propagation regimes can be characterized by the strength parameter Φ and the diffraction parameter Λ, that is, the regimes of different acoustic fluctuation behavior can be defined in terms of regions in $\Lambda - \Phi$ space (Flatté et al., 1979; Esswein and Flatté, 1980; Flatté, 1983). This formulation is based on the results of

Munk and Zachariasen (1976) (e.g., Eq. 4.22) and therefore applies most directly to the ray picture of sound propagation. The propagation parameters Φ and Λ are functions of ray geometry, acoustic frequency, and certain medium parameters. The medium parameters will be defined generally and their definition in terms of internal-wave spectra will be presented in a subsequent section.

The description of ocean acoustic wave propagation regimes begins with the Born (or equivalently the Rytov) approximation (Munk and Zachariasen, 1976; Flatté et al., 1979), and Eq. 4.22 gives the variances of phase and log-amplitude in terms of the acoustic ray path and the ocean medium parameters. Two limits of Eq. 4.22 are of interest here: (1) geometrical acoustics where $|mR_f| \ll 1$; and (2) diffractive acoustics where $|mR_f| \gg 1$. For geometrical acoustics, the cosine term in $G_\pm$ is close to 1 which gives

$$\begin{pmatrix} \Phi^2 \\ \langle \chi^2 \rangle \end{pmatrix} \simeq q_0^2 \int_\Gamma ds \langle \mu^2(z_{ray}) \rangle \, L_p(\theta_{ray}, z_{ray}) \begin{pmatrix} 1 \\ C \frac{\{m^2\}R_f^2(s)}{2\pi} \end{pmatrix}, \qquad (4.32)$$

where C is a constant of order 1 (Flatté (1983) cite $C \simeq \pi/8$). The phase variance in the geometric region is written with the special symbol Φ^2. The term $\{m^2\}$ is a spectral average of the internal-wave vertical wavenumber that will be defined later. For the diffractive acoustics limit of Eq. 4.22, the cosine term in $G_\pm$ has rapid variations as a function of m and does not significantly contribute. Therefore, the simple diffraction limit result is

$$\langle \phi^2 \rangle \simeq \langle \chi^2 \rangle \simeq \frac{\Phi^2}{2}. \qquad (4.33)$$

These two limits can be compactly represented by defining two new parameters Λ, the diffraction parameter, and Φ, the strength parameter, as

$$\Phi^2 = q_0^2 \int_\Gamma ds \, \langle \mu^2(z_{ray}) \rangle \, L_p(\theta_{ray}, z_{ray}), \qquad (4.34)$$

$$\Lambda = \frac{q_0^2}{\Phi^2} \int_\Gamma ds \, \langle \mu^2(z_{ray}) \rangle \, L_p(\theta_{ray}, z_{ray}) \frac{\{m^2\}R_f^2(s)}{2\pi}. \qquad (4.35)$$

The strength parameter Φ is the rms phase in the geometric acoustics limit and is thus directly related to the rms travel time fluctuation $\tau = \Phi/\omega$. Using these two parameters, the two limiting cases become

$$\text{Geometrical Acoustics } (\Lambda \ll 1) \quad \langle \phi^2 \rangle = \Phi^2, \quad \langle \chi^2 \rangle = C\Lambda\Phi^2, \qquad (4.36)$$

$$\text{Large Diffraction } (\Lambda \gg 1) \quad \langle \phi^2 \rangle = \frac{\Phi^2}{2}, \quad \langle \chi^2 \rangle = \frac{\Phi^2}{2}. \qquad (4.37)$$

At full saturation it is found that $\langle\chi^2\rangle = \pi^2/24$, and thus a reasonable condition for the onset of strong scattering can be defined as

$$\sigma_\iota^2 = 4\langle\chi^2\rangle = 1, \tag{4.38}$$

where $\iota = \ln I = 2\chi$ is the log intensity. This value of log-intensity variance is roughly two-thirds of the way to the saturation value of $\pi^2/6 \simeq 1.65$. The threshold value is also close to the boundary where Born theory is expected to break down, namely $\sigma_\iota^2 \simeq 1.2$ (Clifford, 1978). It should also be noted that if a log-normal distribution is assumed, then this threshold corresponds to an unreasonably large scintillation index, namely $SI \simeq e - 1 = 1.7$. However, the validity of the Born approximation is not important in the present discussion; the concern is the onset of strong scattering. Therefore in the two asymptotic cases, the strong scattering boundaries are delineated within factors of 2 by

$$\Lambda \ll 1, \Lambda\Phi^2 = 1, \text{Unsaturated–Partially Saturated Boundary}, \tag{4.39}$$

$$\Lambda \gg 1, \quad \Phi = 1, \quad \text{Unsaturated–Fully Saturated Boundary}. \tag{4.40}$$

These are the boundaries that have been described in the literature going back to the monograph by Flatté et al. (1979).

It is useful, however to be a bit more precise, especially since many of the experiments that have been conducted in the ocean do not fall in the asymptotic regimes. In fact, it is likely that no ocean experiment will ever be conducted in the diffraction limit unless the frequencies are significantly below 10 Hz. Thus the fundamental question is the functional dependence of σ_ι^2 with respect to Φ and Λ. Through extensive numerical experimentation with different sound-speed and buoyancy frequency profiles, as well as different source and receiver depths and Garrett-Munk internal-wave parameters, the following relation has been obtained (Colosi, 2015):

$$\sigma_\iota^2 \simeq \Lambda\Phi^2\left[\left(\frac{1}{\pi} - 0.1\right)e^{-\Lambda/1.5} + 0.1\right], \tag{4.41}$$

which is accurate at the level of 5–30% in the region $0 < \Lambda < 10$. This relation is quite different from the previous one, and it is seen that in the asymptotic limit $\Lambda \ll 1$, $\sigma_\iota^2 \simeq \Lambda\Phi^2/\pi$, not $\sigma_\iota^2 \simeq \Lambda\Phi^2$. Figure 4.10 shows the $\Lambda - \Phi$ diagram with this new boundary between unsaturated propagation and stronger scattering delineated.

A few more words about the interpretation of Φ and Λ is useful here. In Eq. 4.34, the strength parameter, Φ, is expressed as an integral along the unperturbed ray of the depth-dependent fractional sound-speed variance times the correlation length of the sound-speed fluctuations along the direction of the ray. These three factors, unperturbed ray path, depth-dependent fractional sound-speed variance, and correlation length along the ray, represent the three important factors of ocean

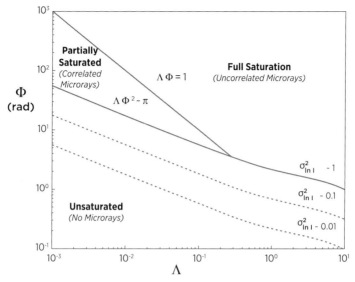

Figure 4.10. Wave propagation regimes as described by the $\Lambda\Phi$ diagram. Borders where the variance of log-intensity is of order $0.01, 0.1$, and 1 are marked. An rms log intensity of one corresponds to an rms intensity of 4.3 dB.

acoustic propagation through internal waves, namely the effects of the background profile, vertical inhomogeneity, and vertical/horizontal anisotropy. The rms phase, Φ, is seen to scale linearly with acoustic frequency and as the square root of propagation range. However, the assumption going into the calculation of L_p that the ray is locally a straight line is not accurate enough for large grazing angle rays. One must take into account the curvature of the ray (Henyey and Macaskill, 1996; Colosi et al., 1999; Flatté and Rovner, 2000). In doing so it is still in general possible to define these local functions μ and L_p, and they are the only aspects of the medium that enter into the definition of Φ (see Chapters 5 and 6).

In Eq. 4.35, the diffraction parameter Λ is a weighted average along the ray of $\{m^2\}R_f^2/2\pi$, where R_f is the vertical Fresnel zone radius. The calculation of Λ thus requires one more aspect of the environment, namely $\{m^2\}$. The quantity m is the vertical wavenumber of the medium sound-speed fluctuation, and the curly brackets indicate an average over the spectrum with the perpendicular wavenumber constraint. This spectral average can be interpreted physically as $\{m^2\} \simeq \hat{L}_z^{-2}$, where $\hat{L}_z$ is an effective vertical correlation length of the sound-speed fluctuations with the perpendicular wavenumber constraint. The diffraction parameter is therefore the weighted average of the ratio of the Fresnel zone divided by the vertical correlation length, that is, a ratio of an acoustic length scale, the Fresnel zone, to a medium length scale, $\hat{L}_z$.[2] It is this ratio that properly quantifies diffraction.

---

[2] This "diffraction" ratio is similar to that for single slit diffraction pattern, namely λ over the slit width.

Based on Eq. 4.22 the expression for $\{m^2\}$ for internal waves is given by

$$\{m^2\} = \left(\frac{\pi N(z)}{N_0 B}\right)^2 \frac{\langle j \rangle}{\langle j^{-1} \rangle}, \tag{4.42}$$

where the WKB relation $m \simeq \pi j N(z)/N_0 B$ has been used. This quantity also varies with depth, giving another form of internal-wave-induced fluctuation inhomogeneity. Note that this spectral average involves the quantity $\langle j \rangle$, which depends on the internal-wave mode cut-off J (see Chapter 6 for further discussion).

Microray Generation and the $\Lambda\Phi > 1$ Condition

Next consider the boundary between the partially saturated and saturated regimes, which only applies for $\Lambda < 1$. In Section 4.1.3 the idea of microray generation was introduced such that sound-speed fluctuations lead to the existence of new Fermat paths. This concept will now be generalized using the strength and diffraction parameters Λ and Φ. Using Eq. 4.3 the travel time difference between an unperturbed ray in the ocean sound channel and a nearby broken ray scattered at x which traverses a region of fluctuating sound-speed is

$$\Delta T(\delta z) = T(\delta z) - T_0 = \frac{T_p}{2} \frac{\delta z^2(x)}{R_f^2(x)} - \frac{1}{c_0} \int_\Gamma \mu(s) \, ds. \tag{4.43}$$

Here $R_f(x)$ is the vertical Fresnel zone and $\delta z(x)$ is the vertical separation between the unperturbed ray and the broken ray at the position x. The second term in Eq. 4.43 represents the contribution to travel time along the broken ray from the random sound-speed changes. The question regarding microray generation is: What values of δz yield extrema for ΔT? To answer this question it is useful to introduce two new dimensionless scaled variables, namely

$$\zeta = \frac{\delta z}{\hat{L}_z}, \quad \text{and} \quad u(\zeta) = -\frac{q_0}{\Phi} \int \mu(s) \, ds, \tag{4.44}$$

where $\hat{L}_z$ is an effective vertical correlation length of μ, $q_0 = \omega/c_0$ is a reference wavenumber, and Φ is the rms phase fluctuation, that is, the strength parameter. Note that in this normalization $u(\zeta) \sim O(1)$. Making this change of variables the resulting phase difference is,

$$\omega \Delta T = \frac{\zeta^2}{2} \frac{\hat{L}_z^2}{R_f^2(x)} + \Phi u(\zeta). \tag{4.45}$$

However, $\hat{L}_z^2/R_f^2(x) \sim O(1/\Lambda)$ and therefore the equation becomes

$$\omega \Delta T \sim \frac{\zeta^2}{2\Lambda} + \Phi u(\zeta). \tag{4.46}$$

The microray condition is that there is an extremum with respect to the scaled vertical deviation ζ so that

$$\frac{d(\omega \Delta T)}{d\zeta}(\zeta_r) = 0 \quad \Rightarrow \quad \zeta_r + \Lambda \Phi \frac{du}{d\zeta}(\zeta_r) = 0. \tag{4.47}$$

For $\Lambda \Phi > 1$ the microray equation (Eq. 4.47) has multiple zero-crossings because the function $u(\zeta)$ is an order one random function whose derivative is as likely to be positive as negative. Furthermore the condition $\Lambda \Phi > 1$ means that the microrays are spread out vertically by a distance greater than $\hat{L}_z$ which then gives rise to uncorrelated microrays needed to get full saturation. Thus $\Lambda \Phi = 1$ is a reasonable, but not unique, condition for the onset of significant microray generation, therefore marking the boundary between the partially and fully saturated regimes.

4.3.2 Acoustic Behavior in Different $\Lambda - \Phi$ Regimes

Figure 4.10 shows the $\Lambda - \Phi$ diagram with the different regions delineated as described in the previous section. In this section various signal characteristics in these regions are described, including phase and intensity variances (and in some cases PDFs of phase and intensity), spatial and temporal coherences, and acoustic pulses time wander, spread, and bias.

The Unsaturated Regime

The unsaturated or weak fluctuation regime is depicted in the $\Lambda - \Phi$ diagram by the empirical condition:

$$\Lambda \Phi^2 \left[\left(\frac{1}{\pi} - 0.1 \right) e^{-\Lambda/1.5} + 0.1 \right] < 1, \tag{4.48}$$

which gives the asymptotic limits $\Lambda \Phi < \pi$ ($\Lambda \ll 1$, geometric acoustics) and $\Phi < 1/\sqrt{2}$ ($\Lambda \gg 1$, diffractive acoustics). If one considers the case in which $\mu = 0$, the primary contributions to the pressure will be from paths within a Fresnel zone of the equilibrium or unperturbed ray, that is, paths whose phase differs by less than half a cycle from that of the unperturbed ray, and are thus not prone to interference or cancellation. The primary concern, however, is the case in which $\mu \neq 0$, and thus the situation is considered where the variables μ, propagation range, and/or frequency are small enough that the regime is weak fluctuations or unsaturated propagation. The physical picture in this regime is that only one ray path is dominant (i.e., no microrays), but this ray and its associated ray tube are perturbed by the sound-speed fluctuations. The intensity fluctuations are therefore caused by the convergence and divergence of the ray tube, while the phase fluctuations are caused by the delay/advance of the energy as it travels through the variable sound-speed structure, as described in Section 4.1.

In this regime the phase and intensity variances are given by Eqs 4.32 and 4.33, and the propagation physics is adequately described by the Born or Rytov approximations. Because the Born and Rytov theories apply here, phase and log-amplitude statistics are directly related to the statistics of the sound-speed fluctuations (more details are in Chapter 6). Because stochastic ocean sound-speed fields are assumed to be consistent with Gaussian PDFs, the PDFs of phase and log-amplitude will also be Gaussian.

The log-normal PDF for the amplitude (or intensity) allows us to obtain a simple expression for the scintillation index in terms of the log-amplitude moments. Taking $\langle \chi \rangle$ and $\langle \chi^2 \rangle$ as the first two moments of log-amplitude, then the PDF is written as:

$$P(\chi) = \sqrt{\frac{1}{2\pi\sigma_\chi^2}} \exp\left[-\frac{(\chi - \langle \chi \rangle)^2}{2\sigma_\chi^2}\right], \tag{4.49}$$

where $\sigma_\chi^2 = \langle \chi^2 \rangle - \langle \chi \rangle^2$. The nth moment of the pressure amplitude is then given by

$$\langle a^2 \rangle = \exp\left[\frac{\langle \chi^2 \rangle n^2}{2} + \langle \chi \rangle n\right]. \tag{4.50}$$

Using these linear moments in the scintillation index equation yields the result,

$$SI = e^{4\langle \chi^2 \rangle} - 1 \simeq 4\langle \chi^2 \rangle, \tag{4.51}$$

where the last step follows from the smallness of the log-amplitude fluctuation in the unsaturated regime. In practice one takes $\langle \chi \rangle = 0$ so that $\langle \chi^2 \rangle$ is the variance of log-amplitude, which is equal to one-quarter the variance of log-intensity.

Other important aspects of the acoustic field fluctuations are the space and time scales of the variability. These space time scales come into play when ocean acoustic data are used for applications such as beamforming, match field processing, communication, or other coherent signal processing. Here consider the acoustic field to be a complex phasor in which rotations of the phasor represent phase variability and changes in the length of the phasor represent amplitude variability. Figure 4.11 shows schematically "typical" phasor behavior in the unsaturated regime. In the geometric regime ($\Lambda \ll 1$), the phasor has an amplitude that is close to a constant, but the phase can vary by large amounts. The rms phase is equal to Φ. The rms log amplitude is roughly equal to $\Lambda^{1/2}\Phi$, which is much smaller than Φ (Eq. 4.36). For $\Phi < 1$, the phasor appears to move on or close to a circular arc in the first and fourth quadrants. In the geometric regime in which $\Phi > 1$, the phasor still has relatively small amplitude fluctuations, and there are no signal fade-outs where the phasor length drops to small values or zero. However, in this case the phasor can occupy all four quadrants. The diffractive regime depicted by $\Lambda > 1$, on the other hand, has similarly sized phase and amplitude fluctuations

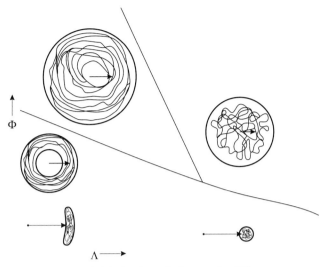

Figure 4.11. Time or spatial behavior of phasors in different wave propagation regimes as described by the $\Lambda\Phi$ diagram.

whose rms variations are given by $\Phi/\sqrt{2}$ (Eq. 4.37). In this diffractive situation the phasor traces out a small circle, again more or less confined to the first and fourth quadrants. The correlation properties of these phasors will be used to derive spatial and temporal acoustic coherence functions.

Lastly the signal characteristics associated with pulse propagation are addressed. Because there are no microrays in the unsaturated regime, the received signal has a temporal shape that is close to the transmitted shape (Figure 4.12). It will be found that in the other regimes the temporal spread of microrays leads to temporal spreading of the received pulse, denoted by τ_0 and a travel-time bias denoted by τ_1. Because there is no microray interference, travel time-fluctuations of the pulse will be simply related to the phase fluctuations, that is, $\tau = \Phi/\omega_c$, where ω_c is the carrier frequency.

Partially Saturated Regime: Approach to Saturation

Now consider cases in which the strength of μ grows, the propagation range increases, or the acoustic frequency increases, and the acoustic pressure therefore has contributions from many ray paths (microrays), each of which, due to the sound-speed fluctuations, is locally an extremum of the travel time. These microrays will each have their own Fresnel zone that will be perturbed by the fluctuations, and the interference of these microrays will lead the signal towards full saturation $SI \rightarrow 1$, as discussed in Section 4.1. The approach to saturation depends strongly on the relationship between the Fresnel zone (R_f), the vertical spatial spread of the microrays (R_{mr}) (such as that exemplified in Figure 4.5), the vertical correlation length of the internal waves (L_z), and the rms phase or

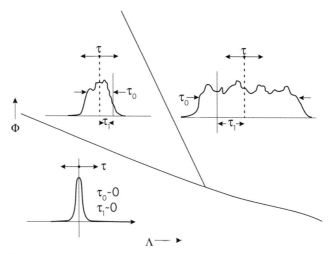

Figure 4.12. Behavior of pulse propagation in different wave propagation regimes as described by the $\Lambda - \Phi$ diagram. Here τ is the rms pulse time wander, τ_0 is the rms pulse time spread, and τ_1 is the pulse travel time bias.

strength parameter (Φ). In the case for which $R_{mr} \gg L_z$, $R_f \ll L_z$, and $\Phi \gg 1$, the microrays will be independent, having traversed independent regions of sound-speed structure. They will also have acquired sufficient phase variation to strongly interfere with one another; this sum of independent randomly phased waves forms the basis of full saturation (Section 4.1). For the case $R_{mr} \ll L_z$, $R_f \ll L_z$, and $\Phi \gg 1$, the large-scale structure of the medium fluctuations changes all the microrays together, so that the received pressure is the sum of several correlated contributions. If there are not too many microrays then there is a high probability that all the microrays can sum constructively to create strong focusing regions; this is the partially saturated regime in which the scintillation index approaches 1 from above . For the case in which $R_f \gg L_z$, there is strong diffraction that destroys the focusing ability of the medium, and thus the approach to saturation has the scintillation index going to 1 from below.

The partially saturated regime can be loosely defined in the $\Lambda - \Phi$ space to occupy the region between the microray generation condition $\Lambda\Phi = 1$ and the strong intensity variability condition $\Lambda\Phi^2 = \pi$, and of course this applies only in the geometric region $\Lambda < 1$. In this regime the microrays are on average spread over a region smaller than L_z. The paths that are separated by less than L_z are partially correlated, while the paths separated by more than a correlation length are statistically independent. It is the correlated paths that join together constructively to generate the strong focuses that are the trademark of the partially saturated regime. In this regime the PDF of intensity has been the subject of much study. It has been found that the generalized Gamma distribution can be fit to a large number of field and numerical experiments (Macaskill and Ewart,

1984; Ewart and Percival, 1986; Ewart, 1989), and this distribution has the log-normal and exponential distributions as asymptotic forms. The generalized Gamma distributions for intensity (I) and log-intensity (ι) are written as (Ewart and Percival, 1986):

$$P(I) = \frac{b\mu^k}{\Gamma(k)} I^{bk-1} e^{-\mu I^b}, \tag{4.52}$$

$$P(\iota) = \frac{b\mu^k}{\Gamma(k)} e^{bk\iota} e^{-\mu \exp(b\iota)}, \tag{4.53}$$

where $\mu = (\Gamma(k+1/b)/\Gamma(k))^b$, Γ is the Gamma function, and k and b are coefficients that must be fit to observations. Convolution PDFs have also proved useful, in particular the log-normal convolved with the exponential distribution has been applied in atmospheric optics (Flatté et al., 1994). In the convolution PDFs, the microrays that are spread beyond L_z obey the exponential PDF, while the rays confined to the region less than L_z obey the log-normal PDF.

Turning to the space–time scales of the phasor motion in the partially saturated regime, Figure 4.11 shows that phase is expected to change much more rapidly than amplitude, thus creating a circular motion to the phasor. Although the amplitude moves more slowly, it eventually covers the entire gamut of possible amplitudes, from zero to infinity giving fade-outs and strong focuses. Because $\Phi > 1$, the phasor occupies all four quadrants.

With regard to pulse propagation in the partially saturated regime, the received signal closely resembles the transmitted pulse with a small amount of microray time broadening, τ_0, and a small travel time bias τ_1 (Figure 4.12). Pulse broadening occurs because the microrays will have a distribution of travel times and the bias means that on average the microrays will have a different travel time from the unperturbed ray: both the spread and bias are nonlinear effects. One may understand the bias as a result of Fermat's Principle. In the presence of the fluctuating sound-speed the acoustic energy can find a quicker over-all route to the receiver, leading to a fast bias. Formulas that predict the change in pulse shape, τ_0, as well as the travel time bias, τ_1, will be given in Chapter 7, and it will be found that the bias can be both positive and negative. Lastly there is the issue of travel time and phase. Because of microray interference, one might expect that travel time and phase would not be simply related in the partially saturated regime. Simulations, however, indicate that this is not the case (Flatté and Rovner, 2000) and rms travel-time fluctuations are accurately given by $\tau = \Phi/\omega_c$.

The Saturated Region

The saturated regime is the regime of the $\Lambda - \Phi$ diagram in which a large number of independent microrays are generated, that is, the unperturbed ray has broken up into many microrays via the triplication mechanism described in Section 4.1.2. In

Figure 4.10 the saturated regime is defined by $\Lambda\Phi > 1$ for $\Lambda < 1$, and the empirical relation for $\Lambda > 1$. In the fully saturated regime, the microrays are on average spread vertically over a region that is larger than the vertical correlation length L_z. Because the rays are spread this way, they are statistically independent, and the pressure field is a sum over a large number of independent, interfering microrays (see Section 4.1.4); that is, one may write

$$p = \sum_{n=1}^{N} x_n + i y_n = X + iY. \tag{4.54}$$

In this case the central limit theorem tells us that as $N \to \infty$ the real and imaginary parts of the pressure, X and Y, tend towards independent zero mean Gaussian random variables. The intensity, which is the sum of the squares of these two Gaussian random variables, then has a Chi-square distribution of degree 2; that is an exponential distribution:

$$P(I = X^2 + Y^2) = \frac{1}{\langle I \rangle} \exp\left[-\frac{I}{\langle I \rangle}\right]. \tag{4.55}$$

For acoustic amplitude $A = I^{1/2}$, the PDF is the well-known Rayleigh distribution (Dyer, 1970):

$$P(A) = 2A \exp(-A^2). \tag{4.56}$$

The exponential distribution has moments $\langle I^n \rangle = n! \langle I \rangle^2$, and in this asymptotic limit $SI = 1$, as described in Section 4.1.4. The moments of the log-intensity can also be obtained, giving

$$\sigma_\iota^2 = \frac{\pi^2}{6}, \qquad \rightarrow \qquad \sigma_\iota^2 (\text{dB})^2 = \frac{\pi^2}{6} \frac{100}{(\ln 10)^2} = (5.6\text{dB})^2. \tag{4.57}$$

Hence in full saturation the rms intensity is 5.6 dB (Dyer, 1970).

With regard to the space–time scales of the phasor evolution, Figure 4.11 shows that in the saturated regime the behavior can be quite complicated indeed. Here the phasor executes movements on the complex plane that are as likely to be in the radial as in the azimuthal direction. While the short-term coherence of the phasor can be calculated, the long-term motion is characteristic of a random walk.

Pulse shape in the saturated regime is shown in Figure 4.12. Here the received pulse is significantly spread in time compared to the transmitted pulse, and there is a large travel-time bias. The amounts of spreading and bias (τ_0 and τ_1) have been calculated using the path-integral method, but due to ray chaos effects, this calculation has proved to be accurate only at ranges less than a few hundred kilometers (Colosi et al., 1999; Flatté and Vera, 2003). Spread calculations at longer ranges have been more successful using ray methods (Beron-Vera and

Brown, 2004, 2009; Flatté and Colosi, 2008). Thus complications exist for estimating spread and bias in the saturated regime. However, the travel time follows the same rules that are found in the unsaturated regime, that is, the wander of the spread pulse follows the relationship $\tau = \Phi/\omega_c$. This occurs because the travel time behavior is dominated by the ocean structure that is large scale in both space and time (Flatté and Vera, 2003). ("Large" is relative in this context, when the statistical component of ocean variability is multiscale, but still small-scale compared with the distances between sources and receivers.) This robust character of travel time means that internal-wave effects on certain types of acoustical remote sensing, like ocean acoustic tomography, are somewhat minimal.

4.3.3 Coherence Functions

The acoustical observable of coherence is a quantity of exceptional importance because acoustic data are often combined either spatially or temporally to increase signal-to-noise levels and to establish incoming signal direction (Beamforming). Acoustic coherence is also an important observable that could be useful in internal-wave tomography (Flatté, 1983; Colosi, 1999, see also Section 7.4.9). Regarding increasing signal to noise or beam forming one can consider adding together data collected at various space and/or time points, that is,

$$P = \sum_{j=1}^{N} p_j + n_j, \qquad (4.58)$$

where p_j and n_j respectively are the signal and the noise in the *jth* sample. With the assumptions that there is no signal-noise or noise-noise correlation ($\langle p_j n_k^* \rangle = 0$ and $\langle n_j n_k^* \rangle = 0$, $j \neq k$), it is found that the expected signal gain from this addition is given by

$$\text{Gain} = 1 + \frac{1}{N} \sum_{j=1}^{N} \sum_{k=1, k \neq j}^{N} \frac{\langle p_j p_k^* \rangle}{\langle |p|^2 \rangle}, \qquad (4.59)$$

where $\langle p_j p_k^* \rangle$ is the acoustic coherence function between samples j and k separated in space or time or both. The gain given by Eq. 4.59 is often termed the white noise gain and it reveals that the expected gain from a beamformer or from some other coherent processing methodology depends critically on the coherence function. If the signal is perfectly coherent (i.e., $\langle p_j p_k^* \rangle / \langle |p|^2 \rangle = 1$) then the gain is N, that is, 3 dB of gain is achieved for every doubling of N.

It is a rather remarkable fact that the coherence function of the pressure p can be estimated by certain simple functions of ocean variability. This fact has been derived in numerous ways, with different assumptions (Flatté, 1983). At relatively short range and high acoustic frequency the normalized coherence between two points p_1 and p_2 is given by (Dashen et al., 1985)

$$\frac{\langle p_1 p_2^* \rangle}{\langle |p|^2 \rangle} = \exp\left(-\frac{D(1,2)}{2}\right), \tag{4.60}$$

whereas at long range the coherence function obeys (Colosi et al., 2005):

$$\frac{\langle p_1 p_2^* \rangle}{\langle |p|^2 \rangle} = \frac{\langle A_1 A_2 \rangle}{\langle |p|^2 \rangle} \exp\left(-\frac{D(1,2)}{2}\right). \tag{4.61}$$

Here p_1 and p_2 are assumed to be normalized with unit mean intensity, and $\langle A_1 A_2 \rangle$ is the amplitude covariance function. In both cases, however, the loss of coherence is primarily driven by the phase structure function $D(1,2)$, which is a function of the space–time separations of two points denoted by $(1,2)$. The specific separations being considered are: time Δt, vertical Δz, horizontal Δy, and acoustic frequency $\Delta \sigma$. Then the phase structure function is defined as (Esswein and Flatté, 1981):

$$D(\Delta) = \left\langle \left(q_0[1 + \frac{\Delta \sigma}{\sigma}] \int_0^R ds\, \mu(\Delta) - q_0[1 - \frac{\Delta \sigma}{\sigma}] \int_0^R ds\, \mu(0) \right)^2 \right\rangle, \tag{4.62}$$

where ds is an integral along the unperturbed ray. Also Δ and 0 denote two unperturbed rays that are separated at the receiver by the appropriate coordinate. For example, if the separation at R involves two receivers separated horizontally by Δy_0, then the separation at a distance x would be $[(x/R)\Delta y_0]$. If the separation of the two rays is in time or frequency, then the two rays are identical in space. The vertical separation of two rays that arrive at two receivers separated by Δz_r is complicated by the ocean waveguide.

The phase structure function can be expressed in terms of medium correlation functions (Flatté et al., 1979; Esswein and Flatté, 1981):

$$D(\Delta) = 2q_0^2 \int_0^R dx \langle \mu^2 \rangle L_p f(\Delta(x); z), \tag{4.63}$$

where f is a normalized correlation function that is zero for zero separation and is unity at large separations where the two μ values in the definition of D are uncorrelated. Calculations using this approach will be discussed in more detail in Chapter 6. A useful conceptual model is to consider the structure function to be quadratic in the separation Δ

$$D(\Delta) = \Delta^2 \, v_\Delta^2, \tag{4.64}$$

where v_Δ is a phase rate for the particular separation being considered. The phase rates can be computed directly from Eq. 4.63, but again a conceptual model for temporal or spatial separations is useful. Here it is found that the phase rate can be simply represented as the rms phase Φ divided by the correlation time (or length scale) of the ocean (Flatté and Stoughton, 1988). Hence the result is

$$v_t \simeq \Phi\{\sigma^2\}^{1/2} = \frac{\Phi}{\hat{T}_{IW}}, \quad v_z \simeq \Phi\{k_z^2\}^{1/2} = \frac{\Phi}{\hat{L}_z}, \quad v_y \simeq \Phi\{k_y^2\}^{1/2} = \frac{\Phi}{\hat{L}_H}, \quad (4.65)$$

where, as discussed in Section 4.3.1, the curly brackets represent spectral averages under the perpendicular wavenumber constraint and thus represent effective internal-wave vertical and horizontal length scales $\hat{L}_z$ and $\hat{L}_H$, as well as internal-wave time scale $\hat{T}_{IW}$ (see Table 3.1 for internal wave correlation values without the wavenumber constraint). These time and length scales are of course functions of the ray angle and depth, but to a first approximation the values above can be considered to be representative of ray upper turning points where the scattering is most pronounced. An order of magnitude estimate of acoustical coherence time or spatial scales is defined as the separation Δ at which the structure function has a value of 1, or equivalently, the coherence function has decayed by $e^{-1/2}$. For the coherence time t_0 as well as vertical z_0 and horizontal y_0 coherence lengths then the simple expressions are

$$t_0 = \frac{\hat{T}_{IW}}{\Phi}, \quad z_0 = \frac{\hat{L}_z}{\Phi}, \quad y_0 = \frac{\hat{L}_H}{\Phi}, \quad (4.66)$$

where the effect of the ocean space–time scales on the acoustical coherence space and time scales can be seen directly. Because the rms phase scales as $\Phi \simeq q_0 \langle \mu^2 \rangle^{1/2} R^{1/2} L_p^{1/2}$ the coherence values decrease as one over the frequency and as one over the square root of range.

PART II
Wave Propagation Theories

5

Ray Theory

5.1 Introduction

A good starting place for the analysis of sound propagation through the stochastic ocean is to consider the ray picture of acoustic transmission. As described in Chapter 2, rays provide a pleasing intuitive geometrical picture of wave propagation, and as such this theory has played a central role in ocean acoustics since the beginning of the field. But, the use of ray theory is not just based on theoretical niceties. Quite the contrary, there are strong observational reasons that compel us to use ray theory. Since the advent of broadband-controlled electronic sources, acoustically navigated moorings, and wide vertical aperture receiver arrays, deep-water observations of time fronts from ranges of 50–5000 km have been unambiguously identified with specific ray paths through the ocean. This is the basis of ocean acoustic tomography, whose history is intertwined with that of sound propagation through the random ocean. Figure 5.1 shows an observed deep-water time front at a range of 3250 km, as well as the time front or ray ID (Worcester et al., 1999). Although the late part of the arrival pattern is a confused interference pattern, the early part of the arrival pattern is ray-like with its separated branches. Upon closer inspection of the arrival pattern, however (Figure 5.2), it is seen that the fronts themselves have a narrow interference pattern with multiple peaks associated with a given time front branch. The arrival finale is a complex interference pattern (Colosi et al., 2001). This behavior is associated with ray chaos, and in this chapter it will be demonstrated that ray theory can provide important insights into the statistics of the acoustic field.

As has been previously described, rays are an asymptotic construction with an array of idiosyncrasies. Important among these is that ray trajectories and amplitudes depend on gradients of the ocean sound-speed structure, and there is therefore sensitivity to small scales. Because the ray is infinitely thin, there is no natural acoustical scale to separate meaningful small-scale ocean structure from irrelevant structure. Invariably the finite wavelength of the sound will set this

187

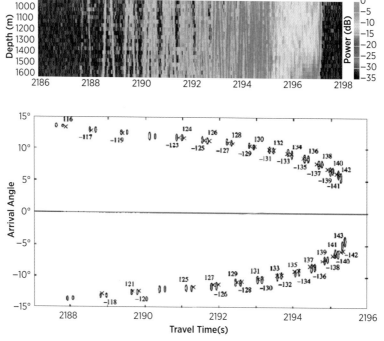

Figure 5.1. Top panel: A 24-hour, incoherently averaged time front from the 3250-km ATOC Acoustic Engineering test. Bottom panel: A demonstration of the identification of over 40 ray paths in time and vertical angle. Measurements are shown with ellipses and ray predictions are shown with crosses. The time front ID is given next to the arrival.
Source: Worcester et al. (1999).

natural separation scale, and this notion has been investigated using the Fresnel zone. Another important issue in ray theory is ray chaos, that is, exponential sensitivity to initial conditions and sound-speed perturbations. A reasonable question is: To what extent do finite wavelength effects "smooth over" complex chaotic ray structures in the wave field and to what extent do they persist? In one form or another, this is the issue in the field of quantum chaos in which one seeks to understand the nature of quantum systems (finite $\hbar$) with chaotic classical dynamics ($\hbar = 0$) (Giannoni et al., 1991; Casati and Chirikov, 1995; Brown et al., 2003). The answer to this question is still a matter of basic research, so one must be careful in the interpretation of some ray results.

Be that as it may, ray theory is still a powerful approach that allows insight into the acoustical field that is not easily obtained in other ways, and it provides an important foundation from which more complicated full wave theories can be understood (e.g., Born approximation/weak fluctuation theory, path integrals, and normal mode theories). In addition many of the idiosyncrasies of ray theory are associated with the ray amplitude, not its phase, and thus ray theories for the

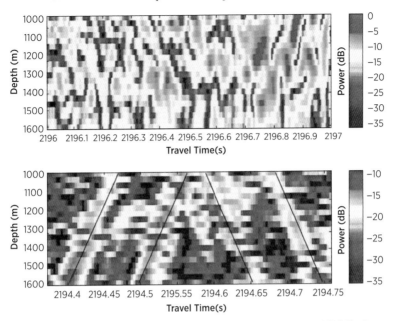

Figure 5.2. Example intensity pattern from the 3250-km ATOC Acoustic Engineering Test. The top panel shows an expanded view of the time front finale region, while the bottom panel shows details of four early time front arrivals. Black lines show ray theory predictions based on climatology.
Source: Colosi et al. (2001).

statistics of phase (or travel time) tend to be quite useful and accurate. Because the normal mode approach is best suited to shallow-water problems, the focus of this chapter will be primarily on deep-water environments.

5.2 Fundamental Equations: Displacement and Current

As was developed in Chapters 2 and 3, ocean internal waves produce vertical displacements $\zeta(\mathbf{r},t)$ that result in sound-speed fluctuations given by Eq. 2.32. However, in certain experimental arrangements, measurable acoustical effects can be produced by internal-wave horizontal currents $\mathbf{u}(\mathbf{r},t)$, which act as a head or tail wind to the sound transmission and can further refract the sound. Indeed, several experiments have measured the relative effects of internal-wave displacements and currents using reciprocal transmission (e.g., Worcester, 1977; Stoughton et al., 1986; Dushaw et al., 1995; Dushaw and Worcester, 1998).

A comprehensive treatment of acoustics in moving and inhomogeneous media can be found in the texts by Pierce (1994) and Ostashev and Wilson (2015), but here we consider a useful but simplified analysis of ray propagation through internal waves in which internal-wave-induced sound-speed perturbations as well as horizontal currents are present. To make the analysis as simple and as physically illuminating as possible, two-dimensional propagation in the (x, z) plane is assumed having sound speed $c(x, z)$ and horizontal current $\mathbf{u} = (u(x, z), 0, 0)$ (Duda, 2005; Colosi, 2006). The vertical component of the internal-wave current is neglected because in a vertically stratified ocean the horizontal component is much larger than the vertical component. Furthermore, the acoustic propagation is primarily in the horizontal direction. The primary effect of currents is to introduce an advective term into the wave equation (Eq. 2.8) (Pierce, 1994),

$$c^2\left(\frac{\partial^2}{\partial x^2} + \frac{\partial^2}{\partial z^2}\right)p = \left(\frac{\partial}{\partial t} + u\frac{\partial}{\partial x}\right)^2 p, \tag{5.1}$$

and the acoustic dispersion relation from Eq. 5.1 is

$$\omega(k_x, k_z, x, z) = c(x, z)\,(k_x^2 + k_z^2)^{1/2} + u(x, z)k_x. \tag{5.2}$$

In this approximation, the horizontal current produces a Doppler shift. The existence of a current also introduces anisotropy into the problem, which means that the velocity of the ray depends on the ray direction in addition to the sound speed and current (Pierce, 1994; Munk et al., 1995; Ostashev and Wilson, 2015).

The ray equations can be derived using the dispersion relation and the Hamiltonian Eqs. 2.38 and 2.39. From Chapter 2, these are written as

$$\frac{d\mathbf{k}}{dt} = -\frac{\partial\omega}{\partial\mathbf{r}} \quad \text{and} \quad \frac{d\mathbf{r}}{dt} = \frac{\partial\omega}{\partial\mathbf{k}}, \tag{5.3}$$

where $\mathbf{r}(t) = (x(t), z(t))$ and $\mathbf{k}(\mathbf{r}(t)) = (k_x(\mathbf{r}(t)), k_z(\mathbf{r}(t)))$. As was seen in Chapter 2, the acoustic frequency ω scales out of these equations, so it is useful to define the "slowness" coordinates $p_x = k_x/\omega$ and $p_z = k_z/\omega$, giving (Colosi, 2006)

$$\frac{dz}{dx} = \frac{p_z c}{(1 - p_z^2(c^2 - u^2))^{1/2}} \equiv \tan\theta, \tag{5.4}$$

$$\frac{dp_z}{dx} = \frac{p_x u - 1}{(1 - p_z^2(c^2 - u^2))^{1/2}}\left(\frac{p_x}{c}\frac{\partial u}{\partial z} + (1 - p_x u)\frac{1}{c^2}\frac{\partial c}{\partial z}\right), \tag{5.5}$$

$$\frac{dT}{dx} = \frac{1}{(c^2 - u^2)}\left(-u + \frac{c}{(1 - p_z^2(c^2 - u^2))^{1/2}}\right), \tag{5.6}$$

where θ is the angle of the ray with respect to the horizontal. Equation 5.4 can be used to give the vertical ray slowness p_z in terms of the ray angle:

$$p_z = \frac{\sin\theta}{(c^2 - u^2\sin^2\theta)^{1/2}}. \tag{5.7}$$

Using the dispersion relation, the horizontal ray slowness p_x can be expressed in terms of the vertical slowness and the local sound speed and current:

$$p_x = \frac{-u + c(1 - (c^2 - u^2)p_z^2)^{1/2}}{(c^2 - u^2)}. \tag{5.8}$$

In the deep ocean, internal-wave-induced horizontal currents are of order 1–10 cm/s, and thus the Mach number for sound transmission through internal-wave currents is extremely small. Expanding the ray equations to first order in Mach number ($u/c \ll 1$) is thus reasonable, and the result is (Colosi, 2006)

$$\frac{dz}{dx} = \frac{p_z c}{(1 - p_z^2 c^2)^{1/2}},\tag{5.9}$$

$$\frac{dp_z}{dx} = -\frac{1}{c^2}\left(\frac{\partial c}{\partial z}\frac{1}{(1 - p_z^2 c^2)^{1/2}} + \frac{\partial u}{\partial z}\right),\tag{5.10}$$

$$\frac{dT}{dx} = \frac{1}{c^2}\left(\frac{c}{(1 - p_z^2 c^2)^{1/2}} - u\right),\tag{5.11}$$

with $p_z = \sin\theta/c$. Here the effects of internal-wave currents are apparent, namely that to first order the current has no effect on the ray angle (Eq. 5.9), but the current shear will cause a first-order refraction effect in much the same way a vertical gradient of sound speed causes a refraction (Eq. 5.10). Regarding ray travel times, the current acts as a head or tail wind (Eq. 5.11). The current-induced refraction and travel time lead to nonreciprocity in acoustic transmission through internal waves (Munk et al., 1981, 1995). Of fundamental importance is the relative size of the current shear and sound-speed gradient terms due to internal waves. Using the Garrett-Munk internal-wave spectrum, it has been shown that (Duda, 2005; Colosi, 2006)

$$\frac{u'_{rms}}{\delta c'_{rms}} = \sqrt{3/2}\frac{N(z)}{(dc/dz)_p(z)},\tag{5.12}$$

where u'_{rms} and c'_{rms} are the rms shear and sound-speed gradients from internal waves, respectively. A bit more insight can be obtained using the Munk canonical sound-speed gradient relation (Eq. 2.24), that is $(dc/dz)_p = c_0 G N^2(z)$. Thus the ratio scales as one over N. Using values of N equal to 3, 1, and 0.5 (cph), and $c_0 G = 1500$ (s), the ratio of rms shear to sound-speed gradient is 0.16, 0.47, and 0.93 respectively, suggesting that current shear is indeed a significant, but not dominant effect. More research is needed in this area. The "traditional" approximation is to neglect shear since much of the scattering takes place at the upper turning point where N is large. This leaves only the horizontal advection of the ray in the travel time equation, which manifests itself as a travel time or phase fluctuation.

Linearization of the Problem for Travel Time

In the traditional approximation in which the current shear is neglected, the ray equations (Eqs. 5.9–5.11) giving the travel time for a sound-speed field $c(x,z) = \bar{c} + \delta c(x,z)$ and current field $u(x,z)$ are nonlinear in the sound-speed perturbation and linear in the current. The equations for the ray path are independent of u. The issue is that the perturbation in sound speed simultaneously changes the ray path and the travel time. A linearization of the equations is thus sought since $|\delta c| \ll \bar{c}$. As with any nonlinear problem, the linearization will eventually break down; this is the ray chaos regime.

The variational approach to ray theory (Section 2.7.1) is helpful here. In that calculation the changes in travel time for paths near a given Fermat path (or ray) have canceling contributions from these nearby paths traveling through (1) slightly different sound-speed structure; and (2) the paths having a different overall length (see Eq. 2.107). This result from Fermat's principle has to be applied with care to the linearization with respect to sound-speed perturbations, because the principle says nothing with regard to medium perturbations. Indeed, the principle is for a fixed medium in which nearby paths are considered to be displaced from the Fermat path. The critical point here is that nearby paths in the fixed ocean can be associated with a path that would have been generated by the sound-speed perturbation δc. In doing so it can be seen that sound-speed and path length changes cancel, giving no contribution to the travel-time variation. This leads to the conclusion that travel-time changes associated with the changing ray path occur at second order due to Fermat's principle. To leading order then one can consider the perturbations as acting along the unperturbed ray. This is the so-called approximation of expanding about the deterministic ray, and as such the linear travel-time equation becomes

$$T \simeq \int_\Gamma \frac{1}{\bar{c}(z)} \, ds - \frac{1}{c_0} \int_\Gamma \mu(x,z) \, ds - \frac{1}{c_0} \int_\Gamma \mu_u(x,z) \cos\theta \, ds, \qquad (5.13)$$

where Γ is the unperturbed ray path, θ is the inclination of the ray from the horizontal, μ is the fractional sound-speed change, and $\mu_u = u/c_0$ is the Mach number. The linearization procedure is also discussed in detail for the tomographic problem by Munk et al. (1995). Note that by linearizing the equations, the mean of the travel-time fluctuations becomes zero, because $\langle \mu \rangle = \langle \mu_u \rangle = 0$. In Section 5.8 (path integrals), the internal-wave travel-time bias will be addressed (which is generally small even at basin-scale ranges). Techniques for computing the relative effects of sound-speed and current on internal-wave-induced travel-time statistics are discussed in the next section.

5.3 Travel-Time Variance

In this section acoustic travel-time variance in the geometrical acoustics approximation is examined. This observable is important because it is directly related to the acoustical phase variance; that is, in the geometrical regime Φ^2 is equal to the travel-time variance times the square of the acoustic frequency (see Chapter 4). Results for internal-wave-induced travel-time variance caused by sound-speed and currents have been presented by Munk et al. (1981) and Stoughton et al. (1986) and the following presentation is based in many ways on those works. The starting point is Eq. 5.13, in which the first term is the unperturbed travel time, T_0, and the last two terms are the travel-time fluctuations due to the internal-wave displacements and currents, respectively. Because it is useful to separate displacement and current effects, a reciprocal transmission experiment is considered in which the travel times between two moorings are observed in both directions $T_{1\to2}$ and $T_{2\to1}$. The separation of displacement and current is then accomplished by writing,

$$\tau_\zeta = \frac{T_{1\to2} + T_{2\to1}}{2} - T_0 = -\frac{1}{c_0} \int_\Gamma \mu(x,z)\, ds, \tag{5.14}$$

$$\tau_u = \frac{T_{1\to2} - T_{2\to1}}{2} = -\frac{1}{c_0} \int_\Gamma \mu_u(x,z)\cos\theta\, ds. \tag{5.15}$$

The travel-time variances associated with displacements and currents are thus

$$\langle \tau_\zeta^2 \rangle = \frac{1}{c_0^2} \int_\Gamma ds_1 \int_\Gamma \langle \mu(s_1)\mu(s_2) \rangle\, ds_2, \tag{5.16}$$

$$\langle \tau_u^2 \rangle = \frac{1}{c_0^2} \int_\Gamma \cos\theta_1 ds_1 \int_\Gamma \cos\theta_2 \langle \mu_u(s_1)\mu_u(s_2) \rangle\, ds_2, \tag{5.17}$$

where the importance of the correlation functions of μ and μ_u is evident. There is also an oceanographically interesting cross-correlation $\langle \tau_\zeta \tau_u \rangle$, that is a measure of the correlation between internal-wave displacements and currents. This quantity can be related to the vertical flux of horizontal momentum and thus can be used to identify sources and sinks of internal waves (Munk et al., 1981). A nonzero value of $\langle \tau_\zeta \tau_u \rangle$ can only occur if the internal-wave spectrum has (1) vertical wavenumber asymmetry (i.e., differences in upwards and downwards propagating energy); and/or (2) horizontal anisotropy. The GM spectrum has neither of these features.

The correlation function is a Fourier transform pair with the spectrum of μ and μ_u, and these function can be written in terms of the GM spectrum. Equation 2.32 shows that the spectrum of μ is directly related to the GM vertical displacement spectrum F_ζ, which from Eq. 3.64 is

$$F_\zeta(\kappa_h, j) = \zeta_0^2 \frac{N_0}{N(z)} H(j) \frac{4}{\pi} \frac{\kappa_h^2}{\hat{\kappa}_j^3 [1 + (\kappa_h/\hat{\kappa}_j)^2]^2}, \tag{5.18}$$

where $\kappa_h = (k^2 + l^2)^{1/2}$ is the total horizontal wavenumber, and $\hat{\kappa}_j = \pi j f / N_0 B$ is the roll-off wavenumber. The GM spectrum for u was presented in Chapter 3

(Eq. 3.70), and so the spectra of μ and μ_u are

$$F_\mu(\kappa_h, j) = F_\zeta(\kappa_h, j) \frac{1}{c_0^2} \left(\frac{d\bar{c}}{dz}\right)_p^2, \tag{5.19}$$

$$F_{\mu_u}(\kappa_h, j) = F_\zeta(\kappa_h, j) \frac{1}{c_0^2} \frac{m^2(j)}{\kappa_h^2} (\sigma^2 \cos^2\vartheta + f^2 \sin^2\vartheta), \tag{5.20}$$

where $\vartheta = \tan^{-1}(l/k)$ is the azimuthal angle, $m(j)$ is the internal-wave vertical wavenumber, and the WKB dispersion relation has been used (Eq. 3.63).

Because the μ_u spectrum depends on the azimuthal angle, the correlation functions of μ and μ_u in the vertical plane are written such that

$$\langle\mu(1)\mu(2)\rangle(\Delta x, \Delta z) = \sum_{j=1}^{\infty} \int_0^\infty d\kappa_h F_\mu(\kappa_h, j) \int_0^{2\pi} \frac{d\vartheta}{2\pi} \cos(m\Delta z)\cos(\kappa_h \cos(\vartheta)\Delta x), \tag{5.21}$$

$$\langle\mu_u(1)\mu_u(2)\rangle(\Delta x, \Delta z) = \sum_{j=1}^{\infty} \int_0^\infty d\kappa_h \int_0^{2\pi} \frac{d\vartheta}{2\pi} F_{\mu_u}(j, k, \vartheta)\cos(m\Delta z)\cos(\kappa_h \cos(\vartheta)\Delta x). \tag{5.22}$$

In both cases the integral over azimuthal angle can be done analytically, yielding

$$\langle\mu(1)\mu(2)\rangle = \langle\mu^2\rangle(\bar{z}) \frac{4}{\pi} \sum_{j=1}^{\infty} H(j)\cos(m(j)\Delta z)$$
$$\times \int_0^\infty \frac{d\kappa_h\, \kappa_h^2}{\hat{\kappa}_j^3} \frac{J_0(\kappa_h\Delta x)}{[1 + (\kappa_h/\hat{\kappa}_j)^2]^2}, \tag{5.23}$$

$$\langle\mu_u(1)\mu_u(2)\rangle = \langle\mu_u^2\rangle(\bar{z}) \frac{4}{3\pi} \sum_{j=1}^{\infty} \left(\frac{\pi j}{N_0 B}\right)^2 H(j)\cos(m(j)\Delta z)$$
$$\times \int_0^\infty \frac{d\kappa_h}{\hat{\kappa}_j^3} \frac{[(\sigma^2 + f^2)J_0(\kappa_h\Delta x) - (\sigma^2 - f^2)J_2(\kappa_h\Delta x)]}{[1 + (\kappa_h/\hat{\kappa}_j)^2]^2}, \tag{5.24}$$

where $\bar{z}$ is the mean depth, and J_0 and J_2 are Bessel functions of the first kind. The factors $\langle\mu^2\rangle$ and $\langle\mu_u^2\rangle$ quantify the variance of fractional sound speed and Mach number as a function of depth and are given by

$$\langle \mu^2 \rangle = \zeta_0^2 \frac{N_0}{N(z)} \frac{1}{c_0^2} \left(\frac{d\bar{c}}{dz} \right)_p^2, \tag{5.25}$$

$$\langle \mu_u^2 \rangle = \frac{u_0^2}{2} \frac{1}{c_0^2} \frac{N(z)}{N_0}, \quad u_0^2 = 3\zeta_0^2 N_0^2, \tag{5.26}$$

where GM values are $N_0 = 3$ cph, $\zeta_0 = 7.3$ m, and u_0^2 is the reference variance of the total horizontal current, thus explaining the factor of $1/2$ in the Mach number variance. Using the GM parameters the rms current in the x-direction is $u_0/\sqrt{2} \simeq$ 5 cm/s, so the Mach number for acoustic propagation through internal waves is indeed small. It must be remembered that the assumptions about the stochastic internal-wave field via the GM spectrum are that the fluctuations are isotropic in the horizontal and that there is up/down symmetry of the waves such that a mode description can be used. It is also important to recall that kinematics dictates that internal-wave fluctuations are inhomogeneous in depth and anisotropic in the depth-horizontal plane. Typical correlation scales are discussed and presented in Chapter 3.

5.3.1 Ray-Tangent Approximation

If the ray path is approximately linear over an internal-wave correlation length in the direction of the ray, Eqs. 5.16 and 5.17 can be simplified significantly. This approach is termed the ray-tangent approximation, and it is most appropriate for small grazing angle rays that have minimal curvature. There are clearly issues with this approach near ray turning points where the ray is not straight at all, but this has been the standard approach for ray-based methods, including weak fluctuation theory and path integrals. Using $ds = dx\sec\theta$, writing the second integrals in Eqs. 5.16 and 5.17 in terms of a relative coordinate $\Delta x = x_2 - x_1$, and making the straight ray approximation $\Delta z = z_2 - z_1 = \Delta x \tan\theta$ one obtains

$$\langle \tau_\zeta^2 \rangle = \frac{1}{c_0^2} \int_\Gamma dx_1 \, \sec^2\theta \int_{-\infty}^{\infty} d\Delta x \, \langle \mu(x_1, z_1)\mu(x_1 + \Delta x, z_1 + \Delta x \tan\theta) \rangle,$$

$$\tag{5.27}$$

$$\langle \tau_u^2 \rangle = \frac{1}{c_0^2} \int_\Gamma dx_1 \int_{-\infty}^{\infty} d\Delta x \, \langle \mu_u(x_1, z_1)\mu_u(x_1 + \Delta x, z_1 + \Delta x \tan\theta) \rangle.$$

$$\tag{5.28}$$

Here the integration limits of the Δx integral are taken out to infinity because at large separations the correlations are essentially zero. It proves insightful to bring Eqs. 5.21 and 5.22 into the travel-time variance equations, and to focus on the angular and Δx integrals. Starting with Eq. 5.27 it is found that

$$\int_0^{2\pi} \frac{d\vartheta}{2\pi} \int_{-\infty}^{\infty} d\Delta x \cos(m\tan\theta\Delta x)\cos(\kappa_h\cos\vartheta\Delta x) = \int_0^{2\pi} \frac{d\vartheta}{2}\Big[\delta(m\tan\theta - \kappa_h\cos\vartheta)$$

$$+\delta(m\tan\theta + \kappa_h\cos\vartheta)\Big],$$

$$= \frac{2}{l}. \tag{5.29}$$

A resonance condition therefore exists such that only waves at the correct azimuth can contribute to the travel-time variance. These are in fact the internal waves that have wavenumbers perpendicular to the ray with slope θ, which is the same result that was described in Chapter 4. For example, in this calculation one of the contributing internal waves has wavenumber $\kappa = (-m\tan\theta, l, m)$, which is perpendicular to the ray vector $\mathbf{r}_{ray} = (1, 0, \tan\theta)$. Because of horizontal isotropy and up-down symmetry there are four possible directions for these resonant waves, thus accounting for the factor of 2 in the above equation. Next, it is useful to do the remaining integrals over the spectrum in mode number/frequency space. Using the WKB dispersion relation, the required expression for the y-component of the internal-wave wavenumber is $l = (\pi j/N_0 B)(\sigma^2 - \sigma_L^2)^{1/2}$ where $\sigma_L^2 = f^2 + N^2\tan^2\theta$ is the cut-off frequency below which the perpendicular wavenumber resonance is forbidden by the dispersion relation. The travel-time variance due to internal-wave displacements is thus

$$\langle\tau_\zeta^2\rangle = \frac{1}{c_0^2}\int_\Gamma ds\,\langle\mu^2\rangle\,L_p(z,\theta). \tag{5.30}$$

Here the fractional sound-speed variance is given by Eq. 5.25 and the function L_p is the correlation length of the sound-speed fluctuations in the direction of the ray (θ), which is given by

$$L_p(z,\theta) = \frac{1}{\langle\mu^2\rangle}\sum_{j=1}^{\infty}\int_{\sigma_L}^N \frac{2N_0 B}{\pi j}\frac{F_\mu(j,\sigma)}{(\sigma^2 - \sigma_L^2)^{1/2}}\,d\sigma. \tag{5.31}$$

For the GM spectrum the correlation length along the ray is

$$L_p(z,\theta) = \langle j^{-1}\rangle\frac{4}{\pi^2}\frac{N_0 B}{f}\left(\frac{1}{1+a^2} + \frac{a^2}{2}\frac{1}{(1+a^2)^{3/2}}\ln\left(\frac{(1+a^2)^{1/2}+1}{(1+a^2)^{1/2}-1}\right)\right), \tag{5.32}$$

where $a = N(z)\tan\theta/f$ and $\langle j^{-1}\rangle = \sum_{j=1}^{\infty} H(j)/j$. (Note that $1 + a^2 = \sigma_L^2/f^2$.)

Next the travel-time fluctuations associated with internal-wave currents is examined. Here the same resonance condition is obtained. Performing the Δx and the angular integral is slightly more complicated, but it is found that

$$\int_0^{2\pi} \frac{d\vartheta}{2\pi}(\sigma^2\cos^2\vartheta + f^2\sin^2\vartheta)\int_{-\infty}^{\infty} d\Delta x\cos(m\tan\theta\Delta x)\cos(\kappa_h\cos\vartheta\Delta x) = \frac{2\sigma_L^2}{l},$$

(5.33)

where the WKB dispersion relation has been used. The result for the current induced travel-time variance is

$$\langle\tau_u^2\rangle = \frac{1}{c_0^2}\int_\Gamma ds\ \cos^2\theta\ \langle\mu_u^2\rangle L_{p_u}(z,\theta).$$

(5.34)

The Mach number variance is given by Eq. 5.26, and the correlation length of the current in the direction of the ray is

$$L_{p_u}(z,\theta) = \frac{1}{\langle\mu_u^2\rangle}\sum_{j=1}^{\infty}\int_{\sigma_L}^{N}\frac{2N_0 B}{\pi j}\frac{\sigma_L^2}{(\sigma^2-\sigma_L^2)^{1/2}}\frac{N^2-\sigma^2}{\sigma^2-f^2}F_\zeta(j,\sigma)\,d\sigma.$$

(5.35)

Here the relation from the internal-wave mode equation has been used such that $m^2/\kappa_h^2 = (N^2-\sigma^2)/(\sigma^2-f^2)$. For the GM spectrum there is an analytical expression for the correlation length of the current along the ray given by (Munk et al., 1981)

$$L_{p_u}(z,\theta) = \langle j^{-1}\rangle\frac{4}{\pi^2}\frac{N_0 B}{f}\left(-1+\frac{1}{2}\frac{2+a^2}{(1+a^2)^{1/2}}\ln\left(\frac{(1+a^2)^{1/2}+1}{(1+a^2)^{1/2}-1}\right)\right).$$

(5.36)

There is an integrable singularity in the equation for L_{p_u} for $a = 0$, a ray turning point. This singularity exists because of the strong contributions at $\theta \simeq 0$ from nearly horizontally propagating inertial waves that have small horizontal wavenumbers. Note that in the WKB dispersion relation $\sigma = f$ corresponds to $\kappa_h = 0$.

Equations 5.32 and 5.36 provide important information about the sensitivity of travel-time fluctuations to the scales of the ocean. Not only is there the perpendicular wavenumber resonance but because of the $\langle j^{-1}\rangle$ term and the integrand of the L_p functions, it is the low-mode and low-frequency waves that provide the greatest contribution to the travel-time variance. These are the large spatial scale waves, and thus this view is consistent with the results from weak fluctuation theory (Chapter 4) that travel time (or phase) is sensitive to the largest scales in the problem.

Calculations by Munk et al. (1981) and Stoughton et al. (1986) using the formalism developed here show that the rms travel-time fluctuation associated with vertical displacements is larger than the current-induced effect by roughly a factor of 10. For conditions relevant to the 300-km, 1983 Reciprocal Transmission

Experiment (RTE83) in the North Atlantic near Bermuda, Stoughton et al. (1986) find values of $\langle \tau_\zeta^2 \rangle^{1/2}$ and $\langle \tau_u^2 \rangle^{1/2}$ of order 4.3 and 0.42 ms for small-angle rays, and 1.8 and 0.28 ms for steep angle rays.

5.3.2 Accuracy of Ray-Tangent Approximation

In many cases of observational interest the angles of the rays are too large for the ray-tangent approximation to be accurate (Colosi et al., 1999), and alternative approaches are needed. If the travel-time variance is considered to be of the form $\int_\Gamma W_\tau ds$ where W_τ is the ray weighting function, then various approximations can be examined. The exact form of W_τ is simply the inner integral of Eq. 5.16, that is,

$$W_\tau(s_2) = \frac{1}{c_0^2} \int_\Gamma ds_1 \langle \mu(s_1)\mu(s_2) \rangle, \tag{5.37}$$

which can be evaluated numerically. The ray-tangent approximation gives $W_\tau = \langle \mu^2 \rangle L_p(\theta, z)/c_0$. An approach to compute L_p at the ray upper turning point is to use a quadratic ray-path approximation instead of a linear one. In this case $\Delta z = C_r \Delta x^2$ where $C_r = (dc/dz)/(2c)$ is an estimate of the ray curvature. The effective value of L_p at the turning point is then (Colosi et al., 1999)

$$L_p(C_r, z^+) = \frac{\pi^2}{2} \sqrt{\frac{N_0 B}{C_r N(z^+)}} \langle j^{-1/2} \rangle. \tag{5.38}$$

The weighting function in this turning point approximation is determined as follows. Over most of the ray path the ray-tangent approximation is used, but when $L_p(\theta, z)$ exceeds the turning point value $L_p(C_r, z^+)$ the value is then capped. Figure 5.3 shows three different calculations of the weighting functions for the case of long-range propagation in the central North Pacific Ocean (Colosi et al., 1999). The ray-tangent approximation is seen to predict a much too narrow and strong weight distribution in the neighborhood of the upper turning point, whereas the exact solution shows a broader and lower distribution. In this particular case, oddly enough, the area under the two curves is roughly the same, that is, the predicted travel-time variances are roughly the same. This is not always the case, and these weighting functions for other observables can lead to dramatically incorrect predictions, such as those associated with travel time bias and pulse spread (Colosi et al., 1999). Importantly, the results shown in Figure 5.3 have a significant physical meaning, that is, they show that the ray-tangent approximation overestimates the anisotropy of the internal-wave-induced scattering, which means that the dependence on ray angle is not as strong as previously estimated.

The exact calculation of the weighting function is somewhat computationally intensive, so approximate empirical relations have been developed in terms of the function L_p (Flatté and Rovner, 2000). Here it is found that to a fair approximation

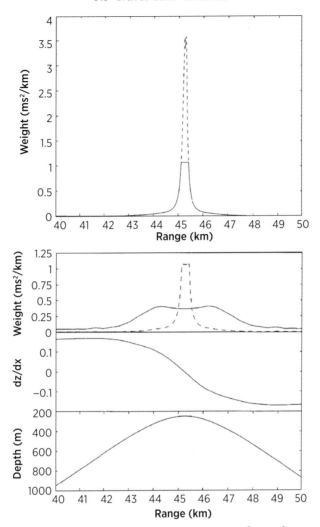

Figure 5.3. Travel-time variance weighting functions $\langle \mu^2 \rangle L_p / c_0^2$ near one upper turning point for a ray path in the ATOC Acoustic Engineering Test (AET). The top panel shows the results for the ray-tangent approximation (dashed), and a crude parabolic upper turning point correction (solid). The bottom panels show the ray path, ray slope, and a comparison between the weighting functions based on a full numerical integration over the ray path (solid), and the parabolic approximation shown in the top panel (dashed).
Source: Colosi et al. (1999).

one may write

$$L_p(z, \theta) = L_{p0} \frac{[1 - \exp(-(\sigma_c/\sigma_{curv})^p)]^q}{1 + (\rho_s/\rho_c)^{q_a}(\sigma_c/\sigma_{curv})^{p_a}}, \tag{5.39}$$

where $\rho_c = 3.5$, $\sigma_c = 0.0204$, $p = 0.385$, $p_a = 0.5$, $q = 1.3$, and $q_a = 2.0$ are dimensionless constants. The reference $L_{p0} = 12.7$ km, and the factor $\rho_s =$

$N(z)\tan\theta/f$. Weighting functions along the ray paths for observables other than travel-time variance will be discussed in more detail in subsequent chapters on weak fluctuation theory (Chapter 6) and path integrals (Chapter 7).

5.3.3 Observations

The travel-time variance is an acoustic fluctuation quantity that is easily calculated and observed, provided that a well-identified time front can be measured. In deep-water propagation, resolved time fronts that are associated with high angle ray paths constitute the early part of the arrival pattern. In shallow-water propagation, resolved time fronts are rarely observed. The focus will therefore be on deep-water experiments, discussing the longest range experiments first and moving on to shorter ranges. Long-range propagation is perhaps the most challenging test of theory, because errors at each ray turning point accumulate. There are also uncertainties about the background profiles.

The longest range experiment that has been analyzed for travel-time variance is the 3250-km range ATOC Acoustic Engineering Test, in which broadband pulses with a center frequency of 75 Hz were transmitted from the sound-channel axis to a 700-m long vertical receiver array (Colosi et al., 1999). At this long range one must account for the range-dependent background sound-speed and buoyancy frequency profiles, as well as the changing Coriolis parameter. The background profiles were estimated using eXpendable BathyThermograph (XBT) and Conductivity, Temperature, Depth (CTD) observations combined with climatology. The XBT and CTD data were also used to estimate internal-wave strength that was found to be roughly half the standard GM level, that is, $\zeta_0 = 7.3/\sqrt{2} = 5.2$ m. Figure 5.4 shows a comparison between observations and two theoretical calculations plotted as a function of ray upper turning point depth. It must be noted here that fluctuations in travel time at these long ranges are composed of two often equal contributions. There is the wandering of the time front that is modeled by the theoretical expressions, and there is the fracturing of the time front (i.e., multiple arrivals for a single branch) that is not in the present models. Comparisons between data and theory therefore require careful removal of the time front fracturing component. Returning to the comparison, the theoretical calculation using Eq. 5.16 with no approximations is seen to be in excellent agreement with the observations, except for the shallowest turning rays. The theoretical estimate using the upper turning point approximation (Eq. 5.38) is shown to vastly underpredict the variation. Oddly enough, as previously mentioned, the ray-tangent approximation gives results for travel-time variance in this case that are quite close to the values from the exact expression.

The next long-range experiment that has been analyzed is the 1000-km range SLICE89 experiment, where pulses with a 250-Hz center frequency were transmitted from the sound channel axis to a 3000-m long vertical receiving array

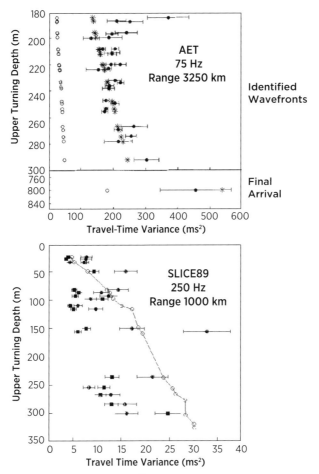

Figure 5.4. Travel-time variance τ^2 as a function of the upper turning point (UTP) depth for the 75-Hz, 3250-km range AET experiment (top panel) and the 250-Hz, 1000-km range SLICE89 experiment (bottom panel). In the AET, theoretical results using the exact weighting function are shown with stars, and those from a parabolic upper turning point approximation are shown with open circles. Observations from the AET are shown with filled circles and error bars. In SLICE89, open circles are theoretical estimates using the empirical approach of Flatté and Rovner (2000), closed circles are results from parabolic equation Monte Carlo simulations, and the closed squares are experimental results for different identifiable time fronts.
Source: Top panel from Colosi et al. (1999).

(Duda et al., 1992). Figure 5.4 shows observations, Monte Carlo simulation results, and a theoretical estimate for travel-time variance. The theoretical calculations for the SLICE89 experiment used the empirical approach of Flatté and Rovner (2000). Both the theory and the parabolic equation Monte Carlo simulations utilized an internal-wave strength that was one-half the GM reference level. Over the UTP depth region of 100–300 m there is reasonable agreement between the numerical simulations and the experimental results, with values between 5 and 25 ms^2. The theoretical calculations differ from the simulations by factors of order 2.

The remainder of the experiments to be considered consist of relatively high-frequency, short-range field efforts conducted in the 1970s and 1980s. In all cases, theoretical calculations were carried out using the ray-tangent approximation. A summary of the results is given in Table 5.1. The observations show no dependence on frequency (as expected), and the theoretical estimates are mostly within a factor of two of the observations. The discrepancy between observed and computed values may hold important oceanographic information concerning the internal-wave spectrum, information that could be obtained from an inversion.

Finally, there is the difficult observation of travel-time fluctuations induced by small-scale internal-wave currents. Only one experiment to date has been analyzed for this quantity, and that is the 400-Hz, 300-km, 1983 Reciprocal Transmission Experiment (RTE83) conducted in the North Atlantic. This measurement is difficult because of the small values of $\langle \tau_u^2 \rangle^{1/2}$, as previously discussed. For this experiment Stoughton et al. (1986) report rms reciprocal travel-time fluctuations to be between 0.90 and 1.8 ms, far exceeding model predictions using the GM spectrum that fall between 0.28 and 0.42 ms rms. The causes of the discrepancies are not presently known, but likely candidates are (1) some degree of nonreciprocity due to source receiver separations, and (2) inaccuracies in the GM spectrum near the inertial period, which provides much of the current responsible for the travel-time fluctuations.

5.4 Other Ray-like Observables

Of course there are other ray fluctuation quantities that can be treated, and here the subject of ray angle fluctuations and ray intensity fluctuations is taken up.

5.4.1 Ray-Angle Variance

Rays moving through the stochastic internal-wave field will not only be advanced and delayed by the fluctuations, but they will also acquire angular deviations in accord with the laws of refraction. The deviation in vertical ray angle ($\delta\theta$) caused by the sound-speed perturbation μ can be calculated from the ray equations in the small-angle approximation (see Eq. 2.111). Here it is found that

$$\frac{d^2z}{dx^2} \simeq \frac{d\delta\theta}{dx} = -\mu',$$ (5.40)

so that the deviation in angle after traveling a distance dx is approximately given by $d\theta \simeq -\mu' dx$, where μ' is the random vertical gradient of fractional sound-speed.

Table 5.1. *Comparison of observed and estimated travel-time variance for a number of experiments.*

Name/Year	Range/Frequency (km/kHz)	ID	τ^2 (Obs) (ms^2)	τ^2 (Theory) (ms^2)	Reference
Cobb/1976	17.2/4	−1	0.16	0.04	Flatté et al. (1979)
	17.2/8	−1	0.16	0.04	
MATE/1977	18.1/2	−1	0.24	0.16	Ewart and Reynolds (1984)
	18.1/4	−1	0.24	0.16	
	18.1/8	−1	0.24	0.16	
	18.1/13	−1	0.24	0.16	
RTE/1983	305/0.4	+8	20.3	8.9	Stoughton et al. (1986)
	,	−8	15.3	8.6	
	,	+9	8.2	6.7	
	,	+11	5.6	17.7	
	,	−11	6.6	14.1	
	,	+12	8.8	16.6	
	,	−12	6.4	16.6	
	,	+13	7.4	19.2	
	,	−13	7.6	16.4	
	,	+14	16.4	18.1	
	,	−14	10.1	18.1	
	,	+15	11.7	20.0	
SLICE89/1989	1000/0.25	±36 to ±45	5–25[a]	5–30	Flatté and Rovner (2000)
Barents Sea/1992	25/0.224	NR[b]	100–900	100–400	Lynch et al. (1996)
AET/1994	3250/0.075	±126 to ±139	180–300[a]	180–250	Colosi et al. (1999)

[a]See Figure 5.4.
[b]For the Barents Sea Experiment multiple arrivals were observed but ray identifiers were not reported.

In the linear approximation the deviation of the ray angle at range R is obtained by integrating over the unperturbed ray path Γ, giving

$$\theta(R) = - \int_\Gamma ds\, \mu'(s). \tag{5.41}$$

In the ray-tangent approximation, the variance in the ray angle after a range R is

$$\langle \delta\theta^2 \rangle = \int_\Gamma L_p'(\theta_r(s), z(s))\, \langle (\mu')^2(z) \rangle\, ds, \tag{5.42}$$

where L_p' is the correlation length of the vertical gradient of the internal waves in the direction of the ray. Calculation of the ray-angle variance therefore requires the spectrum of internal-wave strain, and using the WKB formula for vertical wavenumber the result is

$$F_{\zeta'}(\sigma, j) = \left(\frac{\pi j N(z)}{N_0 B} \right)^2 F_\zeta(\sigma, j). \tag{5.43}$$

The variance of μ' depends on the variance of internal-wave strain (Munk, 1981; Colosi, 2006)

$$\langle (\mu')^2(z) \rangle = \left(\frac{dc}{dz} \right)_p^2 \langle (\zeta')^2 \rangle \simeq \pi^2 \left(\frac{dc}{dz} \right)_p^2 \frac{\zeta_0^2}{B^2} \frac{N}{N_0} \frac{j_* J}{N_j}, \tag{5.44}$$

where N_j is the mode spectrum normalization, J is the maximum vertical mode number, and B is the vertical scale length of the stratification. Recall from Chapter 3 that J is controlled by shear instability (Gregg, 1977; Munk, 1981). The variance of μ' scales as N while the variance of μ itself scales as $1/N$. The analytic expression for L_p' corresponding to Eqs. 5.32 and 5.36 is

$$L_p'(\theta, z) = \langle j \rangle \frac{4}{\pi^2} \frac{N_0 B}{f} \left(\frac{1}{1 + a^2} + \frac{a^2}{2} \frac{1}{(1 + a^2)^{3/2}} \ln\left(\frac{(1 + a^2)^{1/2} + 1}{(1 + a^2)^{1/2} - 1} \right) \right), \tag{5.45}$$

where again, $a = N(z) \tan\theta / f$ and $\langle j \rangle = \sum_{j=1}^J j H(j)$. The cut-off mode number J is particularly important for L_p' given the divergence of $\langle j \rangle$ for $J \to \infty$. In this example it is seen how different ray observables are sensitive to different scales of the internal-wave field. Travel-time variance, for example, is proportional to $\langle j^{-1} \rangle$ and thus is sensitive to large vertical scale internal waves. The angle deviation, however, is proportional to $\langle j \rangle$ and small vertical scale internal waves provide the largest contributions.

5.4.2 Ray-Intensity Variance

Ray theory formulations of intensity variance have received little attention in the ocean acoustics literature because of the aforementioned problems of small-scale cut-offs. However, a brief development here is useful both from a pedagogical standpoint and from the fact that the topic has some potential for research applications. Here the development of Brekhovskikh and Lysanov (1991) is followed and a perturbation approach is adopted that will be useful only in the unsaturated regime. The starting point is the eikonal and ray amplitude equations (Eqs. 2.51 and 2.52) for the pressure $p = a(\mathbf{r}, \omega) \exp(i\Theta(\mathbf{r}, \omega))$, which is expanded in terms of the small quantity μ. Expressions have already been obtained for the ray phase function which to zeroth and first order are

$$\Theta_0 = \omega \int_\Gamma \frac{ds}{\bar{c}(z)}, \quad \Theta_1 = -q_0 \int_\Gamma \mu \, ds. \tag{5.46}$$

The ray amplitude equations that are used are slightly altered from those presented in Chapter 2. By looking at the log-amplitude $\chi = \ln(a/a_{ref})$, where a_{ref} is a reference amplitude, the new equation for χ is

$$2\nabla\Theta \cdot \nabla\chi + \nabla^2\Theta = 0. \tag{5.47}$$

The zeroth- and first-order solutions of this equation are respectively given by

$$2\nabla\Theta_0 \cdot \nabla\chi_0 + \nabla^2\Theta_0 = 0, \tag{5.48}$$

$$2\nabla\Theta_0 \cdot \nabla\chi_1 + 2\nabla\Theta_1 \cdot \nabla\chi_0 + \nabla^2\Theta_1 = 0. \tag{5.49}$$

In the ray approximation $\nabla\Theta = \mathbf{k}$, and thus the first term in the first-order equation is a directional derivative of the amplitude along the unperturbed ray, that is, $2q_0 d\chi_1/ds$. Next the perpendicular wavenumber resonance condition is utilized. This means that all derivatives in the second and third terms to a good approximation can be replaced by operators that differentiate in a perpendicular direction relative to the ray. Thus the result is

$$\frac{d\chi_1}{ds} = -\frac{1}{2q_0}\left[\nabla_\perp^2\Theta_1 + 2\nabla_\perp\Theta_1 \cdot \nabla_\perp\chi_0\right] \simeq -\frac{1}{2q_0}\nabla_\perp^2\Theta_1, \tag{5.50}$$

where the last line follows from the fact that the variations of unperturbed wave amplitude along the wave front are small. Equation 5.50 has a pleasing simple geometric meaning. It says that the rate of change of the ray amplitude fluctuation is proportional to the local curvature of the wave front. In Chapter 4 it was seen that in the unsaturated regime a curved wave front leads to focusing and defocusing.

The intensity variance is easily written down in terms of integrals over the unperturbed ray path. For small-angle sound propagating through the ocean internal-wave field whose vertical scales are much smaller than their horizontal scales, the approximation $\nabla_\perp^2 \simeq d^2/dz^2$ can be made. The resulting equation is thus

$$\langle \chi_1^2 \rangle = \frac{1}{4q_0^2} \int_\Gamma ds_1 \int_\Gamma ds_2 \left\langle \frac{d^2\Theta_1(s_1)}{dz^2} \frac{d^2\Theta_1(s_2)}{dz^2} \right\rangle, \tag{5.51}$$

where the correlation function of the vertical phase curvature is written in terms of the vertical phase spectrum $F_\Theta(k_z)$,

$$\left\langle \frac{d^2\Theta_1(s_1)}{dz^2} \frac{d^2\Theta_1(s_2)}{dz^2} \right\rangle = \int k_z^4 F_\Theta(k_z) \cos(k_z \Delta z) \, dk_z. \tag{5.52}$$

Because the phase spectrum scales as ω^2, it is seen that in the geometrical acoustics approximation the intensity variance is independent of frequency. Furthermore, while the phase variance scales linearly with range, the intensity variance scales as the cube of range. In this cubic scaling, two factors of range come from the integrals over the ray paths and the third comes from the phase spectrum.

Although it is not evident here, in weak fluctuation theory in the geometrical acoustics limit it is found that $\langle \chi^2 \rangle << \Phi^2$ (Chapter 6). It is only in the diffractive regime of weak fluctuations where amplitude and phase fluctuations are of comparable magnitude.

5.5 Scattering Along and Across the Wave Front

Another important aspect of ray propagation in the ocean that can be treated relatively accurately using the linear equations is that rays tend to scatter primarily along the wave front as opposed to across it. This problem has been taken up by Flatté and Colosi (2008) and Godin (2007). The behavior was described in qualitative terms in Chapter 4 (Figure 4.4), but more analysis is needed to provide analytic formulae for predicting the rms deviations along and across the front.

The simplest starting point is to consider the case in which there is no waveguide, and the geometry is as shown in Figure 5.5. Without fluctuations, the ray arrives at $z = 0$ and $x = R$ after time T. With fluctuations in sound speed, the new arrival positions after time T are Z and $R + X$, with deviation Z along the wave front and deviation X perpendicular to the wave front or in the ray direction. Because the unperturbed angle in this case of no sound channel is zero, the symbol θ will be used for the angle at position x caused by the perturbations. The deviation in ray angle caused by the perturbation μ was shown in the previous section to be described by the equation $d\theta/dx = -\mu'$ in the small-angle approximation. This angular deflection can be used to estimate the vertical deflection Z of the ray. Let us say that a ray propagates unperturbed to a location x_1, where it undergoes a scattering event that alters the ray angle by $\delta\theta(x_1)$. At the final range, which is $(R - x_1)$ away from the scattering event, the vertical deviation is then $\delta z(x_1)$. The triangular geometry dictates that $(dz/d\theta)(x_1) = (R - x_1)$. Summing up all scattering events along the propagation path, the total vertical deviation given is by (Flatté

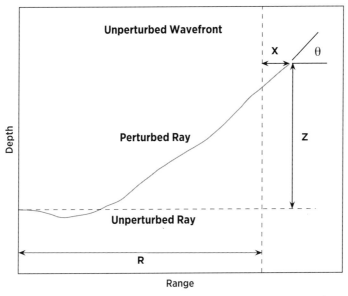

Figure 5.5. Geometry of ray scattering along and across the wave front in the absence of a waveguide.
Source: Flatté and Colosi (2008).

and Colosi, 2008)

$$Z = \int_0^R \frac{dz}{d\theta}(x_1)d\theta(x_1) = -\int_0^R (R - x_1)\mu' \, dx_1. \tag{5.53}$$

This equation has an important physical interpretation. It says that scattering events closer to the source have a larger impact on the vertical deviation because for a fixed scattering angle these events will have a large horizontal distance to wander away from the unperturbed ray. The deviation of the ray across the front is considerably easier to visualize and can be computed by considering that when a ray moves though sound-speed perturbations it will be advanced or delayed by these variations. In the small-angle approximation, this time advance/delay can be associated with the spatial deviation X, so that the result is

$$X \simeq -\int_0^R \mu \, dx. \tag{5.54}$$

Application of these results to the waveguide is relatively straightforward. For the across-the-wave-front deviations, the calculation in terms of the travel-time variance has already been done, which in the ray-tangent approximation yields

$$\langle X^2 \rangle = \int_\Gamma \langle \mu^2 \rangle L_p \, ds. \tag{5.55}$$

The deviation along the wave front is slightly more complicated to compute because an expression for the rate of change of the ray depth at the receiver with respect to the scattering angle is required, that is, $dz/d\theta$. This expression can be obtained using the ray tube equation (Appendix A). The result is that the Z deviation in the waveguide is given by (Flatté and Colosi, 2008)

$$Z = \int_\Gamma \frac{\xi_2(s)}{\xi_2'(R)} \mu' \, ds, \qquad (5.56)$$

where $\xi_2(x)$ is the ray tube function introduced in Chapter 2, and $\xi_2' = d\xi_2/dx$. In the constant background sound-speed case, the ray tube functions yield $\xi_2(x)/\xi_2'(R) = -(R - x)$, consistent with the results from Eq. 5.53. In the ray-tangent approximation the mean square ray deviation along the front is then given by

$$\langle Z^2 \rangle = \left(\frac{1}{\xi_2'(R)}\right)^2 \int_\Gamma \xi_2^2(s) \, L_p'(\theta_r(s), z(s)) \, \langle (\mu')^2(z) \rangle \, ds. \qquad (5.57)$$

As described in Chapter 4, $\langle X^2 \rangle$ and $\langle Z^2 \rangle$ grow in range as R and R^3, respectively. The growth of the along wavefront deviation is much more rapid because of the important contributions near the source as demonstrated by the "moment arm" in Eq. 5.53. Monte Carlo simulations of ray propagation through random realizations of internal-wave-induced sound-speed perturbations have been carried out to test the accuracy of Eqs. 5.42, 5.55, and 5.57 (Figure 5.6). The theory is mostly within a factor of two of the Monte Carlo results, which is a level of accuracy consistent with the ray-tangent approximation for this $7°$ ray. In addition the range scalings of the theory are by and large borne out. At the longest ranges, however, the range growth shifts to $R^{3/2}$ for $\langle X^2 \rangle$. This will turn out to be a manifestation of ray chaos.

5.6 Ray Chaos

In this section the full nonlinearity of the ray equations in the presence of internalwave-induced sound-speed perturbations is confronted. There is a vast literature on this subject from both the classical mechanics side and the ocean acoustics side. The interested reader may start with the review articles by Brown et al. (2003) and Beron-Vera et al. (2003) and utilize the numerous references therein.

This section will focus on several key aspects of the ray equation nonlinearity that are particularly important for underwater acoustics. The discussion starts off with the topic of nonlinear response to forcing by internal waves, and the important issue of nonlinear resonance. This leads to a discussion of the seminal KAM theorem and manifestations of chaos in ray phase space. The discussion is concluded with a treatment of the stochastic aspects of Lyapunov exponents and ray amplitudes.

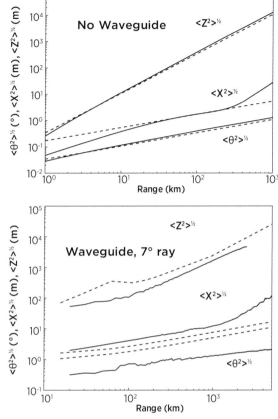

Figure 5.6. RMS deviations of ray displacements along and across the front, as well as ray angular deviations. Displayed are Monte Carlo simulation (solid), and theory from Eqs. 5.55, 5.57, and 5.42 (dash). The top panel shows the statistics for the case of no waveguide, while the bottom panel shows results for a 7° ray in a Munk canonical profile.
Source: Flatté and Colosi (2008).

5.6.1 Nonlinear Response to Forcing

A simplified discussion is an appropriate place to start to understand how nonlinear systems can respond to forcing, such as that provided to ray propagation by the ocean internal-wave field. Consider the ray equations in the parabolic approximation, that is (Eq. 2.111):

$$\frac{d^2z}{dx^2} + U'(z) = -\mu'(x,z), \tag{5.58}$$

where the primes denote derivatives with respect to depth. Here one can view μ to be a small forcing function, and the stochastic aspects of μ will not be addressed just yet. The focus at this point is on ray propagation through a realization of the highly structured internal-wave field. Equation 5.58 is nonlinear because both U' and μ' are nonlinear functions, and the equation is termed *nonautonomous* because μ is a function of the independent variable x (Tabor, 1989). It is important to appreciate that the response of a nonlinear equation to small forcing can be quite different from the response of a linear equation; this strikes at the heart of the notion of chaos and resonances.

First consider the simplified cases in which the unforced equation (e.g., $\mu = 0$) is linear and nonlinear. In the linear case, let $U(z)$ be a quadratic function, so that the ray equation becomes

$$\frac{d^2z}{dx^2} + K_L^2 z = 0. \tag{5.59}$$

The solution to this equation is sinusoidal with horizontal wavenumber $K_L = 2\pi/R_L$ where R_L is the horizontal loop length. However, it is useful to solve this equation using the method of *Integration by Quadrature*. While more laborious, this method provides better insight into the problem and allows solution of an example nonlinear problem. To start, Eq. 5.59 is rewritten as a pair of coupled first-order equations:

$$\frac{dz}{dx} = y, \text{ and } \frac{dy}{dx} = -K_L^2 z. \tag{5.60}$$

Multiplying the first equation by $K_L^2 z$, the second equation by y, and summing, the result is

$$\frac{d}{dx}\left(\frac{y^2}{2} + \frac{K_L^2 z^2}{2}\right) = 0. \tag{5.61}$$

This equation identifies $I_1 = (y^2 + K_L^2 z^2)/2$ as a *constant of the motion* or a *first integral* and is thus computed from the initial conditions.[1] One can therefore write

$$\frac{dz}{dx} = \sqrt{2I_1 - K_L^2 z^2}. \tag{5.62}$$

Integrating both sides of the equation and inverting the result, the solution is

$$z(x) = \frac{\sqrt{2I_1}}{K_L}\sin[K_L(x + I_2)], \tag{5.63}$$

where I_2 is another integration constant established from the initial conditions. The horizontal cycle distance of the ray can be easily read off from the explicit

---

[1] By noting that $y = \tan\theta$, it is seen that the constant I_1 is equal to the Hamiltonian in the parabolic approximation (see Chapter 2) and in some sense could be called the ray *Energy*. The mechanical analogy is the harmonic oscillator for which the independent variable is time. In this case, the first integral is the mechanical energy, kinetic plus potential.

solution, but a method applicable to the nonlinear problem is to utilize the explicit quadrature equation (Eq. 5.62). Here the expected result is obtained, that is,

$$R_L = 2 \int_{-\sqrt{2I_1}/K_L}^{\sqrt{2I_1}/K_L} \frac{dz}{\sqrt{2I_1 - K_L^2 z^2}} = \frac{2\pi}{K_L},$$
(5.64)

where the turning point limits are defined where $dz/dx = 0$.

Next consider the simplest nonlinear problem, that is a quartic profile U, resulting in an unperturbed ray equation of the form

$$\frac{d^2 z}{dx^2} + \beta z^3 = 0.$$
(5.65)

Following the quadrature integration methodology, a first integral is obtained given by

$$I_1 = (2y^2 + \beta z^4)/4.$$
(5.66)

The ray path solution in this case is given in terms of Jacobi elliptic functions and is not of fundamental interest here. The ray loop distance, however, is of interest and is given by

$$R_L = 2 \int_{-(4I_1/\beta)^{1/4}}^{(4I_1/\beta)^{1/4}} \frac{dz}{\sqrt{2(I_1 - \beta z^4/4)}} = \frac{\Gamma^2(1/4)}{I_1^{1/4}} \sqrt{\frac{1}{2\pi\beta^{1/2}}},$$
(5.67)

where Γ is the Gamma function. The loop length is now a function of the first integral I_1; this is the typical case in ocean acoustics, where it is seen that R_L is a function of ray grazing angle or equivalently the Hamiltonian. It will be seen in subsequent discussions that the rate at which the loop length changes with respect to the first integral dictates a great deal about the chaotic evolution of the ray system.

Now the case of $\mu \neq 0$ is considered and the linear and nonlinear responses of these systems are demonstrated. For simplicity, consider a sinusoidal forcing function with spatial frequency K_f so that in the linear case the ray equations are

$$\frac{d^2 z}{dx^2} + K_L^2 z = \epsilon \cos(K_f x),$$
(5.68)

where ϵ is a small forcing amplitude parameter. The well-known solution to this equation is a superposition of oscillations of spatial frequency K_L and K_f given by

$$z(x) = a \sin(K_L x + \delta) + \frac{\epsilon}{K_L^2 - K_f^2} \cos(K_f x).$$
(5.69)

The solution breaks down when the forcing frequency K_f matches the *intrinsic* frequency K_L (i.e., a resonance), but in the neighborhood of that resonance the solution is of the form (Tabor, 1989):

$$z(x) = \hat{a}\sin(K_L x + \hat{\delta}) + \frac{\epsilon}{2K_L}x\sin(K_L x). \tag{5.70}$$

Here linear growth of the solution is seen corresponding to the ray getting pushed further and further from the sound channel axis and having a steeper and steeper grazing angle. This behavior can be contrasted to that of a driven, nonlinear ray equation namely

$$\frac{d^2z}{dx^2} + K_L^2 z + \beta z^3 = \epsilon\cos(K_f x). \tag{5.71}$$

This is the well-known "Duffing" oscillator equation (Lichtenberg and Lieberman, 1983). The solution of the nonlinear homogeneous equation (i.e., $\epsilon = 0$) can be expressed in terms of Jacobi elliptic functions, and it has been shown that the ray loop distances are a function of the first integral. An interesting stability phenomenon occurs when $\epsilon \ll 1$. When the forcing frequency matches the intrinsic frequency, resonances occur, and the solution will start to grow. However, as the solution grows the intrinsic frequency changes because it is a function of the first integral. This takes the system out of resonance, thereby preventing linear growth of the solution. The nonlinearity causes a form of stability sometimes referred to as "nonlinear negative feedback." As the forcing strength grows, however, chaotic behavior in the solution can be found due to more complicated *feedback* mechanisms between the fundamental motion, its harmonics, and the forcing function (Tabor, 1989).

Clearly these are idealized cases but they bring up the important issue of resonances and the variation of ray-intrinsic spatial frequencies throughout the waveguide. But this is for a single wave forcing the ray equations, and in the real ocean there are a large number of internal waves with a rich spectrum of wavenumbers. This means that the potential for resonance goes up considerably, and one might imagine a ray trajectory jumping between resonances. This topic will require a more careful treatment, which will be addressed in some measure in subsequent sections, but before that is done the celebrated Kologorov-Arnold-Moser (KAM) theorem will be discussed.

5.6.2 Features of Chaos and KAM Theory

The KAM theorem is concerned with the stability of periodic solutions to near integrable, nondegenerate (i.e., $K_L' \neq 0$) Hamiltonian systems. It codifies some of the most sacred insights about dynamical systems regarding resonances and stable trajectories (Arnold, 1989; Tabor, 1989; Brown et al., 2003). Here consider a Hamiltonian system in action-angle coordinates of the form

$$H(I, \vartheta, x) = H(I) + \delta h(I, \vartheta, K_f x), \tag{5.72}$$

where δh is a small, periodic perturbation to the integrable Hamiltonian $H(I)$. Because δh is small, there is no loss of generality by considering it to be an additive perturbation. The KAM theorem states that when δh is small enough, many of the initial conditions lead to stable and quasi-periodic (i.e., close to periodic) ray trajectories. The initial conditions that lead to unstable or chaotic trajectories will remain isolated among the stable trajectories, leading to the jargon of "islands" of stability in a "sea" of chaos. The size of the chaotic sea relative to the stable islands of course depends on the strength of δh. But as was seen in Section 5.7.1 the route to chaos or rapid growth/decay of a solution depends critically on resonances between the intrinsic ray spatial frequencies and the forcing spatial frequencies. In fact there will be resonances when the ray loop wavenumber $K_L(I)$ is an integer multiple or fraction of the perturbation wavenumber K_f. The precise resonance condition is

$$K_L(I) = \frac{m}{l} K_f, \tag{5.73}$$

where (l, m) is any integer pair. Here the potentially important role played by resonant harmonics is seen causing broad regions of the phase space to become chaotic, especially when the perturbation strength increases. However, nonlinear negative feedback can knock a ray out of a resonant trajectory and into a nearby stable trajectory. This notion raises the important question of the width of a resonance and whether or not nearby resonant trajectories overlap. If there are multiple nearby resonates and their widths are large enough, then the nonlinear feedback may end up bouncing the rays from resonance to resonance, never finding a stable trajectory. In this case chaotic motion would appear to be well developed. The dispersion of the ray loops, $K_L(I)$, and the strength of the perturbations, δh, will play a large role in modulating this behavior. The perturbation strength dictates how large a change in action, ΔI, a ray will experience from a scattering event, and $K_L(I)$ tells how large a change in the ray wavenumber to expect. To first order, the change in wavenumber is $\Delta K_L \simeq K_L' \Delta I$. But, $\Delta I \propto \delta h$, so scaling arguments tell us that the width of the resonance in wavenumber is

$$\Delta K_L = \left(|K_L' \delta h| \right)^{1/2}. \tag{5.74}$$

If the resonances are isolated, then there are local regions of chaos (positive Lyapunov exponent) mixed with regions of stability (zero Lyapunov exponent and the separation of nearby trajectories is linear with range). As ϵ increases, the fundamental and harmonic resonances grow and overlap, leading to widespread chaos. But the ocean is a power law medium with many forcing wavenumbers, and thus resonances are likely quite numerous and widespread. In this case of broadband forcing the applicability of the KAM theory has been called into question; that is, the assumption that the ocean acoustic phase space can be seen to be partitioned into nonintersecting regular and chaotic regions. By defining an extended phase space, Brown (1998) has shown that the KAM theorem does indeed apply to systems with a finite number N of periodic forcing wavenumbers.

A useful toy ocean acoustic model with a periodic perturbation demonstrates the nature of this mixed phase space, in which there are stable islands and chaotic seas (Brown et al., 1991). Here consider ray propagation in a background bi-linear sound-speed profile, that is, a linear profile above and below the axis but not necessarily with the same slope. Perturbations are introduced such that the slope of the upper ocean profile oscillates periodically in range, with period λ_f. Iterations of this map separated in range by one ray cycle for axial ray angle, θ, and range, r, obey the equations

$$\phi_{n+1} = \phi_n + \epsilon[\sin\rho_n + \sin(\rho_n + \phi_n + \epsilon\sin\rho_n)], \qquad (5.75)$$

$$\rho_{n+1} = \rho_n + \phi_n + \epsilon\sin\rho_n + \gamma\phi_{n+1}, \qquad (5.76)$$

where ϵ is a dimensionless perturbation strength parameter, γ is the ratio of the average upper ocean sound-speed gradient g to the fixed lower ocean gradient, and $\phi_n = (4\pi/g\lambda_f)\theta_n$, and $\rho_n = (2\pi/\lambda_f)r_n$. This mapping is area-preserving, that is,

$$\partial(\phi_{n+1}, \rho_{n+1})/\partial(\phi_n, \rho_n) = 1, \qquad (5.77)$$

which is the discrete version of Liouville's theorem. The phase space here is essentially axial angle and range, as opposed to depth and vertical ray slowness. Figure 5.7 (left panel) shows the phase space stroboscopically (i.e., range modulo of the forcing wavelength), and stable islands as well as chaotic seas are observed[2] Because of the stroboscopic nature of the display, stable ray trajectories trace out closed loops, whereas chaotic seas are a blur of points. For this intermediate value of ϵ, the phase space is perhaps equally partitioned between stable islands and the chaotic seas. For lesser values of ϵ, the islands will dominate the phase space, and small chaotic regions are the result of isolated and narrow resonances. As ϵ increases to large values, the number and width of the resonances increase, leading to a complex mixture of chaotic and stable regions. Also shown in Figure 5.7 is a small section of an eigenray display (right panel), showing initial angle versus range after 250 iterations of the map. The initial conditions for both panels fall on the $r = \lambda_f/2$ horizontal line. The islands that intersect the $\mathrm{mod}(r, \lambda_f) = 1/2$ are

---

[2] The display in Figure 5.7 is not a Poincaré section. A Poincaré section is a two-dimensional projection $(p_z, z, r \bmod \lambda_f)$ of the three-dimensional phase space (p_z, z, r).

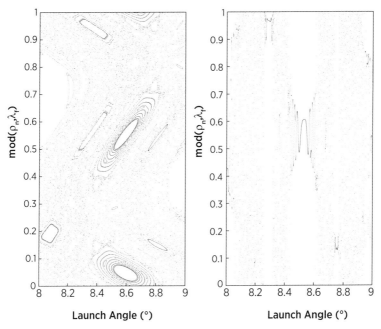

Figure 5.7. Numerical evaluation of the mapping given by Eqs. 5.75 and 5.76 with parameters $g = 1/(30 \text{ km})$, $\lambda_f = 10 \text{ km}$, $\gamma = 4$, and $\epsilon = 0.15$. In both cases, the initial condition corresponds to an axial point source with $r_0 = \lambda_f/2$ and initial angles between $8°$ and $9°$. The left-hand panel shows all points for 500 iterations of the map over 100 launch angles, while the right-hand panel shows the final state of 10,000 rays after 250 iterations of the map.

apparent in both displays, and thus there is a means for identifying "island-like" features in the eigenray plot even if there are no periodic perturbations which of course is the real case in the ocean. This will be important in Section 5.7.7 where ray simulations through random realizations of internal waves are compared to observations.

5.6.3 Levels of Randomness: Ray Statistics

The difficult issue of ray statistics over an ensemble of ocean internal-wave states is now addressed. An important consideration for understanding dynamical systems and their statistical properties is understanding the nature of how the individual trajectories move about in phase space. One aspect of this situation for ray propagation through ocean internal waves has already been seen; that is, the trajectories scatter in such a way that they move primarily along the unperturbed wave front rather than across it. This is a consequence of small-angle forward scattering. Here some of the aspects of a mixed phase space that can strongly affect ray trajectory migration through the phase space are addressed. As one might imagine, a mixed phase space with its combination of regular and chaotic trajectories can have quite odd diffusion properties that do not obey the commonly held rules of Brownian motion or random walks (Brown, 1998; Brown et al., 2003).

The first major constraint is Liouville's theorem, which says that the flow in phase space is incompressible. This means that phase space volume of a group of ray trajectories remains constant as the rays propagate out in range; the volume can be stretched and rotated but not altered. The other key issue here is the inter-mittency of trajectories moving in and out of stable/unstable states. Embedded in the phase space are features termed *Cantori* (Tabor, 1989; Brown et al., 2003) that can inhibit the migration of rays near the boundaries of chaotic seas and islands. Furthermore, island structures by their nature are stable regions, and thus rays that wander into the neighborhood of an island may stay there for quite some time. This is the notion of island boundary *stickiness* (Tabor, 1989; Brown et al., 2003).

Thus a key consideration for determining the statistics of rays is the degree to which there is a mixed phase space with island structures. If there are no islands, then the system may be what is termed an *Anosov system* that is globally unstable, and all trajectories have a positive Lyapunov exponent. Other weaker forms of randomness that are also behaviors of the Anosov system are given the terms: (1) *Ergotic systems*; (2) *Mixed systems*; and (3) *K-systems* (Tabor, 1989). Strong evidence suggests that ray propagation though ocean internal waves is completely chaotic (Wolfson and Tomsovic, 2001; Beron-Vera et al., 2003; Morozov and Colosi, 2004; Beron-Vera and Brown, 2009), which means the phase space is ergotic and strongly mixing. As a result, the canonical variables momentum p_z, and depth z are found to be Gaussian random variables whose standard deviations increase in range diffusively (i.e., as square root of the range). Another consequence of ocean rays being an Anosov system is that all the variational quantities are log-normal variables, meaning that the ray amplitudes are log-normal and the finite-range Lyapunov exponent is a Gaussian random variable (Wolfson and Tomsovic, 2001; Beron-Vera et al., 2003). These results on the distribution functions for the ray quantities come mainly from numerical simulations and theory based on single-scale random media without a waveguide. If the ocean ray system is indeed Anosov, then anomalous diffusion phenomena are not expected (i.e., either super- or sub-diffusion) although the log-normal amplitude distributions may bear some relation to Levy flights (Mandelbrot, 1982).

The influence of these ray statistics on acoustic field travel times and intensity will be discussed further in subsequent sections.

5.7 The α Parameter

Next the important subject of how the background sound-speed profile affects the development of ray chaos and various statistics of the acoustic field is addressed.

The dimensionless parameter,

$$\alpha(I) = \frac{I}{K_L} \frac{dK_L}{dI} = 2\pi \frac{I}{R_L^2} \frac{dR_L}{dH}, \qquad (5.78)$$

provides an important metric that modulates ray fluctuations. Physically, α is the strength of ray loop dispersion as a function of ray action, and the importance of this loop dispersion in estimating the width of resonances has been previously noted.[3] This topic has been extensively developed over the last decade and a half in a series of articles (Beron-Vera and Brown, 2003, 2004; Brown et al., 2003, 2005; Beron-Vera and Brown, 2009) that will be only briefly summarized here. Of key interest here are the effects of α on ray stability (i.e., Lyapunov exponents) and acoustic travel-time statistics. Before going into those topics it is useful to revisit the action-angle representation to better understand the meaning and origin of $\alpha(I)$.

5.7.1 Action Angle Variables Revisited

Action-angle variables from classical mechanics provide a powerful way of looking at Hamiltonian systems (Landau and Lifshitz, 1976; Goldstein, 1980). As was seen in Chapter 2, for $c = c(z)$ the Hamiltonian is a conserved quantity, and there is regular, nonchaotic motion. Systems of this type can be transformed via a canonical transformation such that the Hamiltonian is only a function of one of the conjugate variables, that is,

$$H(p_z, z) \rightarrow H(I, \vartheta) = H(I), \qquad (5.79)$$

where (I, ϑ) are the action-angle variables. The coordinate ϑ that is not in the Hamiltonian is often called a *cyclic coordinate*. Coordinates of this type are associated with conservation laws (e.g., a first integral). Therefore, in this form the conserved quantities of the problem are built in, and so when aspects that break the conservation law are examined (e.g., internal-wave fluctuations), the dynamics become more transparent. The action variable can be written as a function of H and is given by

$$I = \frac{1}{2\pi} \oint p_z \, dz = \frac{2}{2\pi} \int_{z^-}^{z^+} dz \, (c^{-2} - H^2)^{1/2}, \qquad (5.80)$$

where the line integral is understood to be over a closed loop in phase space (p_z, z), and z^+, z^- are the ray upper and lower turning depths. When internal-wave sound-speed structure is introduced, the rays are no longer periodic, and a closed loop cannot be defined to compute the ray action. Equation 5.80 tells us that

---

[3] It should be noted that it has been shown (Brown et al., 2005) that α is closely related to the so-called "waveguide invariant," β (Brekhovskikh and Lysanov, 1991) which is used to describe many aspects of unperturbed ocean acoustic wave fields. In the high-frequency limit $\beta = \alpha$.

rays with small/large grazing angles have small/large values of I. The generating function for this canonical transformation is

$$G(z,I) = \pi I \pm \int_{z^-}^{z} dz' \, p_z(H(I),z'),\tag{5.81}$$

where the $\pm$ sign applies to $\pm p_z$, and

$$p_z = \frac{\partial G}{\partial z}, \text{ and } \vartheta = \frac{\partial G}{\partial I}.\tag{5.82}$$

The Hamiltonian $H(I)$ can be obtained by inverting Eq. 5.80.

Before going into this subject which might seem a bit abstract, it is worth getting a physical picture of action-angle variables by looking at the well-known quantities of ray loop distance R_L and loop time T_L, given by

$$R_L = \int_\Gamma dx = 2 \int_{z^-}^{z^+} \frac{dz}{\tan\theta}, \text{ and } T_L = \int_\Gamma \frac{ds}{c} = 2 \int_{z^-}^{z^+} \frac{dz}{c\sin\theta}.\tag{5.83}$$

For $c = c(z)$ a conserved quantity is the Hamiltonian, and for the Helmholtz equation the hamiltonian is $H = -(c^{-2} - p_z^2)^{1/2} = -\cos\theta/c$. One can therefore write the tangent and sin functions in Eqs. 5.83 in terms of H and get

$$R_L = -2H \int_{z^-}^{z^+} \frac{dz}{(c^{-2} - H^2)^{1/2}}, \text{ and } T_L = 2 \int_{z^-}^{z^+} \frac{dz}{c^2 \, (c^{-2} - H^2)^{1/2}}.\tag{5.84}$$

Taking the derivative of the action (Eq. 5.80) with respect to H, the result is

$$\frac{\partial I}{\partial H} = \frac{2}{2\pi} \int_{z^-}^{z^+} dz \, \frac{\partial}{\partial H}(c^{-2} - H^2)^{1/2} = \frac{R_L(H)}{2\pi} \equiv K_L^{-1}.\tag{5.85}$$

Here the variation of the boundary terms does not contribute because the integrand vanishes there. Next the factor of $1/c^2$ in the ray loop time equation can be written as $1/c^2 - H^2 + H^2$ so that the loop time can be expressed as

$$T_L = 2\pi I - H R_L.\tag{5.86}$$

From this equation and Eq. 5.85 the useful result is

$$\frac{dT_L}{dH} = -H\frac{dR_L}{dH}, \text{ or } H = -\frac{dT_L}{dR_L}.\tag{5.87}$$

Thus the action variable is closely related to the ray loop distance and the loop time. These results correspond directly to the approach presented by Munk et al. (1995).

Now Hamilton's equations for the case $c = c(z)$ can be examined. Here it is found that,

$$\frac{d\vartheta}{dx} = \frac{\partial H}{\partial I} = K_L, \tag{5.88}$$

$$\frac{dI}{dx} = -\frac{\partial H}{\partial \vartheta} = 0, \tag{5.89}$$

$$\frac{dT}{dx} = IK_L - H + \frac{d}{dx}(G - I\vartheta), \tag{5.90}$$

where the last term in the travel time equation is an end-point term that is generally small, except at the shortest ranges. In the literature it has generally been neglected, and this will be done here as well. The solutions of these equations are the simple relations:

$$\vartheta(x) = \vartheta_0 + K_L(I_0)x, \tag{5.91}$$

$$I = I_0, \tag{5.92}$$

$$T(x) = (IK_L - H)x. \tag{5.93}$$

Thus, while the action variable I is closely related to the ray loop distance and time, the angle variable is essentially a phase that increases by 2π for every cycle of the ray.

5.7.2 Ray Stability

With some familiarity with action-angle variables, an analysis of ray stability in that representation can be done by looking at the case of a *nearly* integrable system. For this analysis the Hamiltonian has the form (Brown et al., 2003; Beron-Vera and Brown, 2004)

$$H(I, \theta, x) = H(I) + \delta h(I, \vartheta, x), \tag{5.94}$$

where δh is a small perturbation to the integrable Hamiltonian $H(I)$. With this perturbation, Hamilton's equations have the form

$$\frac{d\vartheta}{dx} = K_L(I) + \frac{\partial \delta h}{\partial I}, \tag{5.95}$$

$$\frac{dI}{dx} = -\frac{\partial \delta h}{\partial \vartheta}, \tag{5.96}$$

$$\frac{dT}{dx} = IK_L - H - \delta h. \tag{5.97}$$

With the added term δh the Hamiltonian is no longer a constant of the motion, and thus the solutions may be chaotic, that is, terms in the stability equations (Eq. 2.61) may grow exponentially yielding significant values of the Lyapunov exponent (see Chapter 2). To examine this chaotic behavior in more detail, the stability equations for the extended phase space of (I, ϑ, T) are required. The result given by Beron-Vera and Brown (2004; see also Chapter 2) is

$$\frac{d}{dx}\begin{pmatrix} \delta I \\ \delta\vartheta \\ \delta T \end{pmatrix} = \begin{pmatrix} 0 & 0 & 0 \\ K_L' & 0 & 0 \\ IK_L' & 0 & 0 \end{pmatrix}\begin{pmatrix} \delta I \\ \delta\vartheta \\ \delta T \end{pmatrix} + \begin{pmatrix} -\dfrac{\partial^2\delta h}{\partial\vartheta\partial I} & -\dfrac{\partial^2\delta h}{\partial\vartheta^2} & 0 \\[2mm] \dfrac{\partial^2\delta h}{\partial I^2} & \dfrac{\partial^2\delta h}{\partial\vartheta\partial I} & 0 \\[2mm] I\dfrac{\partial^2\delta h}{\partial I^2} & I\frac{\partial^2\delta h}{\partial\vartheta\partial I} - \frac{\partial\delta h}{\partial\vartheta} & 0 \end{pmatrix}\begin{pmatrix} \delta I \\ \delta\vartheta \\ \delta T \end{pmatrix}, \quad (5.98)$$

where δI, $\delta\vartheta$, and δT are shorthand for variations with respect to the initial action, and $K_L' = dK_L/dI$. Since the second term in Eq. 5.98 is proportional to the perturbing Hamiltonian, it will be significantly smaller than the first term. Thus one would expect that the dominant growth of the perturbations to the extended phase space $(\delta I, \delta\vartheta, \delta T)$ will be driven by the magnitude of K_L' (and therefore α), which is only a function of the background sound-speed profile $c(z)$. Of course nonzero δh is required to get the instability going, but it is the magnitude of the K_L' term that allows the instability to grow more quickly or more slowly. This effect has been demonstrated in numerous numerical simulation scenarios (Beron-Vera and Brown, 2004, 2009). Physically, large α corresponds to a rapid change in ray loop length as a function of ray grazing angle. This loop dispersion manifests itself as a shearing of the ray Lagrangian manifold (see Figure 2.7), and thus regions of large α are regions of large Lagrangian manifold shear. The analogy to instability in sheared fluid flows is apparent (Beron-Vera and Brown, 2004).

An insightful example of the relationship between α, and the finite-range Lyopunov exponent, ν, is shown in Figure 5.8 (Beron-Vera and Brown, 2003). The calculations for the C89 and S89 sound-speed profiles (derived from the SLICE89 experiment) show a relatively small variation of ν and α across the initial angles. For the canonical profile (C89), there is a clear correspondence between ν and α with larger launch angles showing larger values. The calculations for the C18 and cosh profiles dramatically show the effects of α. The C18 profile is intended to model the Atlantic Sargasso Sea, where an 18°C water mass in the main thermocline creates a depth region of strong sound-speed gradient and therefore large α values for rays that turn in that region. The cosh profile is interesting because it has zero loop dispersion, that is, $\alpha = 0$ for all rays. The simulations show large values of ν for rays that turn in the 18°C water, and for the cosh profile essentially no exponential growth is observed.

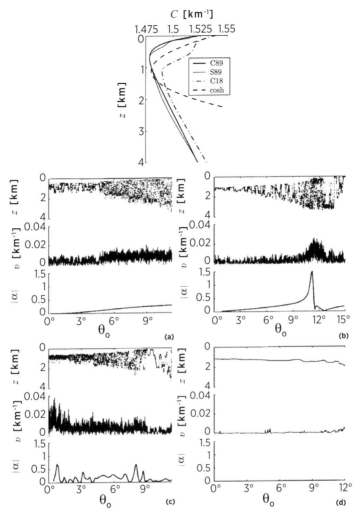

Figure 5.8. Top panel: Four background sound-speed profiles used for ray simulations through random realizations of internal-wave-induced sound-speed perturbations. Bottom four panels each display ray final depth (eigenray plot), finite-range range Lyapunov exponent, ν, and α as a function of initial ray launch angle, for the four background sound-speed profiles. The panels correspond to (a) C89, (b) C18, (c) S89, and (d) cosh waveguides. The range is 1000 km.
Source: Beron-Vera and Brown (2003).

5.7.3 Travel-Time Stability

From the variational equations (Eq. 5.98) in action-angle coordinates the growth of travel time variations, δT, of nearby rays is controlled by K_L'. Similarly, for the unperturbed problem the travel time equation (Eq. 5.93) gives us

$$\frac{dT}{dI} = IK_L'r,\tag{5.99}$$

and thus K'_L is critical even to the range-independent problem. In this discussion of travel-time stability, it is important to distinguish between two cases. In one case the travel time variability in which the launch angle of the ray is fixed is examined, and in the other case the eigenray constraint is imposed (Beron-Vera and Brown, 2004). These two cases are referred to as unconstrained and constrained travel-time fluctuations. These cases have, in fact, been examined before when ray spreading along and across the wave front was computed. The former case is unconstrained, and the latter case is constrained. Loosely, one could write the rms-unconstrained and -constrained travel-time variations as $\langle Z^2 \rangle^{1/2}/c_0$ and $\langle X^2 \rangle^{1/2}/c_0$, respectively. While in Section 5.5 it was shown that the distribution of these rays could be predicted given parameters of the internal-wave field and the background environment, in this section, however the concern is with the relation to α. Figure 5.9 shows Monte Carlo simulations of unconstrained travel-time fluctuations as a function of ray launch angle for two background sound-speed profiles. The correlation with α is clear. Similar correlations to constrained travel times have been shown (Beron-Vera and Brown, 2004).

However, the theoretical treatment of ray scattering along and across the wave front presented in Section 5.5 also applies to the case of a homogeneous background sound-speed (i.e., $c(z) = c_0$) for which $\alpha = 0$. With respect to travel-time fluctuations, there is clearly more going on than simply the relation to α.

5.8 Travel-Time Statistics: Random Walk Model

The theory that has been developed so far can be used to formulate useful statistics for ray travel times in the two important cases of unconstrained and constrained travel times, that is, for a fixed source location and range either the launch angle or the receiver depth can be fixed. In the linear approximation (i.e., integrating over an unperturbed ray), it was seen from Section 5.5 that the variance of rays along the front (unconstrained) scaled as the cube of the range, while variance across the front (constrained) only scaled linearly. The following model allows us to examine nonlinear effects in that the ray path deviates. To be consistent with the literature, travel-time statistics are considered rather than deviations along and across the front.

As shown in Chapters 4 and 5, internal-wave scattering occurs primarily at the upper turning points. Since turning points are separated horizontally by more than an internal-wave correlation length, a useful model is a random walk where the ray trajectory rapidly changes from one value of the Hamiltonian to another (see, for example, Beron-Vera et al., 2003, figure 12). For each turning point (or scattering event), the ray loop distance $R_L(H)$ and time $T_L(H)$ are expanded to second order

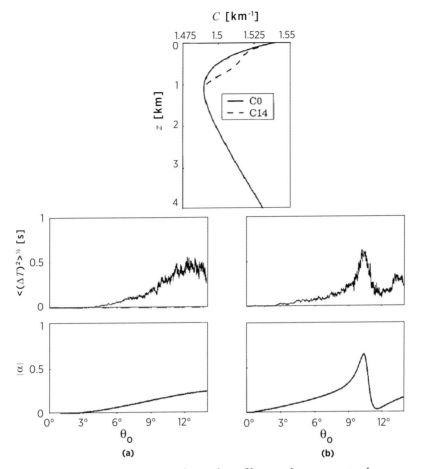

Figure 5.9. Top panel: Two sound-speed profiles used to compute the unconstrained travel-time fluctuation statistics shown in the four bottom panels. The four bottom panels show rms unconstrained travel time spreads and α for the two profiles $C0$ (a) and $C14$ (b) as a function of initial angle. The range is 2000 km. *Source*: Beron-Vera and Brown (2004).

around some reference value of the Hamiltonian H_0 to obtain

$$\delta R_L = 2\pi\left(\delta H \frac{\partial^2 I}{\partial H^2}\bigg|_{H_0} + \frac{\delta H^2}{2}\frac{\partial^3 I}{\partial H^3}\bigg|_{H_0}\right), \qquad (5.100)$$

$$\delta T_L = \frac{H_0}{2\pi}\delta R_L + \frac{K_L'}{K_L^3}\bigg|_{H_0}\frac{\delta H^2}{2}, \qquad (5.101)$$

where Eq. 5.86 has been used and the fact that $\partial^2 I/\partial H^2|_{H_0} = -K_L'/K_L^3$. Now the case for the total travel-time fluctuation occurring after propagation over N loops can be calculated. For a random walk the following definitions are useful:

$$\Delta H_j = \sum_{n=1}^{j} \delta H_n, \tag{5.102}$$

$$\langle \Delta H_j \rangle = \sum_{n=1}^{j} \langle \delta H_n \rangle = 0, \tag{5.103}$$

$$\langle \Delta H_j \Delta H_i \rangle = \sum_{n=1}^{j} \sum_{k=1}^{i} \langle \delta H_n \delta H_k \rangle = \langle \delta H^2 \rangle \min(j,i). \tag{5.104}$$

The first equation says that the perturbation to the Hamiltonian at step j is the sum over all previous steps with step perturbations δH_n (i.e., the random walk). The second equation says that the mean perturbation at step j is zero because the individual steps have a zero mean. Lastly, the third equation gives the correlation between the Hamiltonian perturbations at steps j and i. Because the individual steps are uncorrelated, only steps shared by both terms will contribute. The last equation also tells us that the variance $\langle \Delta H_j^2 \rangle = \langle \delta H^2 \rangle j$ increases linearly with range as is expected from a random walk. The travel-time fluctuation after taking N steps is thus

$$\delta\tau = \sum_{n=1}^{N} \left[\frac{H_0}{2\pi} \delta R_L(n) + \left. \frac{K_L'}{K_L^3} \right|_{H_0} \frac{\delta H_n^2}{2} \right]. \tag{5.105}$$

In order to compute the mean and variance of the fluctuations, the constrained and unconstrained cases must be considered separately.

5.8.1 *Eigenray Constrained Model*

For the case of an eigenray constraint, in the linear approximation, the travel time bias is zero, and the variance increases linearly with range. Here some corrections to these predictions can be made in the regime in which expansion about the unperturbed ray is no longer accurate. The eigenray constraint dictates that the ray must arrive at the final range such that $\sum_{n=1}^{N} \delta R_L(n) = 0$, so that the travel-time fluctuation is given only by the second term in Eq. 5.105. The eigenray constrained travel time bias is then given by

$$\langle \delta\tau \rangle_{eig} = N(N+1) \frac{\langle \delta H^2 \rangle}{4} \left. \frac{K_L'}{K_L^3} \right|_{H_0}, \tag{5.106}$$

where the summation rule $\sum_{n=1}^{N} n = N(N+1)/2$ has been used. The bias is a second-order effect, being proportional to $\langle \delta H^2 \rangle$. This is Fermat's principle. Secondly, the bias is proportional to K_L' (and therefore α), and increased loop dispersion will therefore lead to increased travel time bias. The sign of K'_L is also significant, leading to a fast or slow bias. Thirdly, the bias scales roughly as the square of the range (i.e., N^2). The variance of the travel time is similarly given by

$$\langle \delta \tau^2 \rangle_{eig} - \langle \delta \tau \rangle_{eig}^2 = \frac{\langle \delta H^2 \rangle^2}{4} \left(\frac{K'_L}{K_L^3} \Big|_{H_0} \right)^2 \left(\frac{N(N+1)(2N+1)}{3} \right), \qquad (5.107)$$

where the summation rule $\sum_{n=1}^{N} n^2 = N(N+1)(2N+1)/6$, and the rule of Gaussian statistics $\langle \delta H^4 \rangle = 3 \langle \delta H^2 \rangle^2$ have been used. The variance is seen to scale as the cube of the range, which can be contrasted with the linearized travel-time variance equation (Eq. 5.30) that only grows linearly with range. This cubic growth is shown in Figure 5.6 for the growth of $\langle X^2 \rangle$ when ray chaos effects become large at long range. The cubic growth is present in both the waveguide and the constant background sound-speed cases (where $K'_L = 0$). Clearly the random walk model does not apply in the latter case.

5.8.2 Unconstrained Model

To model the unconstrained travel times, the travel time perturbation (Eq. 5.105) is written using a shorthand notation akin to the original Taylor expansion, giving the result

$$\delta \tau \simeq \sum_{n=1}^{N} \left[T'(H_0) \, \delta H_n + T''(H_0) \frac{\delta H_n^2}{2} \right], \qquad (5.108)$$

where

$$T' = \frac{dT_L}{dH} \Big|_{H_0} = -H_0 \frac{K'_L}{K_L^3} \Big|_{H_0}, \qquad (5.109)$$

$$T'' = \frac{d^2 T_L}{dH^2} \Big|_{H_0} = \frac{K'_L}{K_L^3} \Big|_{H_0} - H_0 \left(\frac{K''_L}{K_L^3} \Big|_{H_0} - \frac{3(K'_L)^2}{K_L^4} \Big|_{H_0} \right). \qquad (5.110)$$

For the unconstrained bias case the first-order term does not contribute because the mean displacement is zero for a random walk. For the bias the result is

$$\langle \tau \rangle_{uncon} = \frac{N(N+1)}{4} T''(H_0) \, \langle \delta H^2 \rangle, \qquad (5.111)$$

which grows as the square of the range. The bias depends on the curvature of the travel-time dependence on H_0, not just the loop dependence as demonstrated in the eigenray constrained bias. The bias can be positive or negative depending on the travel time curvature. A similar calculation for the second moment gives

$$\langle \tau^2 \rangle_{uncon} \simeq \sum_{j=1}^{N} \sum_{n=1}^{N} (T')^2 \langle \delta H_j \delta H_n \rangle = \frac{N(N+1)(2N+1)}{6} \langle \delta H^2 \rangle (T')^2. \qquad (5.112)$$

The spread in travel time $\langle \tau^2 \rangle_{uncon} - \langle \tau \rangle^2_{uncon}$ is dominated by the second moment because $T_L(H)$ is not expected to be a rapidly varying function. The spread therefore is seen to grow as the cube of the range. This is in fact the same range scaling result obtained for the scattering of rays along the front. So unlike the constrained travel times that show a change in range scaling from linear to nonlinear regimes, there is no change in range scaling for the unconstrained travel times.

5.9 Ray Chaos in Observations

In this section ideas from ray chaos are utilized to interpret broadband ocean acoustic transmission observations made at propagation ranges of order 1000 km, that is, ranges where one expects ray chaos to be well developed (Simmen et al., 1997; Colosi, 2001; Beron-Vera et al., 2003). Most of these observations were carried out in conjunction with the Acoustic Thermometry of Ocean Climate (ATOC) program, which utilized broadband low-frequency sources with a center frequency of 75 Hz and navigated large vertical aperture receiving arrays. The large aperture arrays were quite useful for interpreting the observations in terms of rays. Analysis of chaotic ray effects for shorter range experiments would be a worthwhile undertaking, and it is hoped that this chapter may spur some work in that direction.

5.9.1 Nature of Chaotic Rays and Time Fronts

Figure 5.10 compares an observed time front from the 3250-km ATOC Acoustic Engineering Test (AET) and numerical ray simulations with and without internal-wave-induced sound-speed perturbations (Beron-Vera et al., 2003).[4] Perhaps the most noticeable feature in both the observations and the simulation is the partitioning of the arrival pattern in that the early arriving energy shows distinct resolved time fronts, and the late arriving energy appears to be a complex interference pattern of many overlapping arrivals. This behavior has been observed in many experiments dating back to 1989 (Colosi et al., 1994; Worcester et al., 1994, 1999, 2000; Van Uffelen et al., 2009). Some degree of overlap of arrivals in the late region of the pulse is expected because the time front branches get closer together, but the smearing of the arrivals is enhanced in both the observations and

---

[4] Ray simulations carried out with internal-wave-induced sound-speed perturbations are sensitive to small-scale cut-offs in the simulated internal-wave spectra. The simulations discussed here and in Beron-Vera et al. (2003) utilized a realistic cut-off in the horizontal wavenumber, but not so for the vertical mode number cut-off J. For computational reasons values of J between 30 and 50 were used, but a more realistic number is $J \simeq 400$–500. It is an open scientific question as to the extent that the higher modes are irrelevant due to finite wavelength effects. In any case, the ray simulations presented here should be considered to represent a lower bound on the magnitude of ray chaos.

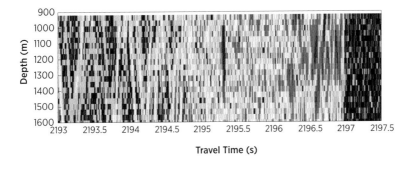

Travel Time (s)

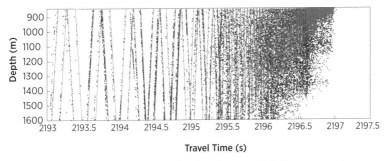

Travel Time (s)

Figure 5.10. Top panel: A measured pulse from a 3252-km transmission in the North Pacific, from the ATOC Acoustic Engineering Test. Bottom panel: Ray simulation calculations with (blue dots) and without (red lines) internal-wave-induced sound-speed perturbations.

the calculation with internal waves. This smearing is associated with constrained ray time spreads, which can grow rapidly once ray chaos is developed (i.e., rms growth as $R^{3/2}$). In addition, significant ray energy in the late region of the pulse is scattered into the shadow zone below the caustics of the unperturbed calculation; this vertical smearing is due to nonlinear effects on unconstrained ray fluctuations. Similar comparisons with observations have been carried out using parabolic equation simulations with random internal-wave structure (Colosi et al., 1994; Worcester et al., 1999). Qualitatively, the simulation with internal waves compares well with the observations, and the cause of the complexity of the ray calculation is ray chaos; that is to say all rays diverge exponentially from their neighbors (Figure 5.11).[5] In addition, the early arriving rays show slower exponential divergence from their neighbors (Beron-Vera et al., 2003), which helps explain the temporal stability of the early time front branches. Further, the large Lyaponov exponents for the late arrivals dictate large nonlinear time and depth spreading thus explaining the smearing of the time fronts and ensonification of the shadow zone. Observations by Dushaw et al. (1999) and later observations and

---

[5] The Lyaponov exponent is a stochastic quantity due to the stochasticity of ocean internal waves. The PDF of v has been shown to be nearly Gaussian (Beron-Vera et al., 2003) with some deviation on the tails, and importantly no values $v = 0$.

Ray Theory

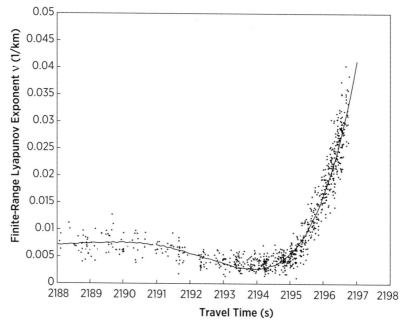

Figure 5.11. Finite-range Lyapunov exponents computed from the ray simulations shown in Figure 5.10. Early arrivals show slower exponential divergence of nearby rays, than later arrivals. The range is 3252 km.

numerical modeling by Van Uffelen et al. (2009) have also revealed internal-wave induced shadow zone depth extension for the early arrivals (see Figure 4.4 for a ray simulation example).

Next the simulation data for the early arriving time fronts are examined in more detail and the issue of the structure of the chaotic rays along the branches of the front is taken up. Figure 5.12 shows the results of a computation of more than 70,000 rays between the launch angles of 7.5° and 11° for the AET environment with internal waves and a much more sparse set of launch angles for the case without internal waves. The bottom panel shows the eigenray plot of ray depth at the receiver range versus ray launch angle. The curve from the calculation with internal waves is quite complicated, and even with 70,000 rays one does not get what should be a smooth unbroken curve. This complicated shape is due to strong folding and stretching of the time front, as is evidenced by the Lagrangian manifold shown in the middle panel. As was shown before, the folding and stretching of the time front occurs primarily along the front (top panel). This behavior that is due to ray chaos leads to an exponential increase in eigenrays and caustics (e.g., where $\partial z/\partial p_0 \propto \partial z/\partial \theta_0 = 0$). Consequently, the interference and phase variability caused by the proliferation of eigenrays and caustics has a tremendous impact on the nature of the acoustic field and its fluctuations. Lastly, there are interesting symmetric structures in the eigenray plot, and expanded views of diagrams of

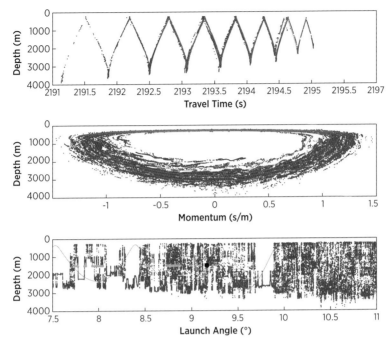

Figure 5.12. Three different ray simulation displays are presented; the time front (top), the Lagrangian manifold (middle), and the eigenray diagram (bottom). Blue dots/red curves are for calculations with/without internal waves. In all panels only results for ray launch angles between 7.5° and 11° are shown. In the eigenray panel for the simulation with internal waves, green dots display eigenray launch angles for time front ID 137 for a receiver at 1495-m depth (218 eigenrays were identified). The single black dot shown in the eigenray display is the launch angle for time front ID 137 in the absence of internal waves. The range is 3252 km.

this sort show smaller and smaller symmetric structures (Beron-Vera et al., 2003). These features are not associated with stable islands because there are no cases of $v = 0$, but they do seem to be associated with smaller values of v.

A critical consideration for acoustic variability is the nature of the eigenrays that make up the pressure field at any given point. For an eigenray plot, if one chooses a specific receiver depth and draws a horizontal line across the display at that depth, points of intersection with that line define eigenray launch angles. Because of the exponential growth of eigenrays and the fact that the Lyaponov exponents are order 1/100 km, several hundred eigenrays can be found computationally for each time front branch at 3250 km. However, because of the strong folding and stretching of the front, the stochastic eigenrays (microrays) do not remain close to the unperturbed ray. In Figure 5.12 eigenrays for time front ID +137 at a receiver depth of 1495 m are shown, and the launch angles for the microrays vary by an enormous amount. For this arrival more than 200 eigenrays were found that had travel times within tens of milliseconds of the unperturbed eigenray for ID

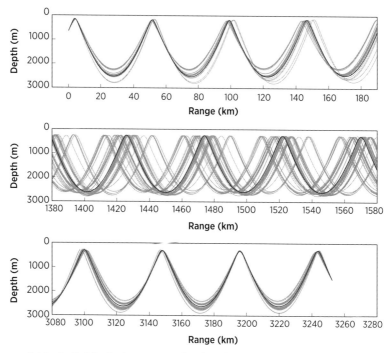

Figure 5.13. Individual eigenray paths for the 218 eigenrays used in the ray synthesis of time front ID 137 with receiver depth 1495 m (gray). The single eigenray path for the calculation without internal waves is shown in black. The three panels give ray paths views, near the source (top), near the receiver (bottom), and in the middle of the 3252-km propagation path (middle).

+137. This geometrical picture differs tremendously from the case considered in Chapter 4 in which the microrays were distributed along a parabolic surface close to the unperturbed ray. Indeed, microrays associated with a particular time front branch are often quite close in launch angle to microrays that make up another branch (Beron-Vera et al., 2003, figure 10). This property has been termed nonlocality, which is due to the fact that the folding and stretching of the time front can extend over multiple branches.

Although there is eigenray nonlocality in launch angle, one aspect of the perturbed and unperturbed rays stays the same, that is, the ray identifier. All microrays falling on a particular wave front branch have the same identifier (ID) as the unperturbed ray on that branch (Simmen et al., 1997; Beron-Vera et al., 2003; Brown et al., 2003). The nonlocality of the microrays and conservation of ID is nicely demonstrated using the microray paths shown in Figure 5.13. Here ray tube broadening is seen from a perspective that differs from the Fresnel zone ideas discussed in the introduction. Because of the eigenray constraint, the perturbed and unperturbed paths are relatively close to one another at the source and receiver, but they differ considerably at the middle range while maintaining the same ID. At the center of the propagation range the ray tube extends to a maximum of about 30 km horizontally but with small vertical separations, while at the source and receiver the horizontal separations are minimal but the vertical variations are of order 50 m. Plots similar to these have been produced by Simmen et al. (1997). Results showing a broadening of the ray tube have strong implications for acoustic fluctuations because the propagation regimes of partially and fully saturated statistics depend critically on the correlation or lack of correlation between microrays. Rays that wander away from one other by more than an internal-wave correlation length become uncorrelated. The broadening of the ray tube both horizontally and vertically also has potentially strong implications for propagation around obstacles such as seamounts, islands, and other geological features. Lastly, this effect has implications for acoustic remote sensing methodologies, such as ocean acoustic tomography, since the path taken by the sound can be altered by internal-wave scattering.

5.9.2 Acoustic Field Statistics

Analysis of long-range propagation data from the ATOC program has revealed four important results related to acoustic field statistics. For the early arriving portion of the time front, pulse time spreads are small (several ms), rms travel-time fluctuations are of order 10–20 ms, and intensity fluctuations obey a log-normal distribution consistent with propagation near the border of the unsaturated and partially saturated regimes (Colosi et al., 1999, 2001). For the late arriving portion of the time front, intensity obeys an exponential distribution consistent with fully saturated propagation (Colosi et al., 1999, 2001). Ray chaos ideas are consistent with all four of these observations, and to address these observations ray estimates of the pressure field are examined.

Figure 5.14 shows a ray synthesis (Eq. 2.65) of 218 microrays for time front ID 137 for a receiver depth of 1495 m drawn from the numerical simulations displayed in Figures 5.12 and 5.13. The ray synthesis requires a set of eigenrays ($\{p_0\}$) for which there are ray amplitudes, travel times, and Maslov indices, but because of the predominantly chaotic nature of the multipaths it is difficult to get a complete set ($\{p_0\}$). However, this limitation is not as severe as one might expect, because standard eigenray finding procedures (e.g., shooting) have no trouble finding the most energetic micro-multipaths. The missing eigenrays are usually of negligible amplitude. Indeed, Figure 5.14 shows that a synthesis using only the 20 most energetic eigenrays represents the total synthesis of 218 eigenrays quite well. More importantly the ray synthesis shows little pulse time spread, in spite of the fact that the microrays are spread themselves over about 20 ms. Although the result shown here is only one case, it is typical of other realizations of the internal-wave field, as well as other receiver depths and wave front IDs in the early arrival section.

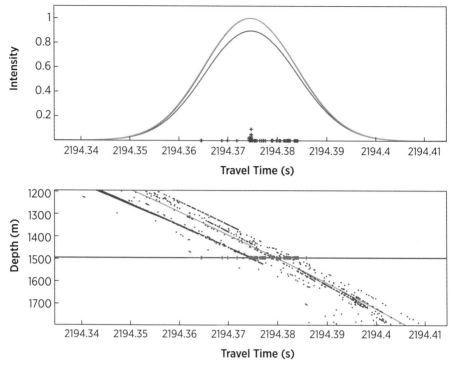

Figure 5.14. A microray synthesis for time front ID 137 and a receiver depth of 1495m. Top panel shows the replica waveform (green), the ray synthesis of all 218 eigenrays (blue), and a ray synthesis with 20 of the largest intensity eigenrays (red). The blue curve cannot be easily seen because it is completely obscured by the replica waveform (green). Pluses show individual eigenray intensity and travel time. Bottom panel displays a section of the time front near the receiver. Blue dots/green line show the rays with/without internal waves. A horizontal blue line marks the receiver depth, and eigenray travel times are marked along this line with red pluses.

These chaotic ray results are thus consistent with the observational results stated in the previous section, namely that (1) there is little pulse time spread; and (2) the microrays are spread over a small time span of order 10 ms rms, thus giving rise to travel-time fluctuations of the same order. There is, however, one unresolved issue with regard to the time spread that can be appreciated by comparing the observed mean pulses from the ATOC AET (Figure 5.15) with the one from the ray synthesis. The observed mean pulses do not show much broadening near the peak, but have a plateau on the tails that is not reproduced by the chaotic ray synthesis. The plateau is not due to ambient noise (Colosi et al., 2001).

The third observed feature of the early arrivals is a log-normal peak intensity PDF (Figure 5.16) (Colosi et al., 2001). As previously stated, work by Wolfson and Tomsovic (2001) and Beron-Vera et al. (2003) has shown that the microray amplitude PDF is log-normal for both the early and late arrivals. Based on the

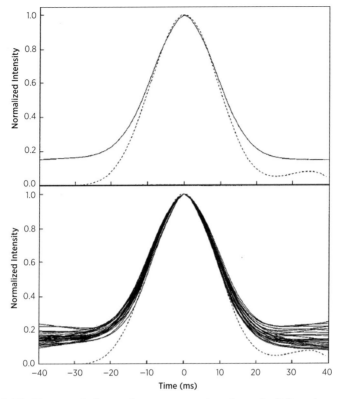

Figure 5.15. Top panel shows the average pulse shape (solid) and transmitted pulse (dash) from the 3250-km range ATOC Acoustical Engineering Test (AET). The bottom panel shows the individual mean pulse shapes for several different time fronts (solid). The pulse spreading is small and is of order several microseconds.
Source: Colosi et al. (1999).

AET numerical simulations discussed here, Figure 5.16 shows an example where the PDF of the variational quantity $|dz/dp_0|$ obeys a log-normal distribution and thus ray amplitudes also obey a log-normal.[6] These results also imply that the finite-range Lyoponov exponent is a Gaussian random variable in which the variance scales as one over range (see center panel of Figure 5.16). Because of the log-normal ray amplitude PDF, in the absence of significant microray travel-time variance, chaotic ray theory predicts a log-normal peak amplitude PDF consistent with the observations. But there are rms travel-time fluctuations of order 10 ms, corresponding to rms phase fluctuations (at 75 Hz) of roughly one cycle, and further there are variations in phase due to Maslov number variations. If the phases were known to be uniformly distributed between 0 and 2π, then the central limit theorem predicts that the intensity field from a large sum of waves

---

[6] All powers of a log-normally distributed variable also obey the log-normal distribution.

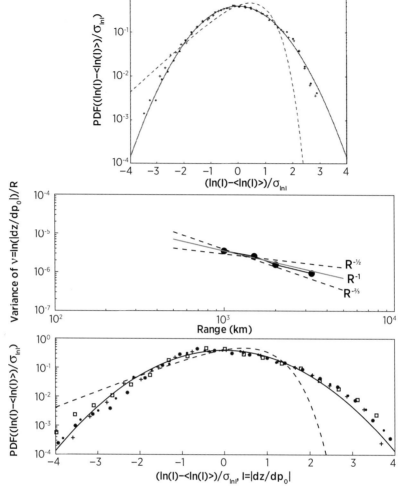

Figure 5.16. Top panel: Observed peak intensity PDFs from time fronts in the ATOC Acoustic Engineering Test. Pluses/circles show PDFs derived from samples with one peak per time front branch and from all peaks. Solid/dashed curves show normal/exponential distributions. Bottom panel: Simulated PDF of $|dz/dp_0|$ at ranges of 1000, 1500, 2000, and 3250 km. Also range scaling of the variance of the finite-range Lyoponov exponent, which is seen to scale as $1/R$. *Source*: Top panel from Colosi et al. (2001).

with log-normal amplitudes and random phases would approach an exponential distribution. Thus the challenge is to reconcile the observed log-normal PDF with the ray picture of an interference pattern of log-normally distributed waves with some degree of phase randomization. Using a simple random phasor sum model and incorporating phase variations consistent with the AET observations and ray simulations, Beron-Vera et al. (2003) show that a log-normal peak intensity PDF is recovered. For larger values of phase randomization (e.g., the rms phase increased by a factor of 2 or 3), the log-normal distribution is not seen. This suggests that the results from the ATOC AET are on the threshold of being in this log-normal regime and not being in the exponential regime that is associated with full saturation.

Lastly, there is the observation of full saturation in the late portion of the pulse (Colosi et al., 2001). This observable is easily reproduced by chaotic ray theory where there is an interference pattern of many overlapping rays with log-normal amplitudes and large phase variation. Here the central limit theorem acts quickly to drive the intensity PDF to the exponential form.

5.10 Wave Chaos

The previous discussion relating chaotic ray predictions to long-range acoustic propagation observations is compelling, but there is a keen, perhaps academic, interest in finding an acoustic observable that shows explosive or exponential growth with range. One such feature that has attracted attention is the behavior of acoustical beams (Tappert, 2003; Morozov and Colosi, 2004; Beron-Vera and Brown, 2009). These beams are an attempt to create a finite-frequency "ray" that then evolves chaotically with range. Figure 5.17 shows an example of a rapidly evolving acoustical beam, and the generation of microrays is quite apparent. Theoretical work by Tappert (2003) for the case of a single-scale random medium and no background sound channel demonstrated that there is competition between exponential increase in the complexity of phase space (i.e., ray chaos) and diffractive smoothing of phase space (i.e., finite wavelength effects). Ray chaos leads to an exponential increase in caustics, and caustics are regions of strong diffraction. Explosive or exponential expansion of a beam will then be seen at shorter ranges where phase space complexity is increasing exponentially, but there are not too many caustics. As the number of caustics proliferates at longer range, phase space will eventually be so full of caustics that diffractive spreading will dominate. Thus one arrives with the somewhat counterintuitive conclusion that some chaotic effects are in fact more apparent at short range, that is at a few e-folding distances of the ray instability. Results for an equivalent problem involving a 2-D electron gas have demonstrated the effect of exponentially expanding beams (Topinka and Westervelt, 2003).

So to test these interesting ideas of Tappert for the ocean acoustic problem, parabolic equation simulations have been carried out by Morozov and Colosi (2004) and Beron-Vera and Brown (2009) for frequencies of 125 and 250 Hz.[7] In

---

[7] Unfortunately, Tappert was not involved in the computational work associated with his theory, which was published posthumously in 2003 (Tappert, 2003).

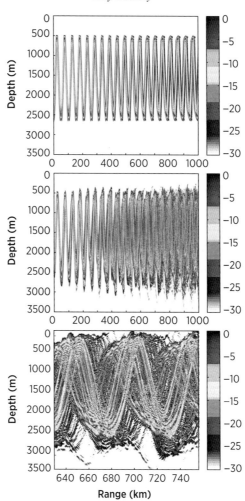

Figure 5.17. Parabolic equation simulations of 250-Hz acoustic beam propagation through a Munk canonical profile with and without internal-wave-induced sound-speed perturbations. Top/middle panels show the unperturbed/perturbed beams. Bottom panel shows an expanded view of the perturbed beam revealing extensive microray generation. The intensity scale is dB referenced to the maximum intensity.

both studies beam spread was shown to evolve diffusively with range (i.e., as the square root) rather than exponentially (Figure 5.18). The null result for exponential beam spread has been discussed extensively by Beron-Vera and Brown (2009). Two important factors need to be considered. First, beam spread is not a direct metric for phase space complexity, because at each range the phase space (p_z, z) is projected onto z to compute the spread. The process of projecting the phase space onto z may destroy the exponential growth, and therefore other phase space distributions like the Wigner or Husimi distributions (Virovlyansky and Zaslavsky,

2000) would be needed to see the exponential behavior. A second factor is the acoustic frequency of the simulations which may be too low, allowing too much diffraction. In order to see the exponential growth over a few e-folding distances higher frequencies may be needed to stave off diffractive smearing. It is important to point out that the reason behind Tappert's prediction of exponential beam growth appears to be essentially correct: it is the details of the ocean environment that seem to be limiting observation of such an effect.

Beam spread is not the only acoustic observable to demonstrate chaotic effects in finite-frequency wave fields. In this regard, Morozov and Colosi (2004) considered the Shannon entropy (Shannon, 1948) of vertical profiles of the acoustical beams. The Shannon entropy, $E(r)$, is an interesting observable because of the close connection between the instability of the ray equations (i.e., Lyapunov exponent) and the rate of KolmogorovSinai (KS) entropy. Recent results have shown that the rate of change of Shannon entropy, $E(t)$, for some idealized dynamical systems is closely related to the KS entropy over an intermediate regime of times (Latora and Baranger, 1999). Thus, the rate of Shannon entropy for the finite-frequency numerical simulations can be directly compared to ray simulation results of average finite-range Lyapunov exponent (a measure of KS entropy). Figure 5.18 shows that after an initial transient, the entropy curves for all the beams considered become linear giving an entropy rate between 0.11 and 0.06 bit/km.[8] Computation of Lyaponov exponents from the ray equations gives an average value of 0.02 bits/km, and this value is associated with the KS entropy rate, that is, $\langle \nu_L \rangle \simeq h_{KS}$. In Figure 5.18 the value of $2h_{KS} = 0.04$ bit/km is compared to the parabolic equation entropy rates, and the order of magnitude agreement suggests that the finite-frequency wave field complexity is indeed increasing exponentially. It should be noted that the factor of 2 added to the ray estimate is because the Shannon entropy of the complex acoustical field includes both phase and amplitude information, whereas the KS entropy rate only has information about the wave energy.

The topic of wave chaos in underwater acoustics is clearly in its infancy, and much work remains to be done.

5.11 Summary

Ray methods are seen to provide a powerful tool for understanding many aspects of the physics of sound propagation through the stochastic internal-wave field. Indeed our basic understanding of acoustic fluctuations draws heavily from the ray notion of microray interference and stability that can be understood in much more detail

---

[8] Conversion from bits/km to 1/km simply involves a change in the base of the logarithms. Thus an entropy rate of 0.02 bits/km would be $0.02(\log_2 e)^{-1}$.

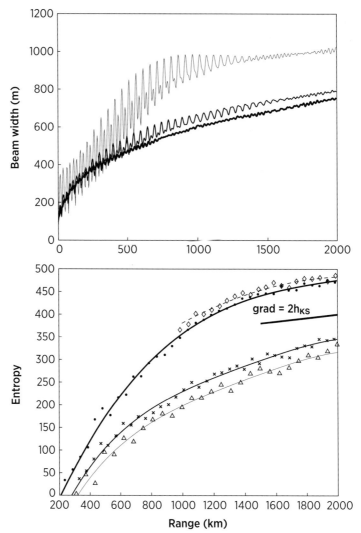

Figure 5.18. Top panel: RMS beam spread computed from parabolic equation simulations of 250-Hz acoustic beam propagation through a Munk canonical profile. Curves correspond to an axial beam (thick), a narrow-angle beam (medium), and a high-angle beam (thin). Bottom panel: Estimated entropy for the 250-Hz beams shown in the top panel, as well as entropy for the high-angle beam at 125 Hz (arbitrary vertical offset, and shown with a dashed curve and diamond symbols). The thick/circle, medium/cross, and light/triangle curves correspond to axial, medium angle, and high-angle beams. Fits in the linear region of the curves yield entropy rates of 0.11, 0.11, and 0.06 bit/km at 250 Hz and 0.06 bit/km at 125 Hz. Also shown in solid is twice the KS entropy, $2h_{KS} = 0.04$ bit/km (arbitrary vertical offset) computed from the ray equations for the axial ray.
Source: Morozov and Colosi (2004).

using the ideas of ray chaos and dynamical systems theory. This development has led to the important realization that the background sound-speed profile plays a critical role in determining acoustic sensitivity to internal waves, and that acoustic scattering in the ocean occurs primarily along the wave front rather than across it.

Furthermore, because acoustic phase is sensitive to the large-scale part of the internal-wave spectrum, ray theory is particularly well suited for the prediction of acoustic observables related to travel time. In subsequent chapters it will be revealed that the basic results of ray theory for phase statistics have a large regime of applicability, most notably for observables such as the rms phase fluctuation and acoustical coherences. However, on the matter of acoustic intensity it is seen that internal-wave ray scattering leads to an exponential proliferation of caustics in the acoustic wave field. Thus there are critical diffractive corrections that need to be made due to the small-scale part of the internal-wave spectrum. These corrections will be discussed at length in the following chapters.

Appendix A Calculation of $dz/d\theta$ in a Waveguide

The material presented here is closely related to the discussion of the Fresnel zone Chapter 2 (Section 2.5.1). Consider a ray in the ocean sound channel that undergoes a small angular deflection $\theta(x')$ at the range position x'. The deflected ray then travels to the final range R, where it has a depth deviation from the unperturbed ray given by $z(R, x')$. The depth deviation of the deflected ray relative to the unperturbed ray is given in terms of the ray tube function ξ, whose evolution along a ray path is given by the ray tube:

$$\frac{\partial^2 \xi}{\partial x^2} + \frac{1}{c_0}\frac{\partial^2 \bar{c}}{\partial z^2}\xi = 0. \tag{A.1}$$

Since Eq. A.1 is second order, two ray tube functions are examined, ξ_1 and ξ_2, whose boundary conditions are

$$\xi_1(0) = 0, \quad \xi_1(R) = 1, \quad \xi_2(0) = 1, \quad \text{and} \quad \xi_2(R) = 0. \tag{A.2}$$

The objective is to construct a ray tube function that is a linear combination:

$$\xi = c_1\xi_1 + c_2\xi_2, \tag{A.3}$$

such that $\xi(R) = z(R, x')$ and $\xi(x') = 0$. This means that at the final range R the depth separation is $z(R, x')$, and at the scattering point x' there is no depth separation. A solution fitting these conditions is

$$\xi(x, x') = z(R, x')\frac{\xi_1(x)\xi_2(x') - \xi_1(x')\xi_2(x)}{\xi_1(R)\xi_2(x')}, \qquad x \geq x', \tag{A.4}$$

$$\xi(x, x') = 0, \qquad\qquad\qquad\qquad\qquad\qquad x < x'. \tag{A.5}$$

The scattering angle $\theta(x')$ is then given by

$$\theta(x') = \frac{\partial \xi}{\partial x}(x', x'),$$

$$= z(R, x') \frac{\xi_1'(x')\xi_2(x') - \xi_1(x')\xi_2'(x')}{\xi_1(R)\xi_2(x')}, \tag{A.6}$$

where $\xi_1' = \partial \xi_1 / \partial x$, etc. Thus the rate of change of final ray depth with respect to scattering angle is

$$\frac{dz}{d\theta}(x') = \frac{\xi_1(R)\xi_2(x')}{\xi_1'(x')\xi_2(x') - \xi_1(x')\xi_2'(x')}. \tag{A.7}$$

The denominator here is the Wronskian and is independent of x', so it is useful to evaluate it at the final range R where ξ_2 is equal to zero. The result is

$$\frac{dz}{d\theta}(x') = -\frac{\xi_2(x')}{\xi_2'(R)}. \tag{A.8}$$

In the case in which there is a constant background sound speed, it is easily found that $\xi_2 = 1 - x/R$, and $\xi_2'(R) = -1/R$ with the result $dz/d\theta = R(1 - x/R)$ as given in Chapter 4 (Section 4.1.3). Returning to the waveguide calculation, a useful approximation to $\xi_2'(R)$ that more clearly reveals the range scaling is

$$\frac{dz}{d\theta}(x') \simeq -\frac{R}{4} \xi_2(x'). \tag{A.9}$$

6

Weak Fluctuation Theory

6.1 Introduction

Geometrical acoustics provides a useful intuitive view of acoustic propagation through ocean internal waves, but there are necessary diffractive corrections that must be made due to the small-scale part of the internal-wave spectrum. These diffractive corrections are primarily associated with amplitude fluctuations. In this chapter the focus is on the effects of diffraction in the regime in which acoustic fluctuations are small, that is, the unsaturated regime (Chapter 4). Here the physical picture to keep in mind is propagation with no microrays, and the acoustic fluctuations are caused by the unperturbed ray and its ray tube being modulated by the internal-wave variations.

Several closely related perturbative methods have been applied to this situation, including the Born approximation (Born and Wolf, 1999), the method of smooth perturbations (Tatarskii, 1971; Chernov, 1975), the super eikonal method (Munk and Zachariasen, 1976), and the Rytov method (Rytov, 1937). Much of this work originated in the atmospheric optics context, in which line of sight propagation through homogeneous, isotropic turbulence was considered.

The seminal contributions in ocean acoustics were provided by Munk and Zachariasen (1976) and Flatté et al. (1979), in which the effects of waveguide propagation and internal-wave inhomogeneity and anisotropy were integrated into theoretical expressions for several observables. In this book the Born approximation is utilized because it provides the simplest conceptual picture, while maintaining a reasonable level of accuracy. For some time the wave propagation in random media community believed that the Rytov approximation, which gives the same equations as the Born approximation, apply under a set of less restrictive conditions. It is now known that this optimistic perspective is unfounded (Clifford, 1978) and that the results of both Born and Rytov are expected to apply for the variance of log-amplitude less than roughly 0.3. The physics behind this issue is that the Rytov method fails to treat consistently multiple scattering, while the Born approximation assumes single scattering. It has been discovered experimentally, however, that in both atmospheric and ocean cases the Born results for phase statistics appear to have broad application in both weak and strong fluctuations where the phase variance is significantly larger than 1 (Clifford, 1978; Flatté, 1983, and Section 6.4). The more restrictive applicability of the Born results for log-amplitude makes the theory relevant to short-range, low-frequency transmissions or situations in which the internal-wave field is weak, such as the Arctic.

6.2 Spectra of Phase and Log-Amplitude: No Waveguide

For simplicity and to gain some insight into the formalism, the case of a constant background sound speed is treated first, in which the ray paths are straight lines. This material may at first glance appear irrelevant to the real ocean situation in which a defined waveguide is present, but this is not the case. The straight ray results of this section will be applied locally along the curved ray path to give useful results for the ocean. A good detailed treatment of this material within the atmospheric optics context is given by Clifford (1978) and Ishimaru (1978).

The observables of interest here are the variances of phase and log-amplitude, and their cross correlation. From the variances, the frequency and wavenumber spectra of phase and log-amplitude naturally follow. These observables give important first-order information on the magnitude of fluctuations and their space–time scales.

6.2.1 Incident Plane Wave

In the first-order Born approximation, introduced in Chapter 2, the pressure field at frequency ω is written as

$$p(\mathbf{r}) = p_0(\mathbf{r})\Big[1 + \varphi_1(\mathbf{r})\Big],$$

$$\varphi_1(\mathbf{r}) = -\int_V d^3\mathbf{r}' G(\mathbf{r}, \mathbf{r}')\Big[2q_0^2\,\mu(\mathbf{r}')\,\frac{p_0(\mathbf{r}')}{p_0(\mathbf{r})}\Big], \qquad (6.1)$$

where p_0 is the unperturbed pressure, and the volume integral gives the relative pressure perturbation, $p_1(\mathbf{r})$, to first order in μ. The real and imaginary parts of φ_1 are understood to be the fluctuation in log-amplitude and phase, respectively (Eq. 2.99). For a constant background sound-speed environment, the Green's function is

$$G(\mathbf{r}, \mathbf{r}') = \frac{1}{4\pi}\frac{e^{iq_0|\mathbf{r}-\mathbf{r}'|}}{|\mathbf{r}-\mathbf{r}'|} \simeq \frac{1}{4\pi}\frac{e^{iq_0(x-x')[1+[(y-y')^2+(z-z')^2]/2(x-x')^2]}}{|x-x'|}, \qquad (6.2)$$

where the last step is termed the Fresnel approximation. The Fresnel approximation follows from the small-angle forward scattering assumption and the condition that the wave is observed at long range, with propagation primarily in the x-direction. Assuming an incident plane wave in the x-direction, $p_0 = e^{iq_0x}$, and an observation plane at $\mathbf{r} = (R, y, z)$, after some manipulation the expression for the relative pressure fluctuation is

$$\varphi_1(\mathbf{r}) = -q_0 \int d^3\mathbf{r}' \, \exp\left(i\pi\frac{(y-y')^2 + (z-z')^2}{\lambda(R-x')}\right) \frac{\mu(\mathbf{r}')}{\lambda(R-x')}. \qquad (6.3)$$

The terms of the form $R_f^2(x) = \lambda(R-x)$ are the plane wave Fresnel zone. Because of the forward scattering approximation, the volume integral in Eq. 6.3 is confined in the x-direction between 0 and the final range R, while the other directions extend to infinite. The Fresnel approximation has transformed the integral equation over the Green's function (Eq. 2.95) to what is called the Fresnel diffraction formula (Eq. 6.3). In the diffraction formula, the exponential term gives the location of interference fringes produced by the interaction of the incident plane wave and the scattered spherical wave originating from the location $\mathbf{r}'$. The physical picture is that the observed phase and amplitude fluctuations are produced by the interference of these two waves.

As shown in Chapter 2, the real and imaginary parts of Eq. 6.3 provide information about the statistics of log-amplitude (χ) and phase (ϕ) (Eq. 2.99). In particular, χ and ϕ are linearly related to the fractional sound-speed fluctuation μ. Therefore, the mean values are $\langle\phi\rangle = \langle\chi\rangle = 0$, because $\langle\mu\rangle = 0$. In the weak fluctuation limit, the acoustic variability is a zero mean process around the unperturbed field. This means there are no biases in phase or intensity. (This is not true when the acoustic variability becomes stronger.) In addition, the probability distribution functions (PDFs) of log-amplitude and phase and are determined by the PDF of μ, after integration over the scattering volume. However, because of the integration, the central limit theorem dictates that regardless of the PDF of μ, the PDFs of log-amplitude and phase will be close to normal distributions.

For sound propagation through a random internal-wave field in the ray approximation it was found that only internal waves that are perpendicular to the ray contribute significantly to the acoustic variability. It is therefore useful to introduce the transverse Fourier transform of μ given by

$$\mu(\mathbf{r}) = \int d\boldsymbol{\kappa}_\perp \, e^{i\boldsymbol{\kappa}_\perp \cdot \mathbf{r}} \, a(x, \boldsymbol{\kappa}_\perp), \qquad (6.4)$$

where $\boldsymbol{\kappa}_\perp = (0, l, m)$ is the internal-wave wavenumber perpendicular to the direction of the incident plane wave. For a random field of internal waves, the wavenumber amplitudes of μ, namely $a(x, \boldsymbol{\kappa}_\perp)$, are statistically independent, simplifying the ensemble averages. Substituting Eq. 6.4 into Eq. 6.3 the z' and y' integrals can be carried out analytically to give

$$\varphi_1(\mathbf{r}) = \int d\boldsymbol{\kappa}_\perp \, e^{i\boldsymbol{\kappa}_\perp \cdot \mathbf{r}} \left[-iq_0 \int_0^R dx' \, \exp\left(-i\frac{|\boldsymbol{\kappa}_\perp^2| |R_f^2(x')|}{4\pi} \right) a(x', \boldsymbol{\kappa}_\perp) \right]. \quad (6.5)$$

Here it is seen that each perpendicular wavenumber contributes some variation to log-amplitude and phase, and that contribution will be determined by an integral over a function along the unperturbed ray path (i.e., the x-axis). In fact, summing over all perpendicular wavenumber contributions can be considered a Fourier synthesis with the random spectral amplitude for each $\boldsymbol{\kappa}_\perp$ given by the quantity in square brackets. For ease of analytical treatment, the following complex quantity is defined:

$$\varphi_1(R, \boldsymbol{\kappa}_\perp) = \chi(R, \boldsymbol{\kappa}_\perp) + i\phi(R, \boldsymbol{\kappa}_\perp) = -iq_0 \int_0^R dx' \, e^{-i|\boldsymbol{\kappa}_\perp|^2 \, R_f^2(x')/(4\pi)} a(x', \boldsymbol{\kappa}_\perp), \quad (6.6)$$

whose real and imaginary parts are the perpendicular Fourier transforms of log-amplitude and phase from Eq. 6.5. This expression will be used for subsequent computations of expectation values.

Of key interest are the spectra of phase and log-amplitude, as well as the cross-spectrum. To obtain these spectra, two correlation functions are defined:

$$B_1 = \langle \varphi_1(R, \boldsymbol{\kappa}_\perp) \varphi_1^*(R, \boldsymbol{\kappa'}_\perp) \rangle, \quad B_2 = \langle \varphi_1(R, \boldsymbol{\kappa}_\perp) \varphi_1(R, \boldsymbol{\kappa'}_\perp) \rangle, \quad (6.7)$$

where $\boldsymbol{\kappa}_\perp$ and $\boldsymbol{\kappa'}_\perp$ are wavenumber vectors in the observation plane (R, y, z). Using these equations, the spectral functions of log-amplitude and phase, as well as the cross-spectrum, are given by,

$$F_\chi(R, \boldsymbol{\kappa}_\perp) = \frac{1}{2} \left[\mathrm{Re}(B_1) + \mathrm{Re}(B_2) \right],$$

$$F_\phi(R, \boldsymbol{\kappa}_\perp) = \frac{1}{2} \left[\mathrm{Re}(B_1) - \mathrm{Re}(B_2) \right],$$

$$F_{\chi\phi}(R, \boldsymbol{\kappa}_\perp) = \frac{1}{2} \left[\mathrm{Im}(B_2) - \mathrm{Im}(B_1) \right]. \quad (6.8)$$

To compute these functions requires information concerning the statistics of the transverse wavenumber amplitudes. In particular, the result is

$$\langle a(x, \boldsymbol{\kappa}_\perp) a^*(x', \boldsymbol{\kappa'}_\perp) \rangle = \delta(\boldsymbol{\kappa}_\perp - \boldsymbol{\kappa'}_\perp) \, C_n(x - x', \boldsymbol{\kappa}_\perp),$$

$$\langle a(x, \boldsymbol{\kappa}_\perp) a(x', \boldsymbol{\kappa'}_\perp) \rangle = \delta(\boldsymbol{\kappa}_\perp + \boldsymbol{\kappa'}_\perp) \, C_n(x - x', \boldsymbol{\kappa}_\perp), \quad (6.9)$$

where the second relation follows because μ is real (i.e., $a(x, \boldsymbol{\kappa}_\perp) = a^*(x, -\boldsymbol{\kappa}_\perp)$). The 2-D spectrum C_n (n for normal) is related to the 3-D spectrum by the Fourier transform relation:

$$F_\mu(k, l, m) = \frac{1}{2\pi} \int_{-\infty}^\infty C_n(x, l, m) \exp(-ikx) \, dx, \quad (6.10)$$

and C_n is therefore seen to be a correlation function in the normal direction of the incident wave.

After some manipulation using Eqs. 6.9, the functions B_1 and B_2 can be written as,

$$B_1(R, \kappa_\perp) = q_0^2 \int_0^R \int_0^R dx' dx'' \, e^{i|\kappa_\perp|^2(x'-x'')/(2q_0)} \, C_n(x'-x'', \kappa_\perp), \qquad (6.11)$$

$$B_2(R, \kappa_\perp) = -q_0^2 \int_0^R \int_0^R dx' dx'' \, e^{-i|\kappa_\perp|^2(2R-x'-x'')/(2q_0)} \, C_n(x'-x'', \kappa_\perp). \qquad (6.12)$$

Using mean and relative coordinates, $\eta = (x'+x'')/2$ and $\xi = x'-x''$, and expressing the correlation function C_n in terms of the 3-D spectrum (inverse of Eq. 6.10) the result is

$$B_1(R, \kappa_\perp) = 2\pi q_0^2 \int_0^R d\eta \, F_\mu(k = -|\kappa_\perp|^2/(2q_0), l, m), \qquad (6.13)$$

$$B_2(R, \kappa_\perp) = -2\pi q_0^2 \int_0^R d\eta \, e^{-i|\kappa_\perp|^2 R_f^2(\eta)/(2\pi)} \, F_\mu(k = 0, l, m). \qquad (6.14)$$

Now the small-angle scattering approximation can be used to simplify the result for B_1, where the issue is the condition $k = -|\kappa_\perp|^2/(2q_0)$. The maximum perpendicular wavenumber is $|\kappa_\perp|^{max} = 2\pi/l_0$, where l_0 is the smallest scale in the internal-wave field. Thus the maximum value for the x-component of the wavenumber is $|k|^{max} = (|\kappa_\perp|^{max}/2)(\lambda/l_0)$. But the scattering angle is small and approximately λ/l_0. Therefore $|k|^{max} \ll |\kappa_\perp|$, and thus $k = 0$ can be used in the equation for B_1. In this approximation, the spectra of interest are

$$F_\chi(R, \kappa_\perp) = F_\mu(0, \kappa_\perp) \left[2\pi q_0^2 \int_0^R \sin^2\!\left(\frac{|\kappa_\perp|^2 R_f^2(x)}{4\pi} \right) dx \right], \qquad (6.15)$$

$$F_\phi(R, \kappa_\perp) = F_\mu(0, \kappa_\perp) \left[2\pi q_0^2 \int_0^R \cos^2\!\left(\frac{|\kappa_\perp|^2 R_f^2(x)}{4\pi} \right) dx \right], \qquad (6.16)$$

$$F_{\chi\phi}(R, \kappa_\perp) = F_\mu(0, \kappa_\perp) \left[\pi q_0^2 \int_0^R \sin\!\left(\frac{|\kappa_\perp|^2 R_f^2(x)}{2\pi} \right) dx \right], \qquad (6.17)$$

where the η variable has been replaced with the x variable. These equations state that the acoustic spectra are simply equal to the ocean spectrum F_μ under the perpendicular wavenumber constraint multiplied by the filter functions that are the quantities in square brackets. These filter functions involve integrals along the x-axis (i.e., the unperturbed straight line ray), and there are nonuniform contributions along this ray. The physical meaning of this nonuniform weighting will be discussed later, in terms of the Fresnel zone and the Fresnel filter that was introduced in Chapter 4.

The variances of log-amplitude and phase, as well as the log-amplitude/phase correlation coefficient, are obtained by integrating the spectra over all perpendicular wavenumbers; that is,

$$\langle \chi^2 \rangle = \int \int d\kappa_\perp F_\chi(R, \kappa_\perp),$$

$$\langle \phi^2 \rangle = \int \int d\kappa_\perp F_\phi(R, \kappa_\perp),$$

$$\langle \chi\phi \rangle = \int \int d\kappa_\perp F_{\chi\phi}(R, \kappa_\perp). \tag{6.18}$$

6.2.2 Point Source

The previous calculation assumed an incident plane wave, but the more relevant oceanographic case is a point source,

$$p_0(\mathbf{r}) = \frac{1}{4\pi} \frac{e^{iq_0|\mathbf{r}|}}{|\mathbf{r}|} \simeq \frac{1}{4\pi x} \exp\left[iq_0\left(x + \frac{y^2 + z^2}{2x}\right)\right], \tag{6.19}$$

where again the last step follows from the small-angle scattering condition for sound traveling primarily in the x-direction. Using this incident field with the free space Green's function (Eq. 6.2) gives

$$\varphi_1(\mathbf{r}) = -q_0 \int d^3\mathbf{r}' \, \exp\left(i\pi \frac{(\gamma y - y')^2 + (\gamma z - z')^2}{\lambda\gamma(R - x')}\right) \frac{\mu(\mathbf{r}')}{\lambda\gamma(R - x')}, \tag{6.20}$$

where $\gamma = x'/R$. Comparing Eqs. 6.3 and 6.20, the spherical wave case can be obtained from the plane wave case by the replacement $(R, y, z) \to (R, \gamma y, \gamma z)$ and $R - x' \to \gamma(R - x')$. For $\gamma = 1$ the plane wave result is recovered. Making this translation, the relative pressure fluctuation function in the Born approximation in terms of the transverse Fourier transform of the sound-speed fluctuations is given by

$$\varphi_1(\mathbf{r}) = -iq_0 \int_0^R dx' \int d\kappa_\perp \, a(x', \kappa_\perp) \, \exp\left(-i\frac{|\kappa_\perp^2| R_f^2(x')}{4\pi}\right) e^{i\kappa_\perp \cdot \gamma \mathbf{r}}, \tag{6.21}$$

which can be compared to Eq. 6.5. The point source Fresnel zone is $R_f^2 = \lambda x(R - x)/R$, as was obtained in Chapter 2. The spectra are thus given by (Ishimaru, 1978;

Ewart et al., 1998; Colosi et al., 2009)

$$F_\chi(R,\boldsymbol{\kappa}_\perp) = 2\pi q_0^2 \int_0^R dx\ F_\mu(\boldsymbol{\kappa}_\perp/\gamma)\ \sin^2\!\left(\frac{|\boldsymbol{\kappa}_\perp|^2 R_f^2(x)}{4\pi\gamma^2}\right), \tag{6.22}$$

$$F_\phi(R,\boldsymbol{\kappa}_\perp) = 2\pi q_0^2 \int_0^R dx\ F_\mu(\boldsymbol{\kappa}_\perp/\gamma)\ \cos^2\!\left(\frac{|\boldsymbol{\kappa}_\perp|^2 R_f^2(x)}{4\pi\gamma^2}\right), \tag{6.23}$$

$$F_{\chi\phi}(R,\boldsymbol{\kappa}_\perp) = \pi q_0^2 \int_0^R dx\ F_\mu(\boldsymbol{\kappa}_\perp/\gamma)\ \sin\!\left(\frac{|\boldsymbol{\kappa}_\perp|^2 R_f^2(x)}{2\pi\gamma^2}\right), \tag{6.24}$$

with $\gamma = x/R$. Here it is seen that the spectral weighting varies not only along the ray path due to the Fresnel zone but also by the dilation of the perpendicular wavenumber $\boldsymbol{\kappa}_\perp/\gamma$.

6.3 Spectra of Phase and Log-Amplitude: Ocean Waveguide

In the ocean waveguide, the unperturbed ray paths are no longer straight lines but distorted sinusoids. As a result, nearby rays can cross, forming caustics and Fresnel zones differ significantly from those in a homogeneous background. In addition, ocean sources are primarily point sources, and so adequate handling of this aspect is critical. These issues in addition to internal-wave inhomogeneity and anisotropy (Chapter 5) must be included in a proper ocean treatment. A comprehensive treatment of weak fluctuation theory for the ocean has not been carried out to the degree seen for ray theory, however. What follows is a treatment of the theoretical tools that are available now, which have seen some success in describing observations.

6.3.1 Local Straight Ray Approximation

Inspired by the seminal paper of Munk and Zachariasen (1976), the traditional approach has been to apply the straight ray results from the previous section locally along the curved ray path Γ. For simplicity the wavenumber dilation effect for a point source is neglected, but point source Fresnel zones are utilized. The physical justification for this approach is that the signal behaves locally like a plane wave as it refracts through the sound channel (i.e., consistent with the ray approximation). The spectra of log-amplitude and phase are thus written (Colosi et al., 2009)[1] as

$$F_{\phi,\chi}(R,l,m) = \pi q_0^2 \int_\Gamma ds\ F_\mu(\boldsymbol{\kappa}_\perp(s);z)\left[1 \pm \cos\!\left(\frac{l^2(\boldsymbol{\kappa}_\perp)\,R_{fy}^2(x) + m^2(\boldsymbol{\kappa}_\perp)R_{fz}^2(x)}{2\pi}\right)\right], \tag{6.25}$$

---

[1] Here the trigonometric identities $\cos^2\theta = (1+\cos(2\theta))/2$ and $\sin^2\theta = (1-\cos(2\theta))/2$ have been used.

where $R_{fy} = \lambda x(R - x)/R$ and R_{fz} are the Fresnel zones in the horizontal (y direction) and vertical (z direction), and $\boldsymbol{\kappa}_{\perp}(s) = (-m\tan\theta, l, m)$ is the local perpendicular wavenumber as a function of the ray angle θ. The plus sign applies to the phase, while the minus sign is for log-amplitude. Equation 6.25 is identical to the result from Munk and Zachariasen (1976), who obtained their expressions from a significantly more detailed derivation. Importantly, there is a direct correspondence between the wavenumbers of the acoustic variability on the left-hand side of Eq. 6.25 in the observation plane (l, m) and the internal-wave wavenumbers on the right-hand side of the equation (Colosi et al., 2009). The spectrum $F_\mu(\boldsymbol{\kappa}_{\perp}(s); z)$ under the perpendicular wavenumber constraint is obtained from Eq. 3.67 by making the substitution $(k, l, m) = (-m\tan\theta, l, m)$.

Because internal-wave horizontal scales are much larger than their vertical scales a good approximation, especially at higher frequencies, is to drop the terms in Eq. 6.25 having to do with the transverse dimension, leaving simply

$$F_{\phi,\chi}(R, l, m) = \pi q_0^2 \int_\Gamma ds \, F_\mu(\boldsymbol{\kappa}_{\perp}(s); z) \left[1 \pm \cos\left(\frac{m^2(\boldsymbol{\kappa}_{\perp})R_{fz}^2(x)}{2\pi} \right) \right] dx.$$

(6.26)

These expressions for the spectra that involve integrals along the ray path do not properly account for ray curvature, as discussed in the previous chapter. On the other hand, these expressions have seen relative success in comparison to data when the scattering is sufficiently weak (Section 6.5). Nonetheless, in the following section a slightly more rigorous derivation of the waveguide result is presented in order to provide the reader a better understanding of the problems involved and to point toward future theoretical work that should be carried out.

6.3.2 A More Rigorous Approach

In this calculation, ray theory is utilized[2] for the Green's function and the unperturbed pressure and so one may write

$$G(\mathbf{r}', \mathbf{r}) = a(\mathbf{r}', \mathbf{r})e^{i\Theta(\mathbf{r}', \mathbf{r})}, \tag{6.27}$$

$$p_0(\mathbf{r}) = a(0, \mathbf{r})e^{i\Theta(0, \mathbf{r})}, \tag{6.28}$$

$$\Theta(\mathbf{r}', \mathbf{r}) = \omega \int_{\mathbf{r}'}^{\mathbf{r}} \frac{ds}{c(z)}, \tag{6.29}$$

where the ray amplitudes, $a(\mathbf{r}', \mathbf{r})$, are obtained from solutions of the ray variational equations as discussed in Chapter 2. Inserting these expressions into the Born

---

[2] Some researchers utilizing the Born approximation to obtain the travel-time sensitivity kernel (Skarsoulis and Cornuelle, 2004) have also used a normal mode approach (Piperakis et al., 2006). Apparently no one has tried to estimate phase and amplitude statistics in the Born approximation using mode representations for the Green's function and unperturbed pressure.

equation the result is

$$\varphi_1(\mathbf{r}) = -2q_0^2 \int_V \frac{a(0,\mathbf{r}')a(\mathbf{r},\mathbf{r}')}{a(0,\mathbf{r})} \mu(\mathbf{r}')e^{i\Delta\Theta(\mathbf{r}',\mathbf{r})}d\mathbf{r}', \tag{6.30}$$

$$\Delta\Theta(\mathbf{r}',\mathbf{r}) = \Theta(0,\mathbf{r}') + \Theta(\mathbf{r}',\mathbf{r}) - \Theta_0(0,\mathbf{r}), \tag{6.31}$$

where $\mathbf{r} = (R,y,z)$ is the observation plane, $\Theta_0(0,\mathbf{r})$ is the phase of the unperturbed ray at the observation point, and $\Delta\Theta$ is the phase difference between the unperturbed ray and the scattered ray emanating from $\mathbf{r}'$. To calculate this phase difference it is necessary to Taylor expand about a ray that travels to the center of the observation plane $(R,0,0)$. This central ray is the unperturbed ray. Because the unperturbed ray is an extremum of the phase the Taylor expansion must be carried out to second order, giving

$$\Delta\Theta(\mathbf{r}',\mathbf{r}) \simeq \frac{\partial^2}{\partial x_i'\partial x_j'}\Big(\Theta(0,\mathbf{r}') + \Theta(\mathbf{r},\mathbf{r}') - \Theta_0(0,\mathbf{r})\Big)\Big|_0 \delta x_i'\delta x_j', \tag{6.32}$$

where the $|_0$ indicates evaluation along the unperturbed ray and implicit summation is understood. Hence all quadratic combinations of ray path deviations from the unperturbed ray must be considered.

Focusing initially on pure z- and y-deviations, it is seen that the pure y-deviation has already been done. Because there is no waveguide in the y-direction, the contribution to the phase difference is

$$(\Delta\Theta(\mathbf{r}',\mathbf{r}))_{yy} = \pi\left[\frac{(\delta y' - \gamma_y y)^2}{R_{fy}^2(x')}\right], \tag{6.33}$$

where $\gamma_y = x'/R$ and $R_{fy}^2(x') = \lambda x'(R-x')/R$ is the point source Fresnel zone for a homogeneous background.

For a pure z-deviation, the result from Chapter 2 is that the phase of a nearby ray can be written to leading order in the parabolic approximation as

$$\Theta = \Theta_{ray} + \frac{q_0}{2}\int_0^R\left[\left(\frac{d\xi}{dx}\right)^2 - \xi^2 V''(z_r)\right]dx, \tag{6.34}$$

where ξ is a vertical displacement away from the central ray that is being expanded around. First consider the calculation of the unperturbed ray phase, that is $\Theta_0(0,\mathbf{r})$. Because the ending depths of the center ray and the unperturbed ray differ, the phase of the unperturbed ray to leading order is

$$\Theta_0 = \Theta_{center} + \frac{q_0}{2}\int_0^R\left[\left(\frac{d\xi}{dx}\right)^2 - \xi^2 V''(z_r)\right]dx, \tag{6.35}$$

where the ray tube function[3] is given by $\xi(x) = z\xi_1(x)$. The integral can be done analytically, giving

$$\Theta_0 = \Theta_{center} + \frac{q_0 z^2}{2} \left[\frac{d\xi_1}{dx}(R) \right]. \qquad (6.36)$$

The phase of the scattered ray, namely $\Theta_s = \Theta(0, \mathbf{r}') + \Theta(\mathbf{r}', \mathbf{r})$, is obtained using Eq. 6.34, giving

$$\Theta_s = \Theta_{center} + \frac{q_0}{2} \left(\int_0^{x'} \left[\left(\frac{d\xi}{dx}\right)^2 - \xi^2 V''(z_r) \right] dx + \int_{x'}^{R} \left[\left(\frac{d\xi}{dx}\right)^2 - \xi^2 V''(z_r) \right] dx \right), \qquad (6.37)$$

where the ray tube equations must satisfy the condition $\xi(x') = \delta z'$ where δz is the vertical separation between the scattered ray and the center ray at x' and $\xi(R) = z$, that is, the scattered ray must end up at the receiver depth z. Ray tube functions that satisfy these criteria and are continuous at x' are (Flatté and Colosi, 2008)

$$\xi(x, x') = \delta z' \frac{\xi_1(x)}{\xi_1(x')}, \qquad\qquad x < x', \qquad (6.38)$$

$$\xi(x, x') = z \left(\frac{\xi_1(x)\xi_2(x') - \xi_1(x')\xi_2(x)}{\xi_2(x')} \right) + \delta z' \frac{\xi_2(x)}{\xi_2(x')} \qquad x > x'. \qquad (6.39)$$

The integrals can again be done analytically, yielding

$$(\Delta\Theta(\mathbf{r}', \mathbf{r}))_{zz} = \pi \left[\frac{(\delta z' - \gamma_z(x')z)^2}{R_{fz}^2(x')} \right], \qquad (6.40)$$

where $R_{fz}^2(x')$ is the vertical Fresnel zone in the waveguide, and $\gamma_z(x') = \xi_1(x')$ is the spherical wave dilation factor. For the case of constant background sound speed, $\xi_1 = x'/R$, and the result given in the previous section is recovered.

The other terms in the Taylor series can be easily shown to be small or identically zero (Munk and Zachariasen, 1976). For small-angle propagation primarily in the x-direction the phase between any two points (x_1, y_1, z_1) and (x_2, y_2, z_2) is

$$\Theta(\mathbf{r}_1, \mathbf{r}_2) \simeq q_0 \left[(x_2 - x_1) + \frac{(y_2 - y_2)^2}{2(x_2 - x_1)} \right] + \Theta(x_1, z_1, x_2, z_2), \qquad (6.41)$$

where the last term represents phase effects from the waveguide. This form means that the variation of the phase with respect to the z- and y-directions is zero, that is, the $\partial^2\Theta/\partial z\partial y = 0$. As a consequence, $\partial^2\Theta/\partial x\partial y = (dz/dx)\partial^2\Theta/\partial z\partial y = 0$. Lastly there are the phase variations with respect to x-direction, and the

---

[3] Recall from Chapter 5 that the boundary conditions on the ray tube functions are $\xi_1(0) = 0$, $\xi_1(R) = 1$, $\xi_2(0) = 1$, and $\xi_2(R) = 0$.

result is $\partial^2\Theta/\partial x^2 = (dz/dx)^2\partial^2\Theta/\partial z^2 = \tan^2(\theta)\partial^2\Theta/\partial z^2 \simeq 0$, and $\partial^2\Theta/\partial x\partial z = (dz/dx)\partial^2\Theta/\partial z^2 = \tan(\theta)\partial^2\Theta/\partial z^2 \simeq 0$.

The Born wave function expanded about the center, unperturbed ray, $(z(x), y(x))$ where the z coordinate origin follows the ray is

$$\varphi_1(\mathbf{r}) = -2q_0^2 \int_0^R dx' \int \int dy' d\delta z' \frac{a(0,\mathbf{r}')a(\mathbf{r},\mathbf{r}')}{a(0,\mathbf{r})} \mu(\mathbf{r}')$$

$$\exp\left[i\pi\left(\frac{(y' - \gamma_y(x')y)^2}{R_{fy}^2(x')} + \frac{(\delta z' - \gamma_z(x')z)^2}{R_{fz}^2(x')}\right)\right]. \tag{6.42}$$

A difficult term here is due to the ray amplitudes. In the previous free space calculations, these terms depended most strongly on x' and the final range R. In fact, it was seen that $a(0,\mathbf{r}')a(\mathbf{r},\mathbf{r}')/a(0,\mathbf{r}) \simeq (2q_0R_f^2(x'))^{-1}$. By analogy the following approximation is made:

$$\frac{a(0,\mathbf{r}')a(\mathbf{r},\mathbf{r}')}{a(0,\mathbf{r})} \simeq \frac{1}{2q_0R_{fy}(x')R_{fz}(x')}. \tag{6.43}$$

This result can also be obtained using the stationary phase approximation (Munk and Zachariasen, 1976). Introducing the transverse Fourier transform of $\mu(\mathbf{r}')$, i.e., $a(x', \boldsymbol{\kappa}_\perp)$, where $\boldsymbol{\kappa}_\perp = (-m\tan\theta, l, m)$ is the internal wave wavenumber perpendicular to the ray, one may write $d\boldsymbol{\kappa}_\perp = dl\,dm/\cos\theta$ (Munk and Zachariasen, 1976), so that $\mu(\mathbf{r}')$ is expressed as

$$\mu(\mathbf{r}') = \int \int \frac{dl\,dm}{\cos\theta} a(x', \boldsymbol{\kappa}_\perp)e^{i\boldsymbol{\kappa}_\perp \cdot \mathbf{r}'}. \tag{6.44}$$

Using Eqs. 6.43 and 6.44, the integrals in the Born wave function over y' and $\delta z'$ can be carried out analytically. The amplitude terms cancel, giving

$$\varphi_1(\mathbf{r}) = -iq_0 \int_0^R dx' \int \int \frac{dl\,dm}{\cos\theta} a(x',l,m)\exp\left[-i\frac{l^2(\boldsymbol{\kappa}_\perp)R_{fy}^2 + m^2(\boldsymbol{\kappa}_\perp)R_{fz}^2}{4\pi}\right]$$

$$e^{i(l\gamma_y y + m\gamma_z z)}e^{-im\tan(\theta)x'}, \tag{6.45}$$

which is close in form to Eq. 6.21. With this expression for φ_1, the homogeneous background point source spectra can be translated to the result for the case of a waveguide, and the result is

$$
F_\chi(R,l,m) = 2\pi q_0^2 \int_0^R \frac{dx}{\cos^2\theta} F_\mu(\kappa_\perp(x,l/\gamma_y,m/\gamma_z))
$$
$$
\sin^2\left[\frac{1}{4\pi}\left(\frac{l^2(\kappa_\perp)R_{fy}^2(x)}{\gamma_y^2} + \frac{m^2(\kappa_\perp)R_{fz}^2(x)}{\gamma_z^2}\right)\right], \tag{6.46}
$$

$$
F_\phi(R,l,m) = 2\pi q_0^2 \int_0^R \frac{dx}{\cos^2\theta} F_\mu(\kappa_\perp(x,l/\gamma_y,m/\gamma_z))
$$
$$
\cos^2\left[\frac{1}{4\pi}\left(\frac{l^2(\kappa_\perp)R_{fy}^2(x)}{\gamma_y^2} + \frac{m^2(\kappa_\perp)R_{fz}^2(x)}{\gamma_z^2}\right)\right], \tag{6.47}
$$

$$
F_{\chi\phi}(R,l,m) = \pi q_0^2 \int_0^R \frac{dx}{\cos^2\theta} F_\mu(\kappa_\perp(x,l/\gamma_y,m/\gamma_z))
$$
$$
\sin\left[\frac{1}{2\pi}\left(\frac{l^2(\kappa_\perp)R_{fy}^2(x)}{\gamma_y^2} + \frac{m^2(\kappa_\perp)R_{fz}^2(x)}{\gamma_z^2}\right)\right]. \tag{6.48}
$$

Equations 6.46–6.48 are new results that have not been studied in detail and have not been applied to the interpretation of observations. They closely resemble the results for a point source with a homogeneous background, but here the y- and z-directions must be treated differently. It is hoped that these results inspire further research into the subject of weak fluctuation theory in a waveguide, but because all observational work has used Eq. 6.26, the rest of the theoretical development will focus on that equation.

6.3.3 Depth/Time Observation Plane

Equation 6.26 gives the phase and log-amplitude wavenumber spectra in the (R,y,z) plane, which is not observationally convenient since measurements are more commonly made along a vertical array. Thus it is necessary to translate the l dependence of $\mathbf{k}_\perp(s)$ into dependence on the internal-wave frequency σ using the WKB internal-wave dispersion relation (Eq. 3.44), and the WBK vertical wavenumber given by $m(j) = \pi j N(z)/N_0 B$. Changing the coordinates of the internal-wave spectrum using the Jacobian function, that is,

$$
F_\mu(\mathbf{k}_\perp(l,m(j)))\,dl\,dj = F_\mu(\mathbf{k}_\perp(\sigma,m))\,d\sigma\,dm, \tag{6.49}
$$

the GM spectrum under the perpendicular wavenumber constraint becomes (see the appendix of Colosi et al., 2009, for details)

$$
F_\mu(\kappa_\perp(\sigma,m);z) = \langle\mu^2(z)\rangle \frac{8}{\pi^3} \frac{m_*}{m(m^2+m_*^2)} \frac{Nf}{\sigma^3}\left(\frac{\sigma^2-f^2}{\sigma^2-\sigma_L^2}\right)^{1/2}, \quad \sigma > \sigma_L \tag{6.50}
$$

where $\sigma_L^2 = f^2 + N^2 \tan^2 \theta$ is the low-frequency internal-wave cut-off imposed by the perpendicular wavenumber resonance condition (see discussion in Chapter 4). This result shows that the GM internal-wave spectrum under the perpendicular wavenumber resonance condition scales like σ^{-3} and m^{-3}, that is, one factor of σ and m different from the spectrum without the resonance condition. In the Born approximation the spectrum of relative sound-speed fluctuations must be integrated along the ray for each frequency σ and each vertical wavenumber m. However, there are forbidden regions in this integral, specifically where $\sigma < \sigma_L$ and $\sigma > N(z)$. To indicate the forbidden regions, Heavyside step functions are added to the spectrum to give

$$
F_{\phi,\chi}(R,\sigma,m) = \pi\, q_0^2 \int_\Gamma ds\, F_\mu(\kappa_\perp(\sigma,m);z) \left[1 \pm \cos\left(\frac{m^2 R_{fz}^2(x)}{2\pi}\right)\right]
$$
$$
H[\sigma - \sigma_L(z_r(x))]\, H[N(z_r(x)) - \sigma]. \qquad (6.51)
$$

If the vertical wavenumber spectrum is sought, the integral of Eq. 6.51 over frequency can be done analytically, yielding

$$
F_{\phi,\chi}(R,m) = \frac{8 q_0^2}{\pi^2} \int_\Gamma ds \langle \mu^2(z) \rangle \frac{m_*}{m(m^2 + m_*^2)} \frac{Nf}{\sigma_L^2} W(N,\sigma_L,f) \left[1 \pm \cos\left(\frac{m^2 R_{fz}^2(x)}{2\pi}\right)\right],
$$
$$
(6.52)
$$

where the function $W(N,\sigma_L,f)$ has the analytic form

$$
W(N,\sigma_L,f) = \int_{\sigma_L}^{N} \frac{\sigma_L^2}{\sigma^3} \sqrt{\frac{(\sigma^2 - f^2)}{(\sigma^2 - \sigma_L^2)}}\, d\sigma,
$$
$$
= \frac{1}{2}\left[\sqrt{\left(1 - \frac{\sigma_L^2}{N^2}\right)\left(1 - \frac{f^2}{N^2}\right)} + \left(\frac{\sigma_L}{f} - \frac{f}{\sigma_L}\right) \ln\left[\frac{\sqrt{1 - \frac{f^2}{N^2}} + \sqrt{\frac{f^2}{\sigma_L^2} - \frac{f^2}{N^2}}}{\sqrt{1 - \frac{f^2}{\sigma_L^2}}} \right] \right].
$$
$$
(6.53)
$$

6.3.4 Variances of Phase and Log-Amplitude

The observationally important variances of log-amplitude and phase are simply the integral of the spectra over all perpendicular wavenumbers. For the GM spectrum, these integrals can be carried out analytically. Here it proves useful to work in frequency, mode-number space where the spectrum under the perpendicular wavenumber constraint is given by[4]

$$F_\mu(\kappa_\perp(\sigma, j); z) = \langle \mu^2(z) \rangle \frac{4}{\pi^2} H(j) \frac{N_0 B}{\pi j} \frac{f}{\sigma^3} \left(\frac{\sigma^2 - f^2}{\sigma^2 - \sigma_L^2} \right)^{1/2}, \quad \sigma > \sigma_L. \quad (6.54)$$

Thus integrating Eq. 6.26 over frequency from σ_L to N, summing over all mode numbers, and utilizing Eq. 5.31 for the correlation length along the ray L_p, the variances of log-amplitude and phase can be written as

$$\langle \phi^2 \rangle, \langle \chi^2 \rangle = q_0^2 \int_{\Gamma_0} ds \, \langle \mu^2(z) \rangle \, L_p(\theta, z) \, G_\pm(R_f, s), \quad (6.55)$$

$$G_\pm(R_f, s) = \frac{1}{2\langle j^{-1} \rangle} \sum_{j=1}^{J} \frac{H(j)}{j} \left[1 \pm \cos\left(\frac{m^2(j) R_f^2(s)}{2\pi} \right) \right], \quad (6.56)$$

where the plus sign is for phase and the minus sign is for log-amplitude. The functions $G_\pm$ which include the Fresnel zone clearly quantify the effects of diffraction. In virtually all cases of oceanographic interest, one can take $G_+ = 1$, and to a high level of accuracy $\langle \phi^2 \rangle = \Phi^2$. On the other hand, in general $\langle \chi^2 \rangle$ must be computed with no approximations to G_-.

6.3.5 An Example Calculation

An example calculation utilizing a deep-water mid-latitude environment is useful here. Computed spectra of phase and log-amplitude are shown in Figure 6.1 for the acoustic frequency, ray geometry, and environmental parameters used in Figure 4.8. In the upper panels, the vertical wavenumber spectra have been evaluated using Eq. 6.52 for an axial ray and an initially up going $10°$ grazing angle ray. In all vertical wavenumber spectra, the large m behavior is m^{-3} and the phase and log-amplitude curves coalesce because in the large m limit the cosine term in the Fresnel filter (i.e., square bracket term in Eq. 6.52) oscillates rapidly, and thus does not contribute. At small m there are significant differences between the spectra of phase and log-amplitude. The phase spectrum shows a nearly m^{-3} behavior throughout, and the log-amplitude spectrum rolls off near a wavenumber of roughly 0.01 cpm. In this case, the Fresnel filter strongly attenuates the large scales for the log-amplitude, but not the phase. Phase is sensitive to the largest scales in the problem, while log-amplitude is sensitive to smaller scales near the Fresnel scale (Chapter 4). Differences between axial and steep rays are not that apparent, although the roll-off of the log-amplitude spectra differs due to the different Fresnel zones of these rays.

In the lower panels of Figure 6.1, the frequency spectra of phase and log-amplitude for the two rays are shown. These spectra were obtained from

---

[4] The apparent extra factor of 2 in Eq. 6.54 comes from the two contributions of $\pm l$.

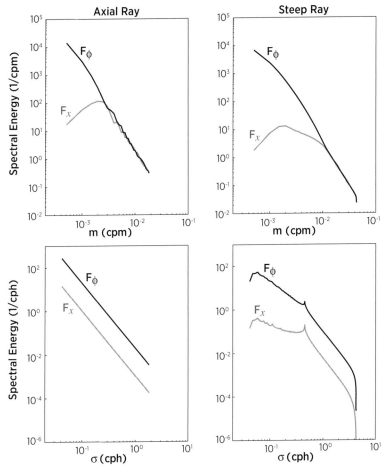

Figure 6.1. Born theory predictions for the vertical wavenumber and frequency spectra for steep (10°) and axial rays using the GM spectrum. The acoustic frequency is 250 Hz, the range is 80 km, and the background is a Munk sound-speed profile and an exponential buoyancy frequency profile (Figure 2.1). The rms phase/intensity fluctuations for the axial and steep rays are 4.9 rad/5.2 dB and 2.8 rad/2.5 dB, respectively. The diffraction parameters, Λ, for the two rays (axial/steep) are 0.20 and 0.16.

Eq. 6.51 by numerically integrating over m. For the axial ray, $\sigma_L = f$ along the whole ray and thus the spectra have a pure σ^{-3} dependence between the cut-off frequencies $N(z_{axis})$ and f. For the steep ray, however, σ_L varies significantly along the ray path (see, for example, Figure 4.8). In particular, when the ray traverses the upper ocean, low frequencies do not contribute to the spectrum; as a result between the frequencies of f and 0.8 cph the steep ray has less low-frequency variation than the axial ray. For frequencies between 0.8 cph and N_{max}, the full internal-wave spectrum contributes and a σ^{-3} dependence results.

This example is for a relatively strong geometric regime, since the phase spectra are larger than the log-amplitude spectra, making $\langle\phi^2\rangle \gg \langle\chi^2\rangle$. In the next section the calculation of the Λ and Φ parameters will be discussed, and the location of this example on the $\Lambda - \Phi$ diagram will be shown.

6.4 $\Lambda - \Phi$ Revisited

In Chapter 4, the $\Lambda - \Phi$ diagram was introduced and used to define the acoustic fluctuation regimes unsaturated, partially saturated, and fully saturated. The parameter Φ is the rms phase variation in the geometrical acoustics limit, and Λ is the weighted average of $\{m^2\}R_f^2(s)/2\pi$ over the ray. The definitions are

$$\Phi^2 = q_0^2 \int_\Gamma ds \, \langle\mu^2(z)\rangle \, L_p(\theta, z), \tag{6.57}$$

$$\Lambda = \frac{q_0^2}{\Phi^2} \int_\Gamma ds \, \langle\mu^2(z)\rangle \, L_p(\theta, z) \frac{\{m^2\}R_f^2(s)}{2\pi}. \tag{6.58}$$

Physically, Φ quantifies the strength of scattering and it increases with increasing frequency, range, and strength of the internal-wave fluctuations. The influences of diffraction are quantified by Λ, which is a path average of the ratio of an acoustical length scale, R_f, to an internal-wave length scale $\{m^2\}^{-1/2}$. Λ decreases with increasing frequency, increases with increasing range, and is independent of internal-wave strength. The diagram provides a useful tool because it is one of the few ways in which to compare experiments and it provides some degree of predictive capability for formulating experiments. In this chapter and the next, the $\Lambda - \Phi$ diagram will be used to interpret observations from experiments, and some further discussion concerning the diagram is needed.

The diagram utilizes weak fluctuation results for the variances of phase and log-amplitude (Eq. 6.55) in the high- and low-frequency limits and can therefore only give general rather than specific information concerning acoustic field statistics in the three regimes. Obtaining specific information concerning an acoustic observable requires the theoretical approaches discussed in Part II of the book: ray methods, weak fluctuation theory, path integrals, and transport theory. Theoretical improvements to the $\Lambda - \Phi$ diagram are discussed below that bring theory into much better agreement with observations (Colosi, 2015). There are four major issues: (1) the calculation of the spectral average $\{m^2\}$ that depends on the small-scale part of the internal-wave spectrum; (2) the calculation of R_f in the parabolic approximation; (3) the sensitivity of R_f to small-scale structure in the background sound-speed profile; and (4) the border between unsaturated propagation and stronger scattering.

There are other issues with the $\Lambda - \Phi$ diagram that require further work. An important issue for high-angle ray paths is the ray-tangent approximation that was discussed at length in Chapter 5. Another issue is long ranges at which ray chaos occurs and the expansion about the unperturbed ray is expected to break down. Figure 5.6 shows that beyond roughly 1000 km, ray chaos effects become evident in the RMS ray scattering across the wave front, that is, $\langle X^2 \rangle^{1/2}$, which is directly related to Φ. Here the scaling goes from $R^{1/2}$ to $R^{3/2}$. There is also further work required in the weak fluctuation theory, to examine the accuracy of Eq. 6.55 compared to the more precise expression that takes into account all the effects of the point source (e.g., phase and log-amplitude variances derived from Eqs. 6.46 and 6.47).

6.4.1 Spectral Average, $\{m^2\}$

The spectral average of the square vertical wavenumber is written as

$$\{m^2\} = \frac{1}{\langle j^{-1} \rangle} \sum_{j=1}^{J} \frac{H(j)}{j} m^2(j) = \left(\frac{\pi N(z)}{N_0 B} \right)^2 \frac{\langle j \rangle}{\langle j^{-1} \rangle}, \qquad (6.59)$$

where the last step follows from use of the WKB vertical wavenumber relation $m(j) = \pi j N(z)/N_0 B$. Because this average depends on $\langle j \rangle = \sum_{j=1}^{J} jH(j)$, the high mode number content of the internal-wave field, in particular the mode number cut-off J, is needed. Using the GM spectrum, $\langle j \rangle$ diverges logarithmically as J becomes large. Although a choice must be made for J, the result is therefore not too sensitive to this choice. In Chapter 3 it was seen that small-scale internal waves are limited by shear instability, and J can therefore be estimated using the inverse Richardson number. The result using the GM spectrum is (Eq. 3.75)

$$J = \frac{(N_0 B)^2}{6\pi \zeta_0^2 N_0 \bar{N} j_*}, \qquad (6.60)$$

where $\bar{N}$ is a value of the buoyancy frequency typical of the main thermocline. Maximum mode numbers computed for the experiments discussed in this chapter range from 200 to 1000 (see Table 6.2), yielding more accurate results for $\{m^2\}$ than previously obtained (Esswein and Flatté, 1980). These results for $\{m^2\}$ are also important for several theoretical expressions from the path integral method.

6.4.2 Fresnel Zone

The issue of the parabolic approximation that has been used in the computation of the Fresnel zone also must be addressed (Colosi, 2015). High-angle methods are needed, especially when ranges longer than a half loop are considered. Differences

Table 6.1. *Observed and predicted phase and intensity statistics for six experiments and for the canonical example (CAN)*

Experiment	Range (km)	Φ (obs/pred) (rad)	Λ	σ_i^2 (obs/pred)	SI (obs/pred)
MATE (2 kHz, ID −1)	18	12.5/6.9	0.028	1.07/0.44	0.73/0.55
MATE (4 kHz, ID −1)	18	25/13.7	0.014	1.65/0.89	1.18/1.43
MATE (8 kHz, ID −1)	18	50/27.4	0.007	1.84/1.77	1.95/4.9
MATE (13 kHz, ID −1)	18	81/44.6	0.0045	1.90/2.84	1.85/NA
AFAR (410 Hz, ID −2)	35	NA/7.4	0.12	1.8/2.2	1.24/NA
AFAR (1010 Hz, ID −2)	35	NA/18.1	0.051	1.7/5.7	1.21/NA
AFAR (4671 Hz, ID −2)	35	NA/83.7	0.013	1.9/NA	1.47/ NA
SD (2.2 kHz, ID −1)	23	NA/15.3	0.011	NA/0.75	1.33/1.11
SD (2.2 kHz, ID +1, early)	23	NA/NA	NA	NA/NA	0.9/NA
SD (2.2 kHz, ID +1, late)	23	NA/19.7	0.020	NA/2.4	1.2/NA
AATE (2 kHz, ID −1)	6.4	NA/0.41	0.43	0.04/0.027	NA/0.027
AATE (4 kHz, ID −1)	6.4	NA/0.83	0.21	0.07/0.064	NA/0.065
AATE (8 kHz, ID −1)	6.4	NA/1.65	0.11	0.12/0.14	NA/0.15
AATE (16 kHz, ID −1)	6.4	NA/3.31	0.053	0.22/0.30	NA/0.35
AET (75 Hz, ID +4)	87	NA/1.23	0.56	0.12[a]/0.20	0.13[a]/0.23
AET (75 Hz, ID −3)	87	NA/0.67	0.30	0.04/0.03	0.044/0.03
PhilSea (284 Hz, ID +5)	107	NA/4.76	0.05	0.34/0.29	0.31/0.33
PhilSea (284 Hz, ID +4)	107	NA/3.36	0.08	0.37/0.22	0.29/0.25
PhilSea (284 Hz, ID −4)	107	NA/3.36	0.08	0.33/0.22	0.30/0.25
PhilSea (284 Hz, ID −3)	107	NA/1.80	0.14	0.24/0.11	0.22/0.12
CAN (250 Hz, Axial)	80	NA/4.3	0.22	NA/1.10	NA/1.99
CAN (250 Hz, ID +3)	80	NA/2.4	0.19	NA/0.28	NA/0.33

[a]The AET ID +4 has been corrected for low-frequency intensity variability below the inertial frequency (see Colosi et al. (2009), figure 10).
The ray *ID* gives the number of ray turning points, and the sign indicates the sign of the initial ray angle (+/− up-surface/down-seafloor). Predictions of log-intensity variance are based on weak fluctuation theory and internal-wave parameters (Table 6.2). In all cases, predicted SI values are computed from $\exp(\sigma_i^2) - 1$. Observed values of σ_i^2 for the MATE experiment were derived from Generalized Gamma Distribution fits to the observed PDFs (Ewart and Percival, 1986). The entry NA means that the value was not reported in the literature or the value from weak fluctuation theory was clearly out of reasonable range.
Source: Adapted from Colosi (2015).

in Fresnel zones computed using the high-angle and small-angle approximations can be quite large (Figure 2.11 and Colosi, 2015). In Chapter 2 it was shown that high-angle ray tube functions can be obtained from the ray variational equations with the Helmholtz equation Hamiltonian (Eq. 2.61). In particular, the quantity of interest is $q_{21} = dz/dp_0$, which is the variation in depth between two rays with an infinitesimal variation in initial launch angle. When the ray equations and variational equations are integrated together with the same Hamiltonian, an internally consistent result is obtained with caustics in the correct locations. This

Table 6.2. *Internal-wave spectral parameters used for the* Λ − Φ *calculations presented in Table 6.1 and Figure 6.4*

Experiment	Latitude	ζ_0 (m)	$N_0 B$ (rad-m/s)	j_*	J	Reference
MATE	45	8.5	6.7	6	563	Ewart and Reynolds (1984)
AFAR	37	13.1	5.2	3	231	Reynolds et al. (1985)
SD	32	7.3	6.4	3	501	Worcester et al. (1981)
AATE	74	1.0	5.4	30	400	Ewart and Reynolds (1993)
AET	30	7.3	9.2	3	1030	Colosi et al. (2009)
PhilSea09	21	8.6	10.2	3	904	Colosi et al. (2013b)
Canonical	30	7.3	8.7	3	909	

has not been the standard procedure in the literature, where Helmholtz rays are combined with parabolic equation ray tube functions.

Using the high-angle ray tube equations to some extent also solves the problem of Fresnel zone sensitivity to small-scale structure in the background sound-speed profiles. Of course, some sensitivity to small scales is expected because the ray tube functions depend on second-order derivatives. However, the parabolic equation ray tube functions are more sensitive to these scales because only the second derivative appears in the equation (Eq. 2.64) while the Helmholtz Hamiltonian gives ray tube equations with both second-order and first-order derivatives (Colosi, 2015). Thus, this approach allows a much more accurate and stable evaluation of the Λ parameter, making the Λ − Φ diagram a more effective tool.

6.4.3 Border of the Unsaturated Regime

A remaining issue is the accurate quantification of the transition region between weak fluctuations (unsaturated propagation) and stronger scattering (partial and fully saturated propagation) in terms of Λ and Φ. In Chapter 4 this transition region was stated to be near the point at which $\sigma_t^2 \simeq 1$, which is roughly two-thirds of the way to the full saturation value of 1.65 and is near the boundary of the region in which the Born approximation is expected to be valid (i.e., $\sigma_t^2 \simeq 1.2$) (Clifford, 1978). With this transition region in mind, how does σ_t^2 vary as a function of Λ and Φ? Through extensive numerical experimentation with different sound-speed and buoyancy-frequency profiles, as well as different source and receiver depths and internal-wave parameters, Colosi (2015) found

$$\sigma_t^2 \simeq \Lambda \Phi^2 \left[\left(\frac{1}{\pi} - 0.1 \right) e^{-\Lambda/1.5} + 0.1 \right], \tag{6.61}$$

which is accurate at the level of 5–30% in the region $0 < \Lambda < 10$, with most of the variability occurring in the larger Λ limit. The small Λ limit is given by $\sigma_t^2 \simeq$

$\Lambda\Phi^2/\pi$. This expression does not include the large Λ limit discussed in Chapter 4 in which $\sigma_t^2 \simeq 2\Phi^2$. This relation is quite different from that used to define the weak fluctuations to strong scattering transition by Flatté (1983), which gave $\sigma_t^2 \simeq \Lambda\Phi^2$ for Λ asymptotically small ($\Lambda \ll 1$). This new boundary between weak fluctuations and stronger scattering is also consistent with the observations that are discussed in the next section.

6.5 Observations

Here comparisons are made between observations and weak fluctuation theory, and the observations are interpreted in terms of the $\Lambda - \Phi$ diagram. When comparing weak fluctuation theory to measurements, the results apply to situations in which a defined acoustic path can be observed. This requires broadband transmissions and temporal resolution of paths free of deterministic path interference. In addition because the concern is weak fluctuations, variations in the acoustic field will be small, and therefore precise source/receiver positioning as well as timing are absolutely necessary. Lastly, the theory requires certain internal-wave parameters, and thus to obtain a truly useful data/theory comparison the experiment must be able to unambiguously estimate these parameters.

To date the experiments that have been analyzed in detail have primarily involved time-series measurements over quite limited vertical apertures. Thus the comparisons have been almost exclusively done with regard to moments of phase and amplitude and the corresponding frequency spectra. One exception was the 75-Hz, 87-km range Acoustic Engineering Test (AET), which had a 700-m aperture receiving array and therefore allowed comparison of the vertical wavenumber spectra. Table 6.1 provides a summary of the four experiments that will be discussed in this chapter. Figures 6.2 and 6.3 show the experimental geometries and some background environmental information. These experiments are (1) the co-located 1971 North Pacific Cobb Seamount experiment (Ewart, 1976) and the 1977 Mid-ocean Acoustic Transmission Experiment (MATE) (Ewart and Reynolds, 1984); (2) the 1975 Azores Fixed Acoustic Range (AFAR) experiment (Reynolds et al., 1985); (3) the 1985 Arctic Acoustic Transmission Experiment (AATE) (Ewart and Reynolds, 1990); and (4) the 1994 Acoustic Thermometry of Ocean Climate (ATOC) Acoustic Engineering Test (AET) (Colosi et al., 2009). Not directly discussed here but also listed in Table 6.1 are the 1978 San Diego (SD) experiment (Worcester et al., 1981) and the 2009 Philippine Sea (PhilSEa09) experiment (White et al., 2013). Figure 6.4 shows where the six experiments and the canonical example from Section 6.3.5 fall on the $\Lambda - \Phi$ diagram (Colosi, 2015). These experiments, which have ranges less than 107 km; acoustic frequencies between 75 Hz and 16 kHz; and locations in the Pacific, Atlantic, and Arctic Oceans, are primarily in the unsaturated regime or in the boundary between the unsaturated and partially saturated regimes. The one

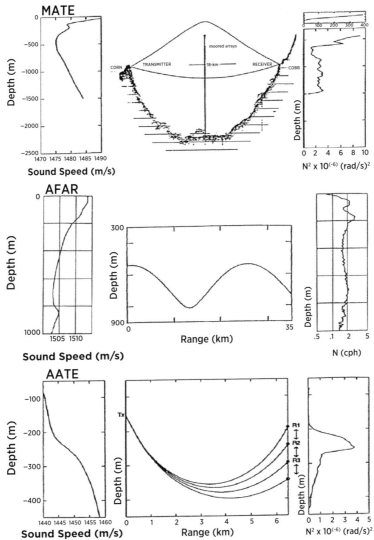

Figure 6.2. Experimental arrangement for the MATE, AFAR, and AATE experiments. The Cobb Seamount experiment was a predecessor to MATE with essentially the same experimental configuration.
Sources: Ewart and Reynolds (1984); Reynolds et al. (1985); Ewart and Reynolds (1993).

exception is the AFAR experiment, which is on the border between the partial and fully saturated regimes.

6.5.1 Cobb Seamount and MATE

Two acoustic paths were observed in the 18-km range Cobb Seamount and MATE experiments. The focus here is on the lower path (Figure 6.2) because the statistics

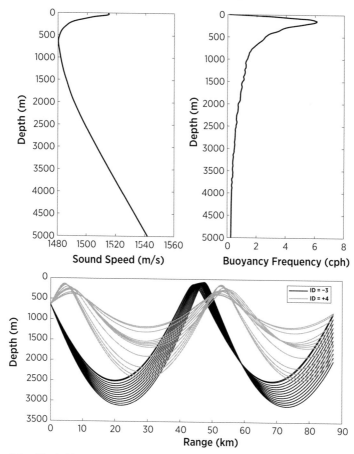

Figure 6.3. (Top) Sound-speed and buoyancy-frequency profiles from the AET 87-km transmission experiment. (Bottom) Resolved ray paths for the AET 700-m long, large-aperture vertical receiving array.
Source: Colosi et al. (2009).

of the lower path are more relevant to weak fluctuation theory and the upper path observations have not been analyzed in detail. The four frequencies between 2 and 13 kHz straddle the boundary between the unsaturated and partially saturated regimes and are in a highly geometric regime, that is, $\Lambda \ll 1$ (Figure 6.4). The predictions of Φ and σ_i^2 are less than the observations (Table 6.1), mostly by a factor of 2, suggesting that the points on the $\Lambda - \Phi$ diagram should be somewhat closer to the partially saturated regime.

Because this experiment is on the boundary where weak fluctuation theory may be expected to break down (i.e., $\sigma_i^2 \sim 1$), one might not anticipate good agreement between observation and theory. However, Figures 6.5 and 6.6 show that the phase spectra for both Cobb and MATE agree well with weak fluctuation theory in the frequency range between f and N. In the case of Cobb, the phase spectrum decays

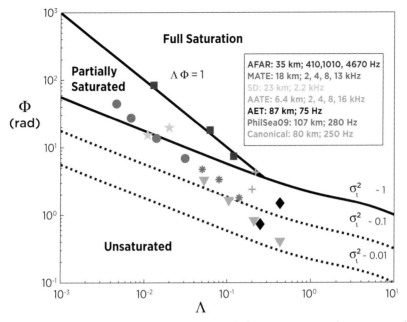

Figure 6.4. Lambda-Phi diagram for several short-range experiments associated with unsaturated and partially saturated propagation regimes. Uncertainty associated with the intensity variance contours due to different background sound-speed and buoyancy-frequency profiles and different source receiver depths are between 5 and 30 percent. An rms intensity fluctuation of 1 corresponds to a rms fluctuation of 4.3 dB.
Source: Adapted from Colosi (2015).

as σ^{-3}, which is precisely what is expected from the perpendicular wavenumber resonance for a near-axial ray for the GM spectrum (e.g., Figure 6.1). With regard to MATE, the internal-wave frequency spectrum was observed to scale as $\sigma^{-1.7}$, and thus the perpendicular wavenumber resonance gives a phase frequency spectrum that scales as $\sigma^{-2.7}$. In accord with weak fluctuation theory, the shape of the phase spectra did not vary across the frequency range from 2 to 13 kHz. The observed MATE internal-wave energy was close to the standard GM level, but the modal bandwidth factor, j_*, was 6 instead of 3 (Table 6.2).

From weak fluctuation theory the frequency spectrum of log-amplitude is expected to scale with frequency in the same way as the phase spectrum (e.g., Figure 6.1). However, the observed Cobb (Figure 6.5) and MATE (not shown) spectra have significantly more energy at high frequencies (Ewart and Reynolds, 1984). The enhancement at high frequencies has been associated with vertical advection of small internal waves by the larger internal waves (Flatté et al., 1980). The effect is more likely caused by microray interference effects that are treated by the method of path integrals (Flatté et al., 1987b) and moment equations (Macaskill and Ewart, 1996).

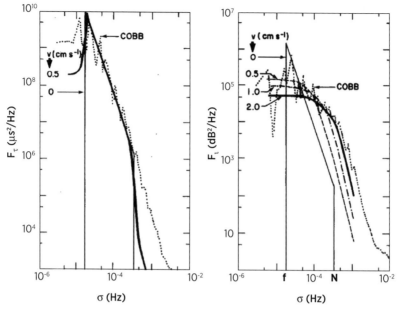

Figure 6.5. Spectra of phase and log-amplitude for the 4 kHz transmissions in the Cobb Seamount experiment. Observed spectra are dotted and predictions are solid, dash-dot, and dash. In the righthand panel, theoretical expressions for the spectra assuming advective background currents of 0.0, 0.5, 1.0, and 2.0 cm/s are shown with light-solid, dash, dash-dot, and heavy solid curves, respectively. In the left panel the light/heavy solid curves are for $v = 0.0$ and 1.0 cm/s.
Source: Flatté et al. (1980).

Several acoustical observables from the Cobb and MATE experiments are consistent with the $\Lambda - \Phi$ prediction that the propagation regime is partially saturated. First, the enhancement of log-amplitude spectral energy at high frequencies is consistent with microray behavior in the partially saturated regime. In addition, the temporal behavior of the acoustic field phasors in MATE are consistent with partial saturation. The phasors at frequencies of 2, 4, and 8 kHz show that phase variations are more rapid than amplitude variations (see Figure 4.11) (Ewart and Reynolds, 1984). The 13-kHz phasors show a pattern tending toward a random walk, which is typical of the border between partial and full saturation. Lastly, the intensity PDFs for MATE are close to a Generalized Gamma Distribution, and the observed shapes of the PDFs show deviations from exponential form that are consistent with partial saturation (Ewart and Percival, 1986).

The conclusion is that the weak fluctuation results for phase appear to have a wider regime of applicability than those for amplitude or intensity. In these experiments the phase variations are in fact quantified using the pulse travel time, and so travel time is a robust observable even down to the scale of internal waves.

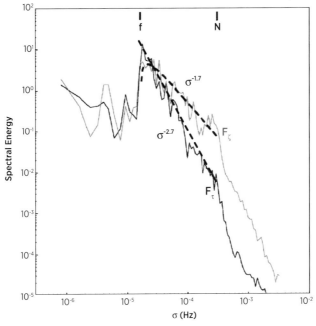

Figure 6.6. Travel time and internal-wave frequency spectra from MATE and predictions from weak fluctuation theory. Spectra are for the MATE lower path, and the travel-time spectrum is for 8-kHz observations. The internal-wave spectrum is normalized to have unit variance. The tides have been removed from the internal-wave and travel-time spectra.
Source: Henyey and Ewart (2006).

6.5.2 AFAR

The 35-km range AFAR experiment, which transmitted pulses at center frequencies of 410, 1010, and 4671 Hz, yielded the most randomized signals that will be considered in this chapter. This experiment took place in a region of rough mid-Atlantic ridge topography where the internal-wave energy was nearly four times the standard GM level (Table 6.2). In addition, the ray path that was resolved propagated in an unusual mid-thermocline sound duct that was formed by the complex intermingling of Atlantic and Mediterranean water masses. This ray path had one lower and one upper turning point (Figure 6.2). The $\Lambda - \Phi$ diagram shown in Figure 6.4 places AFAR on the border between the partial and fully saturated regimes, and this placement is consistent with the observed and predicted intensity moments (Table 6.1). Like Cobb and MATE, the AFAR propagation regime is consistent with the observations of complex temporal phasor behavior and the intensity PDFs (Colosi, 2015). The AFAR log-intensity variance and scintillation index values at 410 and 1010 Hz are unusually high when compared to the MATE and Cobb results, which were at higher frequency but roughly half the range. This anomalous behavior most likely has to do with the unusual environment in which

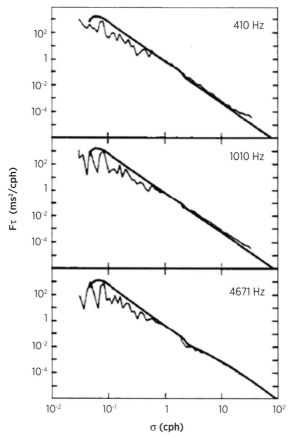

Figure 6.7. Travel-time spectra for AFAR at 410, 1010, and 4871 Hz. Observed spectra are displayed with a light-solid line and predicted spectra are displayed with a heavy solid line.
Source: Flatté (1983).

AFAR was conducted. It is also suspected that energetic internal tides were present, and the effects of these waves can be difficult to disentangle from the intensity records. They show up clearly in AFAR phase spectra as the peak just below 0.1 cph (Figure 6.7).

Like the Cobb and MATE experiments, the AFAR intensity spectra do not agree with weak fluctuation theory, and so only phase spectra from AFAR are shown here (Figure 6.7) (Flatté, 1983; Reynolds et al., 1985). (The intensity spectra will be examined in the next chapter on path integrals.) With regard to the phase spectra, relatively good agreement is obtained with the theory in the frequency range from 1/2 cph to N. However, between f and 1/2 cph the observed spectra are noticeably lower in energy than the theory. This effect has been ascribed to saturation interference effects that render significant differences in the phase from that expected from geometrical acoustics (Flatté, 1983). Another possibility is that the rough topography has attenuated the low-frequency internal waves, and there is some suggestion from the observed internal-wave frequency spectra that this is the case (Reynolds et al., 1985). As expected from weak fluctuation theory, the travel-time spectra are to a good approximation independent of acoustic frequency. The travel-time spectra show nearly power law behavior, which is typical of near-axial ray propagation (Figure 6.1).

6.5.3 AATE

The two experiments to be discussed next (AATE and AET-87) had the smallest variability, and are most appropriately described by weak fluctuation theory. The $\Lambda - \Phi$ diagram shows these observations to be well in the unsaturated regime and the predicted intensity moments agree well with the measurements (Table 6.1).

The 6.4-km range AATE was conducted in the Beaufort Gyre on a drifting ice camp in the spring of 1985. The experiment involved pulse transmissions at frequencies of 2, 4, 8, and 16 kHz (Ewart and Reynolds, 1990, 1993). AATE was in fact a component of the larger AIWEX experiment, whose goal was to characterize Arctic internal waves in the same way that the IWEX experiment characterized mid-latitude internal waves and tested some of the assumptions in the Garrett-Munk spectral model. (Both IWEX and AIWEX are discussed in Chapter 3.) The internal-wave field during AIWEX was found to be nonstationary with strong changes associated with wind events. The acoustic fluctuations were also strongly nonstationary, and the results discussed here are for a limited 57.4-hour period during which the ocean conditions were relatively stable (statistically). Other important considerations are that the internal-wave energy levels were significantly lower than GM and that the frequency and vertical wavenumber spectra show significant departures from GM shape (Figure 3.13). In fact, the internal-wave energy during the stable 57.4-hour period was one-fiftieth of the standard GM level.

The AATE spectra for the 8-kHz transmissions are shown in Figure 6.8. Both the phase and log-amplitude spectra are consistent with weak fluctuation theory. Not surprisingly, spectral predictions using standard GM parameters do not come close to the observations. Spectra at other frequencies (not shown) are also well described by weak fluctuation theory with the AIWEX spectrum. These results suggest that the Born approximation is accurate enough in the Arctic and that perhaps acoustic fluctuations could be used as a tool to measure internal-wave spectra via tomographic methods (Ewart et al., 1998).

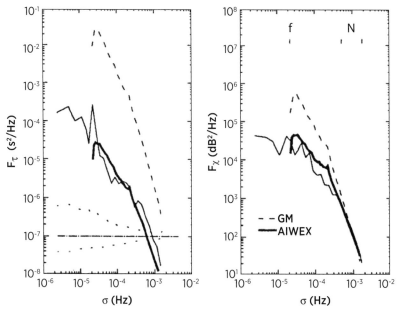

Figure 6.8. Phase and log-amplitude spectra from the AATE at 8 kHz and predictions from weak fluctuation theory. The GM predictions, shown with a dash, use standard mid-latitude parameter values. Predictions for the AATE (heavy solid) use the AIWEX derived internal-wave spectrum (Figure 3.13).
Source: Ewart and Reynolds (1993).

6.5.4 AET-87

Another example of the success of weak fluctuation theory for the intensity spectra are the results from the 75-Hz, 87-km transmissions of the ATOC Acoustic Engineering Test (AET) (Colosi et al., 2009). Like the AATE, the $\Lambda - \Phi$ diagram shows the AET to be well within the unsaturated regime (Figure 6.4) and there is good agreement between the predicted and measured intensity moments (Table 6.1).

In the AET, a 700-m-long vertical receiving array was used so that both frequency and vertical wavenumber spectra can be observed. In addition, two well-resolved time fronts were identified with IDs of +4 and −3 (Figure 6.3). The two arrivals have upper turning points in different parts of the water column and thus sample the internal-wave field quite differently (Colosi et al., 2009). The ray turning depths are between 225 and 350 m for ID +4, and between 90 and 140 m for ID −3. Figure 6.9 shows the frequency spectra of log-amplitude for the two resolved paths, and the agreement between model and observations is excellent. Here the ocean is adequately described by the standard GM spectral model. The smaller grazing angle time front ID +4 has more low-frequency variability than the larger grazing angle ID −3. This is due to the perpendicular wavenumber resonance, which imposes a cut-off frequency σ_L that is larger overall for ID

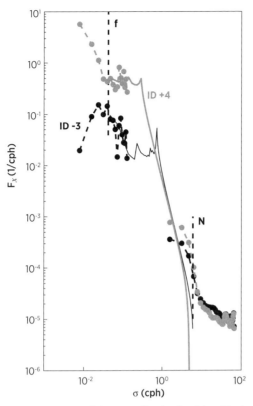

Figure 6.9. Frequency spectra of log-amplitude for identified time fronts with IDs +4 (gray), and −3 (black) from the 75-Hz, Acoustical Engineering Test observations at 87-km range. Observed spectra are displayed with a dashed line and weak fluctuation theory predictions are shown using a solid line. The GM spectrum with standard parameters was used in the predictions. The vertical dashed lines indicate the Coriolis and buoyancy frequencies. The gaps in the spectra are due to 3-h gaps in sampling.
Source: Colosi et al. (2009).

−3 than for ID+4. The frequency spectrum of phase for the AET has not been computed, because the phase tracking method used in the analysis could not overcome multihour gaps in the transmission schedule. Travel-time data were not considered.

The vertical wavenumber spectra from the AET are shown in Figure 6.10. There is again excellent agreement between weak fluctuation theory and observations. In these spectra, as in the examples in Figure 6.1, the effect of the Fresnel filter at low vertical wavenumber can be seen. The log-amplitude spectrum which is not sensitive to large scales, rolls off at small m, but the phase spectrum that is sensitive to large scales keeps growing at small m. At large m, the expected spectral shape of m^{-3} is seen, again due to the perpendicular wavenumber resonance.

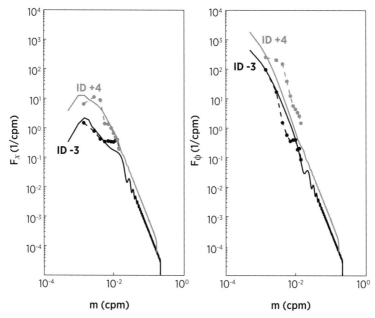

Figure 6.10. Vertical wavenumber spectra of phase (right) and log-amplitude (left) for the AET 87-km transmissions and predictions from weak fluctuation theory with the GM internal-wave model. The observations/predictions are shown with dashed/solid curves, and the two time fronts are in gray (ID +4) and black (ID −3).
Source: Colosi et al. (2009).

6.6 Summary

The results of weak fluctuation theory are quite useful, but there are limitations. Experimental evidence strongly suggests that the results for phase are applicable beyond the unsaturated regime and into the partially saturated regime, where phase and intensity fluctuations are quite large. At the same time, these observations show that the amplitude results break down somewhere between a log-intensity variance/*SI* of 0.51/0.44 (AET, ID +4) and 1.07/0.73 (MATE, 2 kHz) (Table 6.1). From these results it is evident that the limit of weak fluctuation theory as estimated in the atmospheric optic context is overly optimistic for the ocean. In that community the breakdown occurs near a log-intensity variance of roughly 1.2 (Clifford, 1978). Weak fluctuation theory also helps to understand, at first order, the effects of diffraction on both phase and amplitude. The theory also allows us to define the $\Lambda - \Phi$ diagram, which can be used as a guide in comparing different experiments. Clearly new tools are required to treat stronger fluctuations. In the next two chapters two successful techniques that have been developed will be presented, namely path integrals, which are a generalization of the ray-like methods of weak fluctuation theory, and coupled mode transport theory, which is more suitable for single-frequency signals.

7

Path Integral Theory

7.1 Introduction

In the previous chapter it was shown that weak fluctuation theory can adequately describe the phase fluctuations observed in short-range deep-water experiments. However, predicting intensity fluctuations is a much more challenging task, because the perturbation methods break down once significant levels of fluctuations are present. In the partially and fully saturated regimes, therefore, a new analytic tool is needed. The path integral technique is one such tool, and to the extent that the path integral is a complete solution to the parabolic wave equation many of the results are not limited to any regime. While developed within the context of quantum mechanics (Feynman and Hibbs, 1965), the method of path integrals is a powerful theoretical tool in many branches of science, including general relativity, statistical mechanics, diffusion/transport phenomena, and superfluidity (Schulman, 1981). Indeed many problems in wave propagation are well suited to the method as was recognized by Flatté et al. (1979).

The path integral formalism gives the acoustical pressure at frequency ω as

$$p(\mathbf{x}) = N \int D(path) \exp\left(i\Theta(path)\right), \tag{7.1}$$

where the integration is over all paths connecting the source and receiver, Θ is the accumulated phase over the path, and the factor N is a normalization. The path integral can be considered a generalization of the principle of the seventeenth-century Dutch physicist Christiaan Huygens, which states that every point along a wave front can be considered to be a virtual point source of waves that determines the subsequent positions of the wave front (Huygens, 1690). Much of the elegance and physical insight obtained from this method comes from the fact that in the high-frequency limit, the phase factor oscillates rapidly and therefore the primary contribution to the path integral comes from the paths with stationary phase. This is the ray theory limit. It is rather remarkable that the simple path integral equation (Eq. 7.1) involving phase and no amplitude factors is entirely equivalent to the parabolic wave equation. The interested reader is referred to the numerous textbooks discussing this matter within the quantum mechanics context (e.g., Feynman and Hibbs, 1965; Sakurai, 1985; Shankar et al., 1994).

As introduced in Chapter 2, the path integral generalizes the notion that important contributions to the acoustic field come from the neighborhood around a ray path, and that neighborhood can be conceptualized using the idea of a Fresnel zone and the spread of microrays. When $\mu = 0$, the principal contributions to the acoustic pressure come from paths within a Fresnel zone of the unperturbed ray because paths outside of this zone cancel. The primary concern, however, is the case in which $\mu \neq 0$. Consider the situation in which μ is small enough that the regime is weak fluctuations or unsaturated propagation. In the path integral view, only one ray path is dominant, but this ray and its associated ray tube are perturbed by the sound-speed fluctuations. In the unsaturated regime the intensity fluctuations are caused by the convergence and divergence of the ray tube, while the phase fluctuations are caused by the delay/advance of the energy as it travels through the variable sound-speed structure.

As the strength of μ grows or the propagation range increases, the acoustic pressure has contributions from many ray paths (microrays) each of which, due to the sound-speed fluctuations, is locally an extremum of the travel time. Indeed, Figures 5.13 and 5.14 show computed microrays though a numerical realization of internal-wave-induced sound-speed perturbations. Each microray has its own Fresnel zone that is perturbed by the fluctuations. The interference of these microrays leads the signal toward full saturation ($SI \rightarrow 1$), as discussed in Chapter 4.

The approach to saturation depends strongly on the relationship between the frequency dependent Fresnel zone (R_f), the frequency-independent vertical spatial spread of the microrays (R_{mr}), the vertical correlation length of the internal waves (L_z), and the rms phase or strength parameter Φ. In the case in which $R_{mr} \gg L_z$, $R_f \ll L_z$, and $\Phi \gg 1$, the microrays are independent, having traversed independent regions of sound-speed structure. These microrays will also have acquired sufficient phase variation to interfere strongly with one another. This sum of independent randomly phase waves forms the basis of full saturation (Chapter 4). For the case $R_{mr} \ll L_z$, $R_f \ll L_z$, and $\Phi \gg 1$, the large-scale structure of the medium fluctuations affects all the microrays together, so that the received pressure is the sum of correlated contributions. In this case there is a high probability that all the microrays will sum constructively to create strong focusing regions; this is the partially saturated regime in which the scintillation index approaches 1 from above. For the case in which $R_f \gg L_z$, there is strong diffraction that destroys the focusing ability of the medium, and thus the approach to saturation has the scintillation index approaching 1 from below.

There is a close connection between the path integral method and the method of moment equations (Uscinski, 1982) that has had some success in ocean acoustics applications (Codona et al., 1986b; Macaskill and Ewart, 1996). However, the path integral method has been found to be somewhat more general than moment equations because the Markov or ray-tangent approximation is not necessary (though the Markov approximation is generally used in the path integral method to obtain tractable results) (Codona et al., 1986b; Henyey and Macaskill, 1996). When the Markov approximation is made, path integral and moment equation results are identical. A further strength of the path integral is its geometric or intuitive view of the wave propagation physics, which can be compared to the more algebraic moment equation results that are not unlike those discussed in Chapter 8 with regard to mode transport theory. For these reasons the moment equation method will not be treated in this book. In one case, however, moment equation results will be shown that compare favorably to the MATE observations. Those interested in the moment method can refer to the excellent treatments by Uscinski (1982); Codona et al. (1986b); Henyey and Macaskill (1996), and Macaskill and Ewart (1996).

7.2 Path Integral Theory

Here some important aspects of path integrals are reviewed as a brief tutorial on the technique. Further material can be found in Feynman and Hibbs (1965); Sakurai (1985); and Shankar et al. (1994).

The path integral is a representation for the acoustic Green's function for the parabolic equation, and it is sometimes referred to as the propagator. The Green's function, $G(x, y, z; x_0, y_0, z_0)$, takes the PE acoustic pressure field in the plane (y_0, z_0) at range x_0 and propagates it to a new range x. Thus one writes

$$p(x, y, z) = \int dy_0 \, dz_0 \, G(x, y, z; x_0, y_0, z_0) p(x_0, y_0, z_0). \qquad (7.2)$$

In terms of the path integral, the propagator is written as

$$G(x, y, z; x_0, y_0, z_0) = N \int D[y(x), z(x)] \, \exp\left[i\Theta(\text{path}) \right], \qquad (7.3)$$

where the key factor here is the path phase function $\Theta(\text{path})$. If propagation with the Helmhotz equation were to be considered the path phase would be given by $\omega \int ds/c(\mathbf{r})$, where ds is an increment of path arc length. This phase is seen to be frequency times the path travel time. However, for the path integral, the starting point is the parabolic equation. Using the three-dimensional, PE Hamiltonian

(Eq. 2.46)

$$H = \frac{c_0(p_z^2 + p_y^2)}{2} + \frac{U(z) + \mu(\mathbf{r})}{c_0}, \tag{7.4}$$

the PE travel time equation becomes (Eq.2.49)

$$\frac{dT}{dx} = \frac{1}{c_0} \left(\frac{c_0^2 p_z^2}{2} + \frac{c_0^2 p_y^2}{2} - U(z) - \mu(\mathbf{r}) \right). \tag{7.5}$$

Here $p_z = \tan\theta/c_0$, $p_y = \tan\vartheta/c_0$, and $c(\mathbf{r}) = c_0(1 + U(z) + \mu(\mathbf{r}))$, where θ and ϑ are the vertical and transverse path angles relative to the x-direction.[1] Because this is a small-angle approximation, $c_0 p_z = dz/dx$ and $c_0 p_y = dy/dx$ and therefore the path phase is written as

$$\Theta(\text{path}) = \omega \int_{x_0}^{x} dx' L(y(x'), \dot{y}(x'), z(x'), \dot{z}(x')), \tag{7.6}$$

where the Lagrangian density is

$$L(y(x), \dot{y}(x), z(x), \dot{z}(x)) = \frac{1}{c_0} \left[\frac{\dot{y}^2}{2} + \frac{\dot{z}^2}{2} - U_0(z) - \mu(x, y, z) \right]. \tag{7.7}$$

Here the overdot represents differentiation with respect to x. The stationary phase points of the function Θ give the ray trajectories, that is, the Euler-Lagrange equations (Eq. 2.112). For the ocean acoustics problem, one is concerned with fluctuations for a particular time front arrival, so the path integral is analyzed in the neighborhood of a deterministic ray. In addition, the path integral is normalized using the unperturbed pressure from the deterministic path, so that for $\mu = 0$, $p = 1$. Because of these choices, the propagator itself can be used to represent the pressure field, as was written in Eq. 7.1.

   The form of the Lagrangian in the parabolic approximation allows a simplified notation that partitions the path integral into two parts, one having to do with the unperturbed phase along the path and one having to do with the sound-speed fluctuations along the path. The result is

$$p(\mathbf{r}) = N \int D(\text{path}) \exp\left(i\Theta_0(\text{path}) - iq_0 \int_0^R \mu(x, y(x), z(x)) \, dx \right), \tag{7.8}$$

where the unperturbed phase S_0 associated with the path is

$$\Theta_0(\text{path}) = q_0 \int_0^R \left[\frac{\dot{y}^2 + \dot{z}^2}{2} - U_0(z) \right] dx. \tag{7.9}$$

---

[1] Note here that the extra term of $1/c_0$ in Eq. 2.49 has not been included because the path integral is only a solution of the parabolic equation. Also in Eq. 7.5 the result $V = (1 - c_0^2/c^2) \simeq U(z) + \mu(z)$ has been used.

As previously mentioned, the factor $N^{-1} = \int d(path) e^{i\Theta_0(path)}$ is a normalization, so that $p = 1$ for $\mu = 0$.

7.2.1 A Simple Example

An example of the application of the path integral is useful to demonstrate the methodology. Consider propagation in the (x, z) plane with a constant background sound speed. Here it will be seen how the cylindrical free space Green's function is recovered. The path integral for the Green's function is written as

$$G(x, z; x_0, z_0) = N \int D(z(x)) \, \exp\left[i\frac{q_0}{2}\int_{x_0}^{x}\dot{z}^2 dx'\right], \tag{7.10}$$

and the Feynman paths are written as a sum of the ray path and a perturbation away from that path, that is, $z(x') = z_r(x') + \delta z(x')$. The ray path is given by

$$z_r(x') = z_0 + \frac{z - z_0}{x - x_0}(x' - x_0). \tag{7.11}$$

The deviations of the path away from the ray are key to the calculation. It proves useful to write the deviations as a sine series giving

$$\delta z(x') = \sum_{n=1}^{\infty} a_n \sin\left(n\pi\frac{x' - x_0}{x - x_0}\right), \tag{7.12}$$

which satisfies the boundary condition that the path deviation must vanish at the source and receiver (e.g., $\delta z(x_0) = \delta z(x) = 0$). With the perturbation written this way, the integration over all paths can be viewed as an integral over all possible a_n, that is,

$$N\int D(z(x)) = N\int_{-\infty}^{\infty} da_1 \int_{-\infty}^{\infty} da_2..... = N\prod_{n=1}^{\infty}\int da_n, \tag{7.13}$$

where N is the overall normalization factor. The expansion in Eq. 7.13 is an approach that will be used several times in this chapter to evaluate path integrals of interest. Using the expression for $z(x')$, the phase integral in Eq. 7.10 can be done noting that the sine functions are orthogonal with integration over x', yielding

$$\Theta(a_n) = \frac{q_0}{2}\left[\frac{(z - z_0)^2}{x - x_0} + \frac{1}{2}\sum_{n=1}^{\infty}\frac{(n\pi)^2}{x - x_0}a_n^2\right]. \tag{7.14}$$

The path integral then reduces to the form

$$G(x, z; x_0, z_0) = N\prod_{n=1}^{\infty}\int_{-\infty}^{\infty}\exp[i\Theta(a_n)] \, da_n, \tag{7.15}$$

which is an infinite product of Fresnel Integrals (Abramowitz and Stegun, 1964, Chapter 7) that can be evaluated to give

$$G(x,z;x_0,z_0) = N \exp\left(i\frac{q_0}{2}\frac{(z-z_0)^2}{x-x_0}\right)\prod_{n=1}^{\infty}\left(-\frac{iq_0(n\pi)^2}{4\pi(x-x_0)}\right)^{-1/2}. \qquad (7.16)$$

The product term in this equation is not easy to evaluate and there is still the issue of the normalization constant. Thus it is useful to write

$$G(x,z;x_0,z_0) = N'(x-x_0)\exp\left(i\frac{q_0}{2}\frac{(z-z_0)^2}{x-x_0}\right), \qquad (7.17)$$

where by virtue of Eq. 7.16 the new constant N' can only be a function of $x - x_0$. This constant is evaluated using the composition property[2] of the Green's function (Sakurai, 1985), which simply means that the propagation interval from x_0 to x can be divided into two pieces such that

$$\int_{-\infty}^{\infty} dz' G(z,x;z',x')G(z',x',z_0,x_0) = G(z,x;z_0,x_0). \qquad (7.18)$$

Using the composition property the result is that $N'(x-x_0) = \sqrt{q_0/2\pi i(x-x_0)}$. The parabolic equation Green's function is thus

$$G(x,z;x_0,z_0) = \sqrt{\frac{q_0}{2\pi i(x-x_0)}}\exp\left(i\frac{q_0}{2}\frac{(z-z_0)^2}{x-x_0}\right). \qquad (7.19)$$

The skeptical reader can substitute this expression into the parabolic equation and show that a solution has been obtained. Recall that the parabolic equation is for the reduced pressure, that is, a factor of e^{iq_0x} has been removed. Thus to obtain the cylindrical Green's function for the total pressure requires an extra term $e^{iq_0(x-x_0)}$, giving the required result of a cylindrically spreading wave for small-angle propagation primarily in the x-direction.

As the reader can see from this trivial example, the path integral method is simple in concept but rather onerous in execution. Because this monograph is not intended for those with expertise in path integrals, the material will be presented in a self-contained way as much as possible.

7.2.2 A Note on the PE Approximation

While the theoretical development of the path integral method relies upon the parabolic approximation, it is not unreasonable to assert that the results apply equally well for the Helmholtz equation, although no such wide-angle formulation

---

[2] Equation 7.18 is sometimes referred to as the Smoluchowski-Kolmogorov relation from its application to probability and diffusion theory.

has been established (Schulman, 1981). There are many issues associated with writing a path integral solution for the Helmholtz equation (Henyey, 1997), but a main issue in the present topic is the fact that for the Helmholtz equation, the Lagrangian density (and therefore the path integral) does not factor into the sum of deterministic and stochastic parts such as shown in Eq. 7.8. It will be seen in the subsequent development that this additive structure of the path integral is critical for carrying out the ensemble averages to obtain acoustic field statistics. Physically, however, the main approximation with respect to the starting wave equation is that there is weak, small-angle forward scattering. It appears irrelevant whether the ray paths around which the scattering occurs are of small or moderate angle. Therefore, in the following treatment of acoustic statistics the derivations will utilize the parabolic approximation, but actual calculations will be carried out using ray paths and ray-tube functions computed using the Helmholtz equation and related Hamiltonian functions. This approach was also described in Chapter 6 related to the issue of Fresnel zone computations in weak fluctuation theory and the $\Lambda - \Phi$ diagram.

With this background, acoustic fluctuations using path integral theory can be treated. The presentation will start with low-order, relatively easy to solve moments of the field such as the mean pressure and field coherence, and then move onto to higher order more computationally difficult moments such as intensity variance and correlations.

7.3 Mean Pressure

The mean pressure is the simplest moment to observe and is generally obtained with a suitable time average. Using the path integral, the mean pressure is obtained by taking the expectation value of Eq. 7.8. If the reasonable assumption is made that μ is a zero mean Gaussian random variable (ZMGRV), then the expectation values can be moved into the exponent using the well known result $\langle e^{i\alpha} \rangle = e^{-\langle \alpha^2 \rangle /2}$ for α a ZMGRV. The mean pressure is then

$$\langle p(\mathbf{r}) \rangle = N \int D(path) \exp\left(i\Theta_0(path) - \frac{1}{2}\left\langle \left(q_0 \int \mu \, dx \right)^2 \right\rangle \right),$$

$$= e^{-\Phi^2/2} N \int D(path) \exp\left(i\Theta_0(path) \right) = e^{-\Phi^2/2}, \qquad (7.20)$$

where the second line follows from the assumption that the phase variance,

$$\Phi^2 = \left\langle \left(q_0 \int \mu \, dx \right)^2 \right\rangle, \qquad (7.21)$$

is the same for all paths, and the last step comes from the definition of the normalization N. In the partially and fully saturated regimes, $\Phi > 1$, so $\langle p \rangle$ is exponentially small and not particularly interesting.

7.4 Mutual Coherence Functions: Space and Time Separations

An important field quantity, which is a second moment, is the coherence function $\langle p(1)p^*(2) \rangle$, where (1) and (2) represent different space–time endpoints of the paths (1) and (2) at the same frequency. Coherence across frequency and pulse propagation will be discussed in the next section. As described in Chapter 4, the coherence dictates the gain that can be achieved using coherent spatial or temporal processing.

Using Eq. 7.8 the coherence function is seen to be a double-path integral of the form (Dashen et al., 1985)

$$\langle p(1)p^*(2) \rangle = |N|^2 \int D^2(path) \exp\left(i(\Theta_0(1) - \Theta_0(2)) - \frac{V_{12}}{2} \right), \qquad (7.22)$$

$$V_{12} = q_0^2 \left\langle \left(\int_{path_1} \mu \, dx - \int_{path_2} \mu \, dx \right)^2 \right\rangle, \qquad (7.23)$$

where the fact that μ is a ZMGRV has again been used. The quantity V_{12} is seen to be a phase structure function for the two paths. To evaluate this double-path integral, a few important assumptions must be made. First it is assumed that the important contributions to Eq. 7.22 are from microrays that remain somewhat close to the unperturbed ray path. This is done by defining a ray path $z_r(x)$ $(y_r(x) = 0)$ that obeys the ray equations (in the parabolic approximation, Eq. 2.112)

$$\ddot{z}_r + U'(z_r) = 0, \qquad (7.24)$$

where U' is the vertical derivative evaluated along the ray path and the boundary conditions are $z_r(0) = z_s$ and $z_r(R) = (z(1) + z(2))/2$. Here again an overdot means total derivative along the path with respect to range x. This approach of expanding around the unperturbed ray will have some limitations due to ray chaos. Nonetheless, the approach is to define two new path variables, $v_1(x) = z_1(x) - z_r(x)$ and $v_2(x) = z_2(x) - z_r(x)$ that give the vertical separation of paths (1) and (2) relative to the unperturbed ray. Expanding the waveguide variation to second order such that $U(z_1) \simeq U(z_r) + U'v_1 + U''v_1^2/2$ and similarly for $U(z_2)$, the difference in path phases $\Delta\Theta = \Theta_0(1) - \Theta_0(2)$ becomes

$$\Delta\Theta = q_0 \int_0^R dx \, \frac{1}{2} \left[\dot{y}_1^2 - \dot{y}_2^2 + \dot{v}_1^2 - \dot{v}_2^2 + 2\dot{z}_r(\dot{v}_1 - \dot{v}_2) \right]$$

$$- U'(z_r)(v_1 - v_2) - \frac{1}{2}U''(z_r)(v_1^2 - v_2^2). \qquad (7.25)$$

A further change of variables is useful assuming that the fluctuation in path phase $q_0 \int_{path} \mu \, dx$ is a stationary random variable, and thus the structure function V_{12} will only be a function of the difference of the path coordinates. Taking this approach, the new path variables are written as

$$\alpha(x) = y_1 - y_2, \ \beta(x) = (y_1 + y_2)/2, \ u(x) = v_1 - v_2, \ w(x) = (v_1 + v_2)/2, \qquad (7.26)$$

so that the phase difference becomes

$$\Delta\Theta = q_0 \int_0^R \left(\dot{\alpha}\beta + \dot{u}\dot{w} + \dot{z}_r\dot{u} - U'u - U''uw \right) dx. \qquad (7.27)$$

Since V_{12} is assumed not to be a function of the mean coordinates β and w, these two path integrals are given by

$$I_\beta = \int D(\beta)\exp\!\left(iq_0 \int_0^R dx \, \dot{\alpha}\beta\right), \qquad (7.28)$$

$$I_w = \int D(w)\exp\!\left(iq_0 \int_0^R dx \, (\dot{u}\dot{w} - U''(z_r)uw)\right), \qquad (7.29)$$

and can be done analytically (Dashen et al., 1985). Here the notation $\int D(\beta)$, $\int D(w)$ means that only the integrals over the β and w coordinates of the total path integral are being done. The β path integral is done using integration by parts on the x integral in the exponent and by subsuming the boundary terms into the normalization constant. This then gives

$$I_\beta = \int D(\beta)\exp\!\left(-iq_0 \int_0^R dx \, \ddot{\alpha}\beta\right). \qquad (7.30)$$

Because β only appears linearly in the exponent, this path integral is simply a product of delta functions in analogy to the familiar relation (Appendix A and Dashen (1979))

$$\int_{-\infty}^{\infty} e^{iax}dx = 2\pi\delta(a), \qquad (7.31)$$

The w path integral is done in a similar manner, but here some of the terms cancel by virtue of the ray equations. The result is

$$I_w = \int D(w)\exp\!\left(-iq_0 \int_0^R dx \, (\ddot{u} + U''u)w\right), \qquad (7.32)$$

which again is a product of delta functions due to the linear dependence on w. Because of the delta functions, these two path integrals require that the following equations be satisfied

$$\frac{d^2\alpha}{dx^2} = 0, \tag{7.33}$$

$$\frac{d^2u}{dx^2} + U''(z_r)u = 0. \tag{7.34}$$

These equations are equal to zero only along the unperturbed ray. Here the boundary conditions are $\alpha(0) = 0$, $\alpha(R) = y(1) - y(2)$, $u(0) = 0$, and $u(R) = z(1) - z(2)$. Equations 7.33 and 7.34 are simply the raytube equations for the y and z separations of the two rays starting at the source and ending at the two receiver points (1) and (2). As has been seen previously, the depth separation equation is somewhat complicated because of the structure of the sound channel, but the y equation can be solved analytically to yield $\alpha(x) = \alpha(R)x/R$, which is just the separation between two straight-line paths.

Again, because of the delta functions, the β and w path integrals have dictated that the other two path integrals, α and u, are to be evaluated only along the unperturbed ray. Therefore all the path integrals have been done, with the result that

$$\langle p(1)p^*(2)\rangle = \exp\left(-\frac{D(1,2)}{2}\right). \tag{7.35}$$

The task is now to evaluate the structure function $V_{12} = D(1,2)$ along the unperturbed ray path z_r with the appropriate ray separations dictated by Eqs. 7.33 and 7.34. Importantly, there are no amplitude correlation terms in this normalized coherence function (Chapter 4, Eqs. 4.60 and 4.61).

Thus the relationship between coherence and the phase structure function seems to be quite general, where the main approximations have been expansion about the deterministic ray and that V_{12} is not a function of β nor w. For the horizontally isotropic GM spectrum, V_{12} would not be expected to be a function of β, but the inhomogeneity of internal waves in the vertical puts the assumption that V_{12} is not a function of w on much flimsier ground. Furthermore, the existence of ray chaos means that the unperturbed ray path is unstable to internal-wave-induced sound-speed fluctuations, and the growth rate of that instability is of order $1/50$ km to $1/500$ km for deep-water propagation (Chapter 5). Thus the expansion around the unperturbed ray is expected to break down at long ranges of order a few e-folding distances of the growth rate. Presently, little is known of how this breakdown occurs and how the coherence function is modified.

It is also worth noting that the result for the mutual coherence function will be used for higher order moments in the saturated regime, where the pressure is a Gaussian random variable. In this case, the higher order moments are related in a simple way to this second moment.

7.4.1 A Broadband Microray Coherence Theory

It is useful to contrast the path integral coherence result to a kinematic calculation in which the acoustic field is considered to be an interference pattern of many microrays that have some pulse extent. The path integral theory that was just developed did not consider signal bandwidth, and therefore the results from the path integral must be interpreted in terms of narrowband concepts. Here a simple model is presented in which broadband effects can be demonstrated.

As was introduced in Section 4.1.4, consider a demodulated broadband acoustic field to be the sum of N Gaussian pulses of the form (Colosi and Baggeroer, 2004; Colosi et al., 2005) (Eq. 4.8)

$$p = \sum_{k=1}^{N} a_k \exp\left[-\frac{\gamma^2 \delta\Theta_k^2}{2} + i\delta\Theta_k\right]. \tag{7.36}$$

Here N is the number of microrays, a_k is a stochastic microray amplitude, $\gamma = \Delta\omega/\omega_0$ is the fractional bandwidth, and $\delta\Theta_k = \omega_0\delta t_k$ is a zero mean travel-time induced phase fluctuation. It is also understood that the pressure is evaluated at the mean arrival time of the pulses. The coherence function for observations of pressure at some separation Δ is therefore

$$\langle pp^*(\Delta)\rangle = \sum_{k=1}^{N}\sum_{j=1}^{N}\left\langle a_k a_j(\Delta) \exp\left[-\frac{\gamma^2}{2}(\delta\Theta_k^2 + \delta\Theta_j^2(\Delta)) + i(\delta\Theta_k - \delta\theta_j(\Delta))\right]\right\rangle. \tag{7.37}$$

Assuming that the microrays are uncorrelated and that for each microray the amplitude and phase are uncorrelated, the following conditions can be imposed:

$$\langle a_k a_j\rangle = \langle a\rangle^2 \ (k \neq j), \qquad \langle a_k a_k(\Delta)\rangle = \langle aa(\Delta)\rangle,$$
$$\langle \delta\Theta_k \delta\Theta_j\rangle = \Phi^2 \delta_{k,j}, \text{ and } \langle a_k \delta\Theta_j\rangle = 0. \tag{7.38}$$

To move forward analytically, there is no need to impose constraints on the PDF of the microray amplitudes, a_k and $a_k(\Delta)$, but it is necessary to assume a Gaussian joint PDF for the microray phase, $\delta\Theta_k$ and $\delta\Theta_k(\Delta)$, that is,

$$P(\delta\Theta_k, \delta\Theta_k(\Delta); \rho) = \frac{1}{2\pi\Phi^2 \sqrt{1-\rho^2}} \exp\left[-\frac{\delta\Theta_k^2 - 2\rho\delta\Theta_k\delta\Theta_k(\Delta) + \delta\Theta_k(\Delta)^2}{2\Phi^2(1-\rho^2)}\right]. \tag{7.39}$$

Here $\rho(\Delta) = \langle \delta\Theta_k\delta\Theta_k(\Delta)\rangle/\Phi^2$ is the autocorrelation coefficient that is the same for all microrays. Putting it all together, the normalized coherence is found to be (see

Colosi et al., 2005, for details)

$$\frac{\langle pp^*(\Delta)\rangle}{\langle I\rangle} = \frac{1}{\langle I\rangle}\left(\frac{N\langle aa(\Delta)\rangle\, G(\rho)}{\sqrt{(1+\gamma^2\Phi^2)^2 - \gamma^4\Phi^4\rho^2}} + \frac{N(N-1)\langle a\rangle^2 G(0)}{1+\gamma^2\Phi^2}\right), \quad (7.40)$$

$$G(\rho) = \exp\left[-\frac{D(\rho)}{2+\gamma^2 D(\rho)}\right], \quad (7.41)$$

where $D(\rho) = 2\Phi^2(1-\rho)$ is the phase structure function and the mean intensity is given by

$$\langle I\rangle = \frac{N\langle a^2\rangle}{\sqrt{1+2\gamma^2\Phi^2}} + \frac{N(N-1)\langle a\rangle^2 G(0)}{1+\gamma^2\Phi^2}. \quad (7.42)$$

The coherence function depends on the time-bandwidth parameter $\gamma\Phi$, which is a measure of the degree to which the pulses overlap and thus interfere. In the narrowband limit ($\gamma = 0$) and for $\Phi \gg 1$ the familiar result is obtained

$$\frac{\langle pp^*(\Delta)\rangle}{\langle I\rangle} = \frac{\langle aa(\Delta)\rangle}{\langle a^2\rangle}\exp\left[-\frac{D(\rho)}{2}\right]. \quad (7.43)$$

This is close to the path integral result but with amplitude correlation terms, and it is the form that was observed in some long-range, low-frequency experiments (Colosi et al., 2005).

7.4.2 Evaluation of Phase Structure Function

The phase structure function for the two paths Γ_1 and Γ_2 is written as

$$D(1,2) = q_0^2\left(\langle(\int_{\Gamma_1}\mu\, ds)^2\rangle + \langle(\int_{\Gamma_2}\mu\, ds)^2\rangle - 2\int_{\Gamma_1}\int_{\Gamma_2}\langle\mu(s_1,t_1)\mu(s_2,t_2)\rangle\, ds_1\, ds_2\right),$$

$$\simeq 2\Phi^2 - 2q_0^2\int_{\Gamma_1}\int_{\Gamma_2}\langle\mu(s_1,t_1)\mu(s_2,t_2)\rangle\, ds_1\, ds_2, \quad (7.44)$$

where the phase variances over the two paths are both assumed equal to Φ^2 and $\langle\mu(s_1,t_1)\mu(s_2,t_2)\rangle$ is the spatial and temporal correlation function of the fractional sound speed. The second term in Eq. 7.44 is twice the phase correlation function. As discussed in Chapter 5, the correlation function of μ can be expressed in terms of the internal-wave spectrum F_μ giving

$$\langle\mu\mu(\Delta x, \Delta y, \Delta z, \Delta t)\rangle = \sum_{j=1}^{J}\int_0^\infty d\kappa_h\int_0^{2\pi}\frac{d\varphi}{2\pi}F_\mu(j,\kappa_h)\cos(\sigma_j(\kappa_h)\Delta t)\cos(m(j)\Delta z)$$

$$\cos(\kappa_h\cos\varphi\Delta x)\cos(\kappa_h\sin\varphi\Delta y), \quad (7.45)$$

where $\sigma_j(\kappa_h)$ is the internal-wave dispersion relation, and it is understood that the spatial separation variables Δx, Δy, Δz are range-dependent separations between the rays to the two receivers.

The most accurate evaluation of the phase structure function involves carrying out the double integrals in Eq. 7.44. However, the ray-tangent approximation and consequential perpendicular wavenumber resonance condition leads to some important insight. As previously mentioned, the ray-tangent approximation is often referred to as the Markov approximation. Several authors have identified the small parameter

$$\epsilon = q_0^2 \langle \mu^2 \rangle L_p^2 / 4 = \frac{\Phi^2 L_p}{4R} \ll 1, \tag{7.46}$$

as being a sufficient condition for the validity of the Markov approximation (Codona et al., 1986b; Rytov et al., 1989), but there is some feeling that this condition is too restrictive (Henyey and Ewart, 2006). A more useful condition is

$$\frac{dL_p}{dx} \ll 1. \tag{7.47}$$

In any case, using the geometrical acoustics results for the phase spectrum $F_\phi(j,\sigma)$ from Chapter 6, the phase structure function in the Markov approximation can be written using the fact that the spectrum and correlation function are cosine transform pairs. The result is

$$D(1,2) = 2\pi q_0^2 \int_\Gamma ds \sum_{j=1}^{J} \int_{\sigma_L}^{N} d\sigma F_\mu(\kappa_\perp(\sigma, j)) \Big[1 - \cos(\sigma_j(\kappa_h)\Delta t)$$

$$\cos(m(j)\Delta z)\cos(l\Delta y) \Big], \tag{7.48}$$

where $(k,l,m) = (-m\tan\theta, l, m)$ are the perpendicular wavenumbers, and l is derived from the dispersion relation. Here the single-path integration over the unperturbed ray path Γ is from the source to the mean of the receiver locations. The spectrum under the perpendicular-wavenumber constraint is

$$F_\mu(\kappa_\perp(\sigma, j); z) = \langle \mu^2(z) \rangle \frac{4N_0 B \, H(j)}{\pi^2} \frac{f}{j} \frac{\sigma^2 - f^2}{\sigma^3} \left(\frac{\sigma^2 - f^2}{\sigma^2 - \sigma_L^2} \right)^{1/2}, \quad \sigma_L < \sigma < N \tag{7.49}$$

where $\sigma_L = \sqrt{f^2 + N^2 \tan^2 \theta}$ is the low-frequency cut-off.

An equivalent form of the structure function, which is common in the literature (Flatté et al., 1979; Esswein and Flatté, 1981) but is a little more cumbersome, utilizes the L_p notation and is given by

$$D(1,2) = 2q_0^2 \int_\Gamma ds \langle \mu^2 \rangle L_p(s)\{1 - \cos(\sigma_j(\kappa_h)\Delta t)\cos(m(j)\Delta z)\cos(l\Delta y)\}. \quad (7.50)$$

The curly brackets denote averages over the internal-wave spectrum, but the average is only over those waves whose wavenumbers are perpendicular to the local ray slope (Esswein and Flatté, 1981). Explicitly the spectral averages are given by

$$\{a(\Delta;\theta,z)\} = \frac{\sum_{j=1}^J \int_{\sigma_L}^N d\sigma \, \frac{F_\mu(j,\sigma)}{j(\sigma^2-\sigma_L^2)^{1/2}} \, a(\Delta;\theta,z)}{\sum_{j=1}^J \int_{\sigma_L}^N d\sigma \, \frac{F_\mu(j,\sigma)}{j(\sigma^2-\sigma_L^2)^{1/2}}}, \quad (7.51)$$

where $F_\mu(j,\sigma)$ is the internal-wave spectrum in terms of mode number j and frequency σ, and σ_L and N are as defined in Chapter 6. Physical insight into the behavior of the structure function under the ray-tangent approximation for various receiver separations is best obtained by examining the time, depth, and horizontal separations independently.

7.4.3 Depth Separations

Here consider the case in which $\Delta t = \Delta y = 0$. For receivers separated in depth, the range-dependent separation of the two rays is given by the solution of Eq. 7.34 so that $\Delta z(s) = \Delta z(R)\xi_1(x)$ (see Chapter 2, Appendix A). Of course the coherence function could be obtained by integrating Eq. 7.50 for a collection of receiver separations but important analytic insight is obtained by making some further approximations. If the ray separation remains small relative to the vertical correlation length of the internal waves, $\hat{L}_z$, over the propagation range then the cosine in Eq. 7.50 can be expanded, yielding a quadratic structure function $D(1,2) \simeq (\Delta z(R)/z_0)^2$. The resulting expression for the vertical coherence length z_0 is then (Flatté and Stoughton, 1988)

$$\frac{1}{z_0^2} = q_0^2 \int_\Gamma ds \langle \mu^2(s) \rangle \, L_p(s) \, \{m^2(j)\}\xi_1^2(s), \quad (7.52)$$

where the WKB form $m(j) \simeq \pi j N(z)/N_0 B$ is usually used with the GM spectrum to give

$$\{m^2\} = \left(\frac{\pi N}{N_0 B}\right)^2 \frac{\langle j \rangle}{\langle j^{-1} \rangle}. \quad (7.53)$$

This term is recognized from the definition of Λ. The quantity z_0 is the vertical coherence length since a receiver separation of z_0 yields a decay of the coherence function by $e^{-1/2}$. Note that z_0^{-2} senses the internal-wave field with the weight

$\langle\mu^2\rangle L_p$; that is, the biggest contributions to de-coherence occur near the upper turning points. In addition, there is the weight $\{m^2(j)\}\xi_1^2(s) \simeq \xi_1^2(s)/\hat{L}_z^2$, which determines if the nearby rays are spread vertically by more than the vertical correlation length of the internal waves, $\hat{L}_z$. Equation 7.52 reveals that the vertical coherence length z_0 scales as $1/\sqrt{R}$ and $1/\omega$. It can also be seen from this result that problems exist when the receiver is near a caustic, that is, when ξ_1 is undefined (see Chapter 2).

Another important consideration here is that the quantity $\{m^2(j)\}$ depends on the small-scale structure of the internal-wave spectrum, because of the dependence on $\langle j\rangle$ which is divergent if an infinite number of internal-wave modes are considered. A similar situation occurs for the rms ray angle discussed in Chapter 5; the diffraction parameter discussed in Chapter 6; and, as shall be seen, the horizontal coherence length.

Examination of an axial straight line ray is instructive. For such a ray the result is $\xi_1(x) = \sin(K_a x)/\sin(K_a R)$, and $L_p(\theta = 0, z_a) = 4N_0 B\langle j^{-1}\rangle/(\pi^2 f)$. Here $K_a = 2\pi/R_L$ is the wavenumber of an axial sinusoidal ray, where for the Munk canonical profile $K_a = \sqrt{2\gamma_a/B}$. So the vertical coherence length is given by

$$z_0 = \frac{1}{q_0}\left[\langle\mu^2\rangle\frac{\langle j\rangle N^2}{fN_0 B}\frac{K_a R - \sin(K_a R)\cos(K_a R)}{K_a \sin^2(K_a R)}\right]^{-1/2} \simeq \frac{1}{q_0}\left[\langle\mu^2\rangle\frac{2\langle j\rangle N^2 R}{fN_0 B}\right]^{-1/2},$$

(7.54)

where it is understood that the depth-dependent quantities are evaluated at the depth of the sound-channel axis. In the last line of Eq. 7.54 the oscillation in z_0 is removed due to the periodic caustic structure, and a long-range approximation has been made (e.g., $K_a R \gg 1$). Some representative coherences can be computed using typical values for the mid-latitude ocean at the sound-channel axis, that is, take $\langle\mu^2\rangle = 2.0 \times 10^{-8}$, $N = 2$ cph, $N_0 B = 8.73$ rad m/s, $J = 500$, $K_a = 2\pi/41.8$ rad/km, and a latitude of $30°$. Using a frequency of 250 Hz and a range of 100 km, the result is $z_0 = 28$ m, which can easily be re-scaled for different frequencies, ranges, and other parameter changes. Note in this case $\Phi = 10$ rad, so using the rule of thumb from Chapter 4 where $z_0 \simeq \hat{L}_z/\Phi$ with $\hat{L}_z \simeq 100$ then $z_0 \simeq 10$ m which is the correct order of magnitude.

7.4.4 Time Separations

Next consider time separations alone so that $\Delta z = \Delta y = 0$. Again the cosine function in Eq. 7.50 is expanded, yielding a quadratic structure function $D(1, 2) \simeq (\Delta t/t_0)^2$ with the coherence time t_0 given by (Flatté and Stoughton, 1988)

$$\frac{1}{t_0^2} = q_0^2 \int_\Gamma ds\langle\mu^2(s)\rangle L_p(s)\{\sigma^2\},$$

(7.55)

where as usual the WKB form of the dispersion relation is used. As with the vertical coherence length, t_0^{-2} senses the internal-wave field with the weight $\langle\mu^2\rangle L_p$, and in this case the internal-wave correlation time $\{\sigma^2\} \simeq 1/\hat{T}_{iw}^2$ dictates the loss of acoustic coherence (see Chapter 4). Using the GM spectrum, analytic formulas exist for the quantity $\{\sigma^2\}$ (Esswein and Flatté, 1981) and these cumbersome formulas are given in Appendix C.

The result for an axial ray is instructive. In this limit, the spectrum averaged squared frequency is $\{\sigma^2\} = 2N^2 f^2 \ln(N/f)/(N^2 - f^2)$ so the coherence time is[3]

$$t_0 = \frac{1}{q_0}\left[\langle\mu^2\rangle\frac{8}{\pi^2}N_0 B\langle j^{-1}\rangle\frac{N^2 f}{N^2 - f^2}\ln\left(\frac{N}{f}\right)R\right]^{-1/2}. \qquad (7.56)$$

As with the depth coherence the time coherence scales as $1/\sqrt{R}$ and $1/\omega$. Time coherence, like rms phase Φ, depends on $\langle j^{-1}\rangle$ and is therefore sensitive to low mode internal waves. Using the typical mid-latitude, axial ocean values quoted in the previous section, the result is $t_0 = 12.6$ min for 250 Hz at 100-km range.

7.4.5 Horizontal Separations

Finally the case of horizontal separations transverse to the acoustic path with $\Delta z = \Delta t = 0$ is treated. Making the small separation approximation again yields a structure function quadratic in the separation (i.e., $D(\Delta y(R)) \simeq \Delta y^2(R)/y_0^2$). The horizontal coherence length y_0 is given by (Colosi, 2013)

$$\frac{1}{y_0^2} = q_0^2 \int_\Gamma ds\langle\mu^2\rangle L_p(s)\{l^2\}(x/R)^2. \qquad (7.57)$$

As with the other coherences, y_0^{-2} senses the internal-wave field with the weight $\langle\mu^2\rangle L_p$, and in this case the internal-wave horizontal correlation length $\{l^2\} \simeq 1/\hat{L}_H^2$ dictates the loss of coherence (see Chapter 4). The geometric term x/R reflects the fact that the rays separate in y linearly with range. A formula for $\{l^2\}$ using the GM spectrum is given in Appendix C, where it is shown that $\{l^2\} \propto \langle j\rangle$. It is seen then that y_0, like z_0, will be sensitive to small vertical scale internal waves.

The case for an axial ray is again done easily with the result (Colosi, 2013)[4]

$$y_0 = \frac{1}{q_0}\left[\langle\mu^2\rangle\frac{8}{3}\frac{f\langle j\rangle}{N_0 B}\left(\frac{N^2}{N^2 - f^2}\ln\left(\frac{N}{f}\right) - \frac{1}{2}\right)R\right]^{-1/2}. \qquad (7.58)$$

Here it is seen yet again that the coherence scales as $1/\sqrt{R}$ and $1/\omega$. Using typical mid-latitude axial ocean values quoted above the result is that $y_0 = 745$ m for a frequency of 250 Hz and a range of 100 km.

---

[3] In most cases the term $N^2/(N^2 - f^2)$ can be well approximated by 1.

[4] There is a missing factor of $\langle j^{-1}\rangle$ in Colosi (2013).

It is important to mention that an empirical approach that has been widely used was described by Flatté and Stoughton (1988), where the spectral average is approximated by

$$\{1 - \cos(k_y \Delta y(s))\} \simeq \left(\frac{\Delta y(s)}{(\Delta y^2(s) + y_h^2)^{1/2}}\right)^{3/2}, \quad y_h = \frac{L_h f}{\sigma_L}. \tag{7.59}$$

Here $\Delta y(s) = \Delta y(R) x/R$ is the $y-$ separation of the rays along the ray paths, and y_h is an empirical variable with $L_h = 12$ km. At small separations (i.e., $\Delta y(s) \ll y_h$) evaluation of Eq. 7.59 gives $D(\Delta y(R)) \simeq (\Delta y(R)/y_0)^{3/2}$ with

$$\frac{1}{y_0^{3/2}} = 2q_0^2 \int ds \langle \mu^2 \rangle L_p(s) \left(\frac{x}{R}\right)^{3/2} y_h^{-3/2}. \tag{7.60}$$

This equation has the drawback that the frequency and range scaling of y_0 are $\omega^{-4/3}$ and $R^{-2/3}$, which are markedly different from the scalings of the other coherences(Colosi, 2013). In particular, there is no good reason to expect that coherence scales as anything but one over frequency and one over square root range. In addition, the scaling with other environmental parameters is different from Eq. 7.57.

7.4.6 Relation to Signals with Multiple Deterministic Paths

Observationally in many cases one measures signals that are an interference pattern of many deterministic paths, each of which is perturbed by the internal waves. This is the case for single-frequency transmissions, but it is also the situation for broadband transmissions in shallow-water (Figure 1.9) and in the deep-water finale (Figure 5.10). There is a rather simple relationship that relates this case to the results for individual paths that have just been treated.

Consider the acoustic field to be an interference of M fluctuating deterministic paths given by

$$p = \sum_{m=1}^{M} a_m \exp(i(\bar{\Theta}_m + \delta\Theta_m)), \tag{7.61}$$

where $\bar{\Theta}_m$ and $\delta\Theta_m$ are the mean and phase fluctuation of the mth path, and a_m is a stochastic amplitude. The coherence function is then written as

$$\langle p(1)p^*(2)\rangle = \sum_m \sum_j \langle a_{m1}a_{j2}\rangle \exp\left[-\frac{\langle(\delta\Theta_{m1} - \delta\Theta_{j2})^2\rangle}{2}\right]\exp(i(\bar{\Theta}_{m1} - \bar{\Theta}_{j2})), \tag{7.62}$$

where it has been assumed that the phase and amplitude are uncorrelated, and the phase fluctuations are ZMGRVs. If it is further assumed, quite reasonably, that the deterministic path fluctuations are uncorrelated then the result is

$$\langle p(1)p^*(2)\rangle = \sum_m \langle a_m^2\rangle \exp\left(-\frac{D(1,2)_m}{2}\right)$$

$$+ \sum_m \sum_{j,j\neq m} \langle a_m\rangle\langle a_j\rangle \exp\left(-\frac{\Phi_m^2 + \Phi_j^2}{2}\right) \exp(i(\bar{\Theta}_{m1} - \bar{\Theta}_{j2})).$$

$$(7.63)$$

Here $D(1,2)_m$ and Φ_m^2 are the phase structure function and phase variance for the mth path. The second term will likely be small because the sum over the average phases will produce cancellation, but if the phase variances are large then the second term is indeed negligible. In this case the coherence function is seen to be an average over the coherence functions of the deterministic paths, that is,

$$\langle p(1)p^*(2)\rangle \simeq \sum_m \langle a_m^2\rangle \exp\left(-\frac{D(1,2)_m}{2}\right). \tag{7.64}$$

A similar result was obtained by Dyson et al. (1976).

Another important consequence of Eq. 7.64 is that when a quadratic approximation to the structure function is used, that is $D(\Delta) = \Delta^2 v_\Lambda^2$, then the coherence function is seen to be a Gaussian. Here v_Λ is the phase rate or inverse of the coherence time or length for the particular separation being considered. In this case the spectrum of the real and imaginary parts of the acoustic pressure will also be Gaussian.

7.4.7 Observations: Single Path

The literature on observations of ocean acoustic coherence leaves quite a bit to be desired. Table 7.1 gives a summary of the various experiments to be considered here. The only experiment that has been analyzed in detail for the coherence properties of individual ray paths is the AFAR experiment, where the focus was exclusively on time behavior. For AFAR, direct comparisons were made between observations and predictions from Eqs. 7.35 and 7.50. Other experiments, such as Cobb, San Diego, and Bermuda, have reported single-path coherence estimates based on Eqs. 7.35, but the fundamental form of the coherence and phase structure function were not investigated. These experiments, therefore provide the best comparisons between theory and observation within the expected regime of validity of the theory. Somewhat more oblique comparisons, mostly at ranges exceeding 1000 km, are given in the next section.

Tables 6.1 and 7.3 give the Λ and Φ values for the experiments considered here, and it is seen that they are all in the partially and fully saturated regimes.

Table 7.1. *Coherence times and lengths for experiments that resolved individual ray paths and interfering ray paths, such as would be observed for a CW transmission or in the finale of a long-range pulse transmission*

Experiment	Range (km)	t_0 (obs/pred) (min)	z_0 (obs/pred) (m)	y_0 (obs/pred) (m)	Notes
COBB (4 kHz; ID −1)	18	6.6/4.6	NA/14	NA/ 270	SP
COBB (8 kHz; ID −1)	18	3.3/2.3	NA /7.2	NA/135	SP
AFAR (410 Hz; ID −2)	35	9.4/9.0	NA/21	NA/550	SP
AFAR (1010 Hz; ID −2)	35	3.8/3.7	NA /8.6	NA/220	SP
AFAR (4671 Hz; ID −2)	35	0.82/0.79	NA/1.9	NA/48	SP
SD (2.2 kHz; ID −1)	23	5.9/4.9	NA/21	NA/350	SP
SD (2.2 kHz; ID +1, late)	23	2.6/2.0	NA/12	NA/165	SP
Bermuda (220 Hz; ID 31)	900	5.8/4.5	NA/79	NA/ 450	SP
Bermuda (220 Hz; ID −30)	900	6.0/4.6	NA/66	NA/460	SP
Bermuda (220 Hz; ID −27)	900	10/5.1	NA/98	NA/530	SP
MIMI (406 Hz; ID 21,18)	550	6.0/(3.1,3.6)	NA/(59,88)	NA/(325,430)	MP
MIMI (406 Hz; ID 45, 39)	1250	4.2/(2.1,2.4)	NA/(39,55)	NA/(226,270)	MP
AET (75 Hz; ID 125 142)	3250	12^b/(8.1,12.4)	350^d/(145,150)	NA/(785,1200)	SP
NPAL Billboard (75 Hz; ID 147)	3900	NA/5.3	NA/ 86	2000^c/525	SP
NPAL Billboard (75 Hz; Axial, 147)	3900	NA/(8.9,5.3)	NA/ (49,85)	460^d/(840,525)	MP
NPAL SOSUS O (75 Hz; ID 94)	2500	NA/6.6	NA/135	410/670	SP
NPAL SOSUS N (75 Hz; ID 94)	2500	NA/6.6	NA/135	530/670	SP

(*continued*)

Table 7.1. (cont.)

Experiment	Range (km)	t_0 (obs/pred) (min)	z_0 (obs/pred) (m)	y_0 (obs/pred) (m)	Notes
ATOC (75 Hz; Near Axial)	5200	NA/7.9	30/50	NA/800	MP
ATOC (75 Hz; Near Axial)	3500	NA/9.4	50/58	NA/900	MP
AST (28 Hz; Near Axial)	5200	NA/21	125/130	NA/2200	MP
AST (28 Hz; Near Axial)	3500	NA/25	150/150	NA/2400	MP
ASIAEX (300 Hz, ID 13)	18.9	15.1–2.1/14	NA/60	NA/620	MP
ASIAEX (500 Hz, ID 13)	18.9	5.9–1.0/8.5	NA /36	NA/370	MP
SW06 (224 Hz, IDs 38 34)	30	NA/(14,12)	NA/(53,52)	66/(350,320)	MP
SW06 (400 Hz, IDs 38 34)	30	NA/(7.6,6.8)	NA /(30,29)	37/(200,180)	MP

Predictions are based on Eqs. 7.55, 7.52, and 7.57 using range-independent climatological profiles and parameters from Tables 6.2 and 7.2. The label NA indicates no value available, and SP/MP denotes single/multiple path nature of the data. Primary references are COBB (Flatté, 1983), SD (Worcester et al., 1981), AFAR (Reynolds et al., 1985), Bermuda (Spiesberger and Worcester, 1981), MIMI (Dyson et al., 1976), AET (Worcester et al., 1999), NPAL (Andrew et al., 2005; Voronovich et al., 2005; Dzieciuch and Vera, 2006; ATOC/AST (Colosi et al., 2005), ASIAEX (Mignerey and Orr, 2004), and SW06 (Duda et al., 2012).

[a]Based on beamformer gain.
[b]Based on temporal coherent processing gain.
[c]Dzieciuch and Vera (2006).
[d]Voronovich et al. (2005).

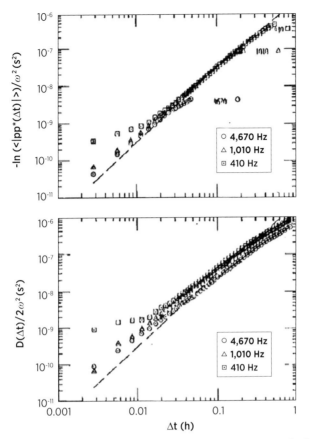

Figure 7.1. Temporal field coherence for the AFAR experiment at the frequencies of 410, 1010, and 4671 Hz. The top panel is the logarithm of Eq. 7.35. The bottom panel shows a direct comparison of the phase structure function. Predictions based on the path integral relation and the quadratic structure function (Eq. 7.55) are shown with dashed lines.
Source: Reynolds et al. (1985).

AFAR, Cobb, San Diego, and Bermuda

The most important experiment in this group is AFAR, whose primary objective was to examine acoustic fluctuations in time. As was seen from Chapter 6 this experiment is on the border between partial and full saturation. Figure 7.1 shows comparisons between theory and experiment related to time coherence (Reynolds et al., 1985). Taking the logarithm of Eq. 7.35 and scaling by the square of the acoustic frequency gives

$$-\frac{\ln|\langle pp^*\rangle|}{\omega^2} = \frac{D(\Delta t)}{2\omega^2}. \qquad (7.65)$$

By path integral theory, the right-hand side of this equation is independent of frequency. To test this result, the top panel of Figure 7.1 shows the left-hand side of the equation for the three frequencies in AFAR. The three curves nearly fall on top of one another, validating the frequency dependence of the coherence function. In addition, the slope of the curves nearly follows a Δt^2 form, which validates the quadratic approximation to the structure function. Some of the deviation from the Δt^2 form at small lags may be due to amplitude effects (i.e., Eq. 7.43). The figure also shows the right-hand side of the equation as predicted using Eq. 7.55 for environmental parameters from AFAR (Reynolds et al., 1985). The comparison in this case is quite good.

The lower panel of Figure 7.1 is the same comparison, but for the observed phase structure function, $\langle(\phi(\Delta t) - \phi(0))^2\rangle$. Even though the AFAR is on the border between the partially and fully saturated regimes, the phase is behaving in a way typical of geometrical acoustics. This result shows that in this border regime, the phase is still strongly influenced by larger scale internal waves and rather than by microray interference effects and smaller scale internal waves.

Finally, Table 7.1 compares the observed AFAR coherence times t_0 to those predicted by Eq. 7.55 using climatological profiles and parameters from the experiment (Table 7.2). The agreement between the path integral prediction and the observations is quite good. The Cobb Seamount (Flatté, 1983) and San Diego (Worcester et al., 1981) experiments, which are qualitatively similar to AFAR (i.e., high frequency, short range), show reasonable agreement with the path integral results also (Table 7.1). Lastly, the 220 Hz, 900-km Bermuda experiment (Spiesberger and Worcester, 1981), which resolved several ray arrivals, reveals a good comparison to the path integral estimates for all paths.

7.4.8 Observations: Multiple Paths, Long Range, and Shallow Water

As previously mentioned, other experimental results provide a somewhat less rigorous test of path integral theory. In the AET, the observed coherence times and lengths were based on temporal gain and beam former performance. In some other cases, like MIMI, NPAL, ATOC, AST, and the Shallow Water 2006 experiment, the data were not form a single ray path but from a complex combination of rays. In this case, the observations give a weighted average of the coherence functions of the individual paths (Eq. 7.64). In addition, several of the observations discussed here are for long ranges in which ray chaos effects are expected to become important, though our understanding of ray chaos influences on the coherences function is nonexistent. Table 7.1 gives a summary of the experiments considered here.

MIMI

In the single-frequency 400-Hz, 550-, and 1250-km range MIMI experiment the signal statistics are dominated by the interference of deterministic as well as

Table 7.2. *Parameters used for calculations presented in Tables 7.1 and 7.3*

Experiment	Latitude	ζ_0 (m)	$N_0 B$ (rad-m/s)	j_*	J	Reference
Bermuda	30	7.3	9.0	3	980	Spiesberger and Worcester (1981)
MIMI	25	7.3	9.9	3	1176	Dyson et al. (1976)
AET (R = 3250 km)	25	5.2	9.3	3	2044	Colosi et al. (1999)
NPAL	30	7.3	8.9	3	954	Andrew et al. (2005)
ATOC/AST (HVLA)	25	7.3	9.0	3	977	Colosi et al. (2005)
ATOC/AST (KVLA)	20	7.3	9.8	3	1171	Colosi et al. (2005)
ASIAEX	21	7.3	1.95	1	25	–
SW06	40	2.5	1.26	1	44	Colosi et al. (2012)

scattering induced microrays (Dyson et al., 1976; Flatté, 1983). As such the signals are in the fully saturated regime. Here the quadratic form of the structure function yields a Gaussian coherence function and thus a Gaussian frequency spectrum, which can be compared to the observed spectra of the real and imaginary parts of the demodulated complex pressure. Figure 7.2 shows the theory/observation comparisons (Dyson et al., 1976; Flatté, 1983) and the agreement is quite good (Note in the figure $\nu = 1/t_0$). It should be remarked, however, that in this case of deterministic multipath, each multipath will have a slightly different value of t_0, and thus theoretical comparisons that utilize an average t_0 value cannot be strictly correct. Table 7.1 shows a range to coherence time estimates relevant to MIMI and the agreement between observations and theory is reasonably good.

NPAL:ATOC

Between 1994 and 2011 several deep-water long-range experiments were conducted in the North Pacific ocean under the name the "North Pacific Acoustic Laboratory" (NPAL) (Worcester and Spindel, 2005; Worcester et al., 2013). Figure 7.3 shows the acoustic transmission paths for several of these experiments. From 1994 to 1997 the measurements were part of the Acoustic Thermometry of Ocean Climate (ATOC) program that included the AET, the Alternate Source Test (AST), and ATOC receptions at Navy SOSUS stations (Figure 1.7). In 1998–1999 there was the NPAL Billboard Array that gave observations of horizontal coherence. There are other more recent experiments associated with NPAL, but none of them have been analyzed for coherence except for the 2005 LOAPEX that was analyzed for normal mode coherence (Chapter 8).

The NPAL experiments discussed here were conducted at ranges in excess of 2000 km and at frequencies less than 100 Hz. In these experiments pulses were transmitted so that individual acoustic paths could be resolved, and in many cases

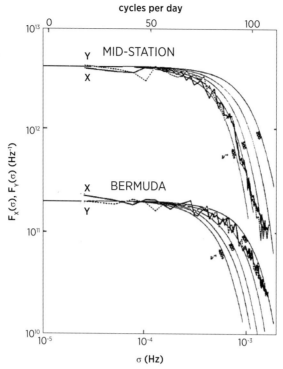

Figure 7.2. Spectra of the real (X) and imaginary (Y) parts of the complex pressure for the 406-Hz, single-frequency MIMI transmissions. Mid-station and Bermuda are at ranges of roughly 550 and 1250 km. The theoretical curves are for full saturation and thus use $\langle pp^*(\Delta t) \rangle \simeq \exp(-(\Delta t)^2/2t_0^2)$ for the coherence function, which when Fourier transformed gives the spectrum. *Source*: Dyson et al. (1976).

large numbers of paths were identified. All the arrivals are identified to be in the saturated regime (Table 7.3).

The earliest ATOC experiment was the 75-Hz, 3250-km AET that has been discussed previously. Coherence times for individual ray paths were estimated to be roughly 12 minutes, based on a calculation of coherent processing gain, and the vertical coherence was estimated to be roughly 350 m based on beam forming gain (Worcester et al., 1999). Path integral estimates from Eqs. 7.52 and 7.55 give values of 8–12 minutes and 150 m for these coherences (see Table 7.1). Remarkably, the temporal coherence estimate is quite good. The vertical coherence estimate, on the other hand, appears too low, but given the rough estimate based on beam former gain the difference may not be unreasonable.

NPAL: Billboard Array

The ATOC receptions at Navy SOSUS stations and the NPAL Billboard array study have provided estimates of transverse coherence. Voronovich et al. (2005)

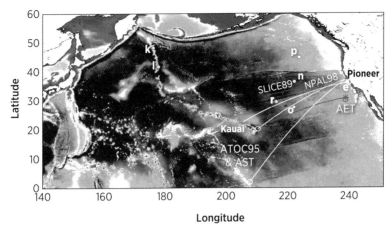

Figure 7.3. Map showing several experiments conducted in the Eastern North Pacific Ocean as part of the NPAL effort and the SLICE89 experiment. The letters k, l, p, r, o, e, and f denote fictive locations of SOSUS receivers. Acoustic sources were located at Pioneer and Kauai. The SLICE89 source was located to the west. *Source*: Adapted from Worcester and Spindel (2005).

estimated transverse coherence using the NPAL billboard receiving array for 75-Hz transmissions at 3900-km range. The coherence value was obtained by cross-correlating the entire reception across hydrophones at roughly equal depth and therefore does not represent the coherence for a particular path. Here the observed value of 460 m is compared to the path integral estimate for a near-axial ray of 840 m and a high-angle ray of 525 m (Table 7.1). The observed value is somewhat below the range of the axial and high-angle predictions.

Observational work by Dzieciuch and Vera (2006) and Monte Carlo numerical studies by Vera (2007) examined individual ray arrivals on the Billboard array and compared them to the results of Voronovich et al. (2005) (Figure 7.4). In this analysis the transverse coherence of the individual paths was shown to be much larger than the estimates obtained by cross-correlating the entire reception. The observations show a coherence length of about 2000 m that is larger than the Monte Carlo estimate of roughly 1000 m. The observations and the Monte Carlo result can be compared to the path integral prediction of 525 m for a high-angle ray. The source of this large discrepancy is not known.

NPAL: SOSUS Receptions

Further work by Andrew et al. (2005) quantified horizontal coherence functions of single paths from 75-Hz observations using Navy SOSUS receivers (Figure 7.4). For a range of roughly 2500 km, coherence lengths y_0 of 410–530 m were obtained, compared to the path integral prediction of 670 m (Table 7.1). In contrast to the NPAL billboard array results, the path integral prediction is reasonably good. It should be noted that path integral predictions for the observations presented in

Table 7.3. Λ *and* Φ *parameters computed for the experiments in Table 7.1*

Experiment	Range (km)	Φ (rad)	Λ
Bermuda (220 Hz, ID 31)	900	9.4	1.4
Bermuda (220 Hz, ID -30)	900	9.1	1.9
Bermuda (220 Hz, ID -27)	900	5.4	0.69
MIMI (406 Hz, ID 39)	1250	10.5	0.61
MIMI (406 Hz, ID 45)	1250	21.4	0.50
MIMI (406 Hz, ID 18)	550	7.1	0.30
MIMI (406 Hz, ID 21)	550	14.6	0.25
AET (75 Hz, ID 125)	3250	4.1	5.5
AET (75 Hz, ID 142)	3250	3.8	1.8
NPAL (75 Hz, ID 94)	2500	4.7	5.6
NPAL (75 Hz, ID 147)	3900	5.8	8.8
NPAL (75 Hz, Near Axial)	3900	7.7	1.7
ATOC (75 Hz, Near Axial)	3500	7.6	6.8
ATOC (75 Hz, Near Axial)	5200	13	3.4
AST (28 Hz, Near Axial)	3500	2.8	18
AST (28 Hz, Near Axial)	5200	4.8	9.0
ASIAEX (300 Hz, ID 13)	18.9	2.0	1.50
ASIAEX (500 Hz, ID 13)	18.9	3.3	0.87
SW06 (224 Hz, ID 38)	30	1.7	0.86
SW06 (224 Hz, ID 34)	30	1.8	1.0
SW06 (400 Hz, ID 38)	30	3.1	0.48
SW06 (400 Hz, ID 34)	30	3.3	0.57

Figure 7.4 were obtained using the empirical relation Eq. 7.60. These predictions give coherence lengths of 414 m and 367 m for the SOSUS N and SOSUS O receptions that compare more favorably to the observations than the 670 m prediction from Eq. 7.57. For reasons given in Section 7.4.5, the predictions based on Eq. 7.57 are the most accurate.

The lower coherences seen in the NPAL Billboard and SOSUS observations may be due to bottom interactions near the receivers. The observations for individual paths described by Dzieciuch and Vera (2006) have the least bottom interaction. Clearly other factors such as ray chaos may be important as well.

NPAL: ATOC/AST

Observations of vertical coherence in the finale of the long-range NPAL transmission data for frequencies between 28 and 84 Hz (Colosi et al., 2005) are discussed. Figure 7.5 shows the coherence functions for seven different experiments of varying frequencies and ranges, and Table 7.1 gives the observed and predicted coherences at frequencies of 28 and 75 Hz. Because the observations are in the finale region, the acoustic field is an interference pattern of many

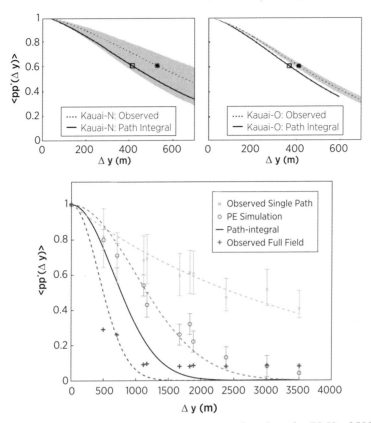

Figure 7.4. Top two panels: Horizontal coherence functions for 75-Hz, 2500-km transmissions to two different SOSUS stations in the North Pacific Ocean. Path integral estimates using the empirical relation (Eq. 7.60) are shown with solid lines. Symbols indicate the $e^{-1/2}$ points. Bottom panel: Horizontal coherence functions for 75-Hz transmission to a Billboard receiving array at 3500-km range (Dzieciuch and Vera, 2006). Observations for individual ray paths (Refracted Arrivals) are compared to Monte Carlo simulations (Vera, 2007) (PE Simulation), the results from Voronovich et al. (2005) where coherence was estimated by correlating the entire arrival pattern (Full Field), and predictions from path integral theory (Eq. 7.60). Dashed curves are fit through the various data points. *Source*: Top panels from Andrew et al. (2005).

rays with low grazing angles. The observations and predictions are remarkably consistent to better than a factor of 2.

SW06/ASIAEX

The Shallow Water 2006 (SW06) experiment was conducted on the New Jersey continental shelf during the month of August (Tang et al., 2007). One of the goals of this field effort, was to quantify the anisotropy of shallow-water propagation in the presence of significant nonlinear internal-wave variability. Using an L-shaped array and acoustic frequencies between 85 and 450 Hz, Duda et al. (2012)

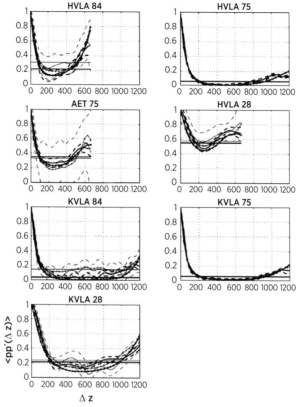

Figure 7.5. Observed vertical coherence functions in the finale of the pulse transmissions for various long-range experiments in the North Pacific Ocean. The AET had a range of 3250 km and a center frequency of 75 Hz. ATOC transmissions to HVLA and KVLA had ranges of 3500 and 5200 km with a center frequency of 75 Hz. The Alternate Source Test (AST), which also transmitted to HVLA and KVLA, has similar ranges but the center frequencies were 28 and 84 Hz. Small dashed curves are uncertainty estimates, and the remaining curves that are close to one another use various methods of estimating the coherence function. The square of the mean field is shown with horizontal lines.
Source: Colosi et al. (2005).

were able to quantify horizontal coherences in the along-shore and across-shore directions at propagation ranges of 19 and 30 km, respectively. In the along-shore direction, sound travels primarily along the crests of the nonlinear internal-wave packets. Horizontal ducting and Lloyds mirror effects can therefore occur (Lin et al., 2009; Lynch et al., 2010, 2006). In the across-shore direction, however, the localized nonlinear wave packets occupy a small fraction of the propagation path, and random linear internal waves are a critical factor (Colosi et al., 2011). In this experiment individual ray paths could not be identified and the coherence estimates are associated with a collection of paths. In addition the background sound-speed profile for SW06 had a small duct that trapped some acoustic energy

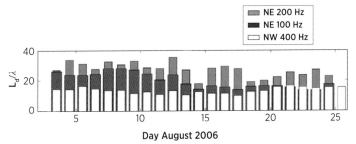

Figure 7.6. Daily averaged SW06 observed horizontal coherence length L_d scaled by wavelength for 100, 200, and 400 Hz, and propagation along shelf (NE, 19 km range) and across shelf (NW, 30 km range). The along-shelf coherence lengths vary considerably in time due to the propagation along the crests of nonlinear internal-wave packets. The coherence across shelf is much more stable and is driven primarily by random linear internal waves. For the NW path only the 400-Hz results are plotted because the 224-Hz data are almost identical, indicating a coherence function that scales as one over frequency. L_d varies between 11 and 16 λ. The coherence length L_d is defined as the e^{-1} point of the coherence function, so this value should be reduced by a factor of $\sqrt{2}$ for comparisons to path integral results.
Source: Duda et al. (2012).

entirely in the water column (Figure 2.3). Thus, to some extent the acoustic fields have a significant component of non-bottom interacting energy for which the path integral results may apply.

Figure 7.6 shows the temporal evolution of the daily averaged horizontal coherence length (L_d) divided by the acoustic wavelength. The across-shelf coherences, denoted by NW, are seen to be relatively stable over time, scale inversely with frequency, and have values between 11 and 16 λ. The along-shore coherences, denoted by NE, on the other hand are quite variable in time and do not scale inversely with frequency as the across-shore coherences do. Because the observed coherence lengths are based on an e^{-1} metric, the values to be compared to the path integral are reduced by $\sqrt{2}$, giving 8 and 11 λ. Path integral estimates of the horizontal coherence using the GM spectrum with some small parameter modifications appropriate for shallow water (Colosi et al., 2012) are much larger than the observations (Table 7.1). Several factors not included in the theoretical calculation could account for the discrepancy, including anisotropy and horizontal inhomogeneity of the internal-wave spectrum, rough bottom and surface scattering, and attenuation of high-angle energy from the seafloor.

At this point it is also appropriate to mention the work of Carey (1998), who examined transverse coherence measurements for many shallow-water experiments. In this analysis, transverse coherence lengths varied between 20 and 40 λ, generally independent of ranges that were between 5 and 40 km. The results from SW06 clearly fall below this range, and the path integral predictions at roughly 50 λ exceed Carey's range.

Lastly, the 2001 ASIAEX South China Sea experiment has provided some observations of shallow-water temporal coherence at 300 and 500 Hz (Mignerey and Orr, 2004). In this case, data was analyzed for an 18.9-km, along shelf acoustic path in which energetic nonlinear internal waves propagated through the region with their crests aligned with the path. Like the SW06 along shelf results, the ASIAEX observations show that the coherences are highly diminished when nonlinear waves are present in the path. At 300/500 Hz the base line coherence times are 15.1/5.9 minutes but when nonlinear waves are present the values drop to 2.1/1.0 minutes. These baseline coherence times are roughly consistent with the path integral estimates of 14 and 8.5 minutes at 300 and 500 Hz (Table 7.1); however, the ASIAEX background sound-speed profile is strongly downward refracting leading to strong bottom interaction.

The topic of shallow-water coherences will be revisited in Chapter 8 where the topic of coupled mode transport theory is introduced, which more naturally treats the shallow-water environment and its bottom interactions.

7.4.9 Internal-Wave Tomography

The path integral formalism discussed thus far suggests the interesting possibility of using acoustic fluctuations to infer aspects of the ocean internal-wave field, that is, internal-wave tomography. As discussed in Chapter 3, there is limited dynamical understanding behind the GM spectrum, including sources and sinks of energy. The basic question concerning the sources of ocean internal waves was articulated by Wunsch (1976) quite some time ago, who wrote

Many sources for the internal wave field have been suggested; but it has not yet been possible to make the kind of statement that can be made about surface waves: namely when the wind blows, surface waves are generated, and the larger the fetch and duration, the larger the waves.

Similar questions exist for the sinks of internal waves and their connection to ocean mixing.

The tool of acoustic propagation would seem to have some potential for a new view into this problem, given that many of the issues have not been resolved using conventional methods, that is, point or moored measurements of currents, temperature, and salinity. A combination of both conventional and acoustic methods would likely be the best. The acoustic methods to be outlined here have several advantages such as spatial averaging, a rapid and repeatable measurement, no calibration issues, and the information increases with the square of the number of moorings. Furthermore, the methods can be combined with Ocean Acoustic Tomography and ocean models to estimate the large-scale structure in which the internal waves are embedded. The objective is to measure deviations in internal-wave spectral energy and/or spectral form over a small three-dimensional region of the ocean and over time. Significant deviations in the spectra would therefore suggest sources and sinks, and more detailed studies could be carried out to establish fluxes into or out of a specific area.

The methodology of internal-wave tomography via the path integral borrows heavily from the field of ocean acoustic tomography that was pioneered by Munk and Wunsch (1979), and has been further described in the monograph Munk et al. (1995). The tomography problem defined by Munk and Wunsch starts with the expression for the travel time of a pulse through a mesoscale field of sound-speed variability:

$$T(t) = \int_\Gamma \frac{ds}{c(\mathbf{r},t)}, \tag{7.66}$$

where ds is along an acoustic ray path, Γ, between the source and the receiver at range R. Figure 7.7 gives an illustration of mesoscale activity in the ocean, with a set of sources and receivers such as might be used in an acoustic tomography experiment. The mathematical apparatus of tomography is used to deduce $c(\mathbf{r},t)$ *between the moorings* from the observations of $T(t)$ for a large number of ray paths, and some *a priori* information. For the observation of mesoscale eddies, the changes in $T(t)$ are observed over times of order a few days, and the mean of the sound speed is determined over small regions in $(\mathbf{r})$, where the sizes of the regions are determined by the geometry of the crossing ray paths.

For internal-wave observables such as those described by ray and path integral theory, the form of the forward problem is identical to Eq. 7.66. For example, the travel-time variance from Chapter 5 is

$$\langle \tau^2 \rangle = q_0^2 \int_\Gamma \langle \mu^2(s) \rangle L_p(s) \, ds. \tag{7.67}$$

So, as a specific example, the travel time $T(t)$ can be observed every few minutes over several months. Instead of extracting the mean as is done in mesoscale tomography, one can extract the variance of the travel time in the internal-wave band for comparison with a prediction based on internal-wave effects. This variance depends on specific characteristics of the internal-wave field, such as the spectral strength and form. Using several paths, the internal-wave spectral strength and form could be determined as a function of geophysical time and over some region of the ocean. If the receivers were vertical line arrays, then even more information could be extracted concerning the large-scale and internal-wave variability. Importantly, if the internal-wave spectra could be determined simultaneously with the large-scale, mesoscale structure in a region of the ocean, then the correlation between these two fields could be investigated to answer several questions. How is the strength of internal waves correlated with warm or cold eddies or fronts? How rapidly does the internal-wave field respond

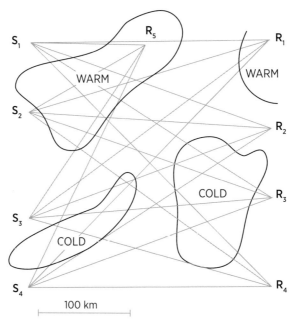

Figure 7.7. A schematic of an internal-wave tomography experiment with background mesoscale ocean structure and a set of ray paths. Each line between a source and a receiver actually represents a vertical plane, in which many rays travel between the two points. If each receiver is a vertical array, then the number of rays is even greater.

to changes in atmospheric forcing? Are internal-wave spectral levels related to the tides or spring/neap cycles?

Such an ambitious experiment has not yet been carried out. However, more modest experiments have been done that have a single source and a vertical line array of receivers. An important early theoretical analysis that helped understand the strengths and weaknesses of these experiments was that of Flatté and Stoughton (1986), who showed that one can determine the internal-wave strength as a function of depth within a small region in the upper ocean, averaged over the region between the source and the receiver array. The first realization of this approach was the work of Stoughton et al. (1986) applied to a 1983 reciprocal-transmission experiment east of Bermuda with a range of 305 km (Howe et al., 1987).

Other observables will add additional constraints to the internal-wave tomography problem. Here it is wise to utilize observables strongly associated with phase given that the forward problem for intensity statistics is on slightly flimsier ground than those for phase. It is also wise to avoid observables associated with internal-wave currents (Munk et al., 1981) because the current effects are just too small to observe with any precision. An example of two observables with some promise are the temporal and vertical coherence functions that constrain the frequency and vertical wavenumber content of the spectrum. As given here in this chapter they are written as

$$\frac{1}{t_0^2} = q_0^2 \int_\Gamma ds \; \langle \mu^2(s) \rangle \; L_p(s) \; \{\sigma^2\}, \tag{7.68}$$

$$\frac{1}{z_0^2} = q_0^2 \int_\Gamma ds \; \langle \mu^2(s) \rangle \; L_p(s) \; \{m^2(j)\}\xi_1^2(s), \tag{7.69}$$

where $\{\sigma^2\}$ and $\{m^2(j)\}$ are important weighting functions giving spectral averages of internal-wave frequency and vertical wavenumber. The ray-tube function ζ_1 gives the vertical separation between two nearby rays. Because of internal-wave anisotropy and inhomogeneity, all of the observables here are strongly sensitive to internal-wave structure in the neighborhood of the ray upper turning point.

Another interesting approach is to use the weak fluctuation results from Chapter 6 to carry out the tomography problem. This approach has been developed by Ewart et al. (1998) and the methodology was tested using Monte Carlo simulations for a variety of experimental configurations and for the case of no ocean sound channel. Here the observables considered are the phase and log-amplitude frequency and vertical wavenumber spectra, which bear a close resemblance to the previously discussed observables of travel-time variance and temporal and vertical coherence. This approach, however, can be applied only in the unsaturated regime. Tomographic methods related to the highly successful normal mode methods described in Chapter 8, have yet to be developed. Here interesting possibilities exist due to the resonant scattering mechanisms of mode coupling by internal waves.

Internal-wave tomography represents an interesting and underdeveloped area of ocean acoustics and physical oceanography, in which many opportunities exist for future research.

7.5 Mutual Coherence Functions: Frequency Separations

The coherence function across frequency, $\langle p(\omega_1)p^*(\omega_2) \rangle$, is critical for describing pulse propagation. The Fourier transform of this coherence function provides a prediction of the mean or Ensemble Average Pulse (EAP). Writing the acoustic pressure time series in terms of its Fourier transform,

$$p(\tau) = \frac{1}{(2\pi)^{1/2}} \int_{-\infty}^{\infty} d\omega X(\omega) p(\omega) \exp(i\omega\tau), \tag{7.70}$$

it is then easy to show that the EAP is given by

$$\langle I(\tau) \rangle = \langle |p(\tau)|^2 \rangle = \int_{-\infty}^{\infty} T_X(\Delta\omega)\langle pp^*(\Delta\omega) \rangle \; \exp(i\Delta\omega\tau) \, d\Delta\omega, \tag{7.71}$$

with

$$T_X(\Delta\omega) = \frac{1}{2\pi} \int_{-\infty}^{\infty} X(\omega + \Delta\omega/2)X^*(\omega - \Delta\omega/2)\,d\omega, \qquad (7.72)$$

where $X(\omega)$ is the normalized amplitude spectrum of the transmitted signal (Worcester et al., 1981). The coherence across frequency is also needed to predict other acoustic observables such as the intensity covariance function (next section).

Acoustic scattering by internal waves causes three things to happen to the EAP. First, travel-time fluctuations, such as those predicted by Eq. 5.27, cause individual realizations of the pulse to wander in time, thus broadening the mean pulse. This is often called Doppler spread. Second, the mean travel time of the pulses in the presence of internal waves may be different from the unperturbed travel time, an effect termed the travel-time bias. One may consider this a consequence of Fermat's principle, though as will be seen, the bias can be both positive and negative. Lastly, loss of coherence across the bandwidth of the signal also causes mean pulse broadening; the amount of broadening in this case is inversely related to the coherent bandwidth of the signal.

The coherence between acoustic pressure fields at frequencies ω_1 and ω_2 is given by a double-path integral of the form

$$\langle p(\omega_1)p^*(\omega_2)\rangle = |N|^2 \int D^2(path)\exp\!\left(i(\Theta_0(1;\omega_1) - \Theta_0(2;\omega_2)) - \frac{V_{12}}{2}\right),$$
$$(7.73)$$

$$V_{12} = \left\langle\!\left(q_1 \int_{path_1} \mu\,dx - q_2 \int_{path_2} \mu\,dx\right)^{\!2}\right\rangle, \qquad (7.74)$$

where $q_1 = \omega_1/c_0$ and $q_2 = \omega_2/c_0$. To evaluate this path integral, the treatment of Dashen et al. (1985) is followed. It proves useful to define a new quantity called the acoustic path length $S(path)$ that is independent of acoustic frequency and is related to the path phase by $\Theta(path) = qS(path)$. As was the case for the coherence functions with separations in space and time, one first examines the unperturbed phase by expanding about the unperturbed ray and carrying out the change of variables used for Eq. 7.27. After a fair amount of algebra, this gives the phase difference $\Delta\Theta = \Theta_0(1;\omega_1) - \Theta_0(2;\omega_2)$ as

$$\Delta\Theta \simeq \bar{q}(S_0(1) - S_0(2)) + \frac{\Delta q}{2} \int_0^R \Big[\dot{\beta}^2 + \dot{w}^2 + \frac{1}{4}(\dot{\alpha}^2 + \dot{u}^2) + 2\dot{w}\dot{z}_r + \dot{z}_r^2$$
$$-2U(z_r) - 2U'w - U''w^2 - \frac{1}{4}U''u^2\Big]\,dx, \qquad (7.75)$$

where $\bar{q} = (q_1 + q_2)/2$, $\Delta q = q_1 - q_2$, and $S_0(1) - S_0(2)$ is given by Eq. 7.27 without the factor of q_0. The key assumption is that V_{12} is not a strong function of β

or w, that is, the internal waves are assumed to be horizontally and vertically homogeneous. This allows the β and w path integrals to be done by using Eq. 7.75.

The β integral is done most readily, but here delta functions are not obtained as before due to the quadratic dependence on β. The path integral is given by

$$I_\beta = \int D(\beta) \exp\left[\frac{i\Delta q}{2} \int_0^R \left(\dot{\beta}^2 + \frac{2\bar{q}}{\Delta q} \dot{\alpha}\dot{\beta} \right) dx \right].$$

(7.76)

Completing the square in the exponent allows the path integral to be done, giving

$$I_\beta = \exp\left[-\frac{i\bar{q}^2}{2\Delta q} \int_0^R \dot{\alpha}^2 \, dx \right].$$

(7.77)

In this expression several constant factors from the path integral have been subsumed into the normalization.

The w path integral is simplified as before using integration by parts, the parabolic equation ray equations, and subsuming boundary terms into the normalization to get

$$I_w = \int D(w) \exp\left[-\frac{i\delta q}{2} \int_0^R \left(wLw + \frac{2\bar{q}}{\Delta q} wLu \right) dx \right],$$

(7.78)

where $L = -d^2/dx^2 - U''$ is an operator associated with the ray tube equation. Completing the square and moving constant terms into the normalization gives

$$I_w = \exp\left(-\frac{i\bar{q}^2}{2\Delta q} \int_0^R uLu \, dx \right).$$

(7.79)

Bringing together all the terms, what is left is a double-path integral over the transverse separation variable α and the vertical separation variable u

$$\langle p(\omega_1) p^*(\omega_2) \rangle = N \int D(\alpha) D(u) \exp\left[-\frac{i\bar{q}^2}{2\Delta q} \int_0^R \left(\dot{\alpha}^2 + uLu \right) dx - \frac{V_{12}}{2} \right].$$

(7.80)

Now, the internal-wave perturbations along the path are examined. Expanding the V_{12} term with $q_1 = \bar{q} + \Delta q/2$ and $q_1 = \bar{q} - \Delta q/2$ gives three contributions

$$V_{12} = \bar{q}^2 \left\langle \left(\int_{path_1} \mu \, dx - \int_{path_2} \mu \, dx \right)^2 \right\rangle$$

$$+ \bar{q}\Delta q \left[\left\langle \left(\int_{path_1} \mu \, dx \right)^2 \right\rangle - \left\langle \left(\int_{path_2} \mu \, dx \right)^2 \right\rangle \right]$$

$$+ \frac{\Delta q^2}{4} \left\langle \left(\int_{path_1} \mu \, dx + \int_{path_2} \mu \, dx \right)^2 \right\rangle.$$

(7.81)

The first term is the phase structure function at the mean frequency, while the second and third terms are proportional to the difference and sum of the phase variances for the two paths. If the paths are not too far apart, the second term can be neglected and the third term can be written as $(\Delta q/\bar{q})^2\Phi^2$, giving

$$V_{12} = \left(\frac{\Delta q}{\bar{q}}\right)^2 \Phi^2 + 2\bar{q}^2 \int_0^R \langle \mu^2 \rangle L_p \{1 - \cos(l\alpha)\cos(mu)\}\, dx. \qquad (7.82)$$

Here the phase structure function has been written in the ray tangent or Markov approximation.

With these simplifications the cross frequency coherence can be compactly written

$$\langle pp^*(\Delta\omega) \rangle = \exp\left[-\frac{1}{2}\left(\frac{\Delta q}{\bar{q}}\right)^2 \Phi^2\right] Q(\Delta q), \qquad (7.83)$$

with the double-path integral

$$Q(\Delta q) = N \int D(\alpha) D(u) \exp\left[-\frac{i\bar{q}^2}{2\Delta q} \int_0^R \left(\dot{\alpha}^2 + uLu\right) dx \right.$$
$$\left. - \bar{q}^2 \int_0^R \langle \mu^2 \rangle L_p \{1 - \cos(l\alpha)\cos(mu)\}\, dx\right]. \qquad (7.84)$$

The term $Q(\Delta q)$ is often referred to as the microray bandwidth function. This path integral is normalized such that $Q = 1$ for $\mu = 0$. The first term in Eq. 7.83 is the Doppler smearing term that causes pulse broadening from the time wander of the pulse. The exponent can in fact be written as $-0.5\Delta\omega^2\langle\tau^2\rangle$, where it is evident that there is no dependence on the mean frequency. Physically, this term is the result of microrays that stay quite close to the unperturbed ray. The microray bandwidth function is therefore involved with the two other effects on the EAP, that is, travel time bias and scattering induced pulse time spread. These effects come from microrays that meander further from the unperturbed ray.

At this point two further simplifications are made. The first is to neglect the transverse term in the phase structure function, namely $\cos(l\alpha)$, and the second is to Taylor expand the other cosine. Ignoring the transverse term is a reasonable assumption because for internal waves the energy carrying waves have $|l| \ll |m|$; this is equivalent to ignoring the transverse Fresnel zone terms in weak fluctuation theory. Taylor expanding the cosine allows solution of a simpler Gaussian (quadratic) path integral. Thus, the final result is a single-path integral in terms of vertical separations only,

$$Q(\Delta q) = N \int D(u) \exp\left[-\frac{i\bar{q}^2}{2\Delta q} \int_0^R uMu\, dx\right], \qquad (7.85)$$

where

$$M(x) = L - i\Delta q \langle \mu^2 \rangle L_p \{m^2\}. \tag{7.86}$$

7.5.1 Evaluation of $Q(\Delta q)$

Dashen et al. (1985) demonstrated two different methods to evaluate the path integral in Eq. 7.85. In one solution the path integral is evaluated using a differential equation (Feynman and Hibbs, 1965). An approximate solution to that equation for small Δq is

$$Q(\Delta q) \simeq \exp[i\Delta\omega\tau_1 - \frac{1}{2}(\Delta\omega\tau_0)^2]. \tag{7.87}$$

As previously noted, Q in Eq. 7.87 accounts for the displacement of the microrays from the unperturbed ray; τ_1 represents a shift of the mean arrival time of the pulse, and τ_0 describes the pulse spreading due to the differences in arrival times of the different microrays. The reciprocal of τ_0 is the coherent bandwidth. The quantities τ_1 and τ_0 of Eq. 7.87 are calculated as follows (Dashen et al., 1985; Flatté and Stoughton, 1988):

$$\tau_1 = \frac{1}{2c_0} \int_0^R \langle \mu^2 \rangle L_p \{m^2\} g(x,x) \, dx, \tag{7.88}$$

$$\tau_0^2 = \frac{1}{4c_0^2} \int_0^R \langle \mu^2 \rangle L_p \{m^2\} \, dx \int_0^R \langle \mu^2 \rangle L_p \{m^2\} g^2(x,x') \, dx'. \tag{7.89}$$

The raytube green's function $g(x,x')$ is defined in Chapter 2, and a positive/negative bias means a longer/shorter travel time. These expressions show how the bias and spread scale with various parameters of the problem. In keeping with the ray-theoretic nature of the path integral, the time bias and spread are independent of frequency. Because the Greens function scales linearly with range, τ_1 and τ_0 scale as the square of range (Dashen et al., 1985). This is the result found using the random walk model presented in Chapter 5. Finally the bias and spread depend on $\{m^2\}$, and thus these quantities are sensitive to the small-scale part of the internal-wave spectrum.

A useful example is to consider the case in which the transmitted pulse is a Gaussian of the form $X(\omega) \propto \exp(-(\omega - \omega_c)^2\alpha^2/2)$, where ω_c is the center frequency and α is the rms time width of the pulse. Using the Gaussian pulse shape with Eqs. 7.87 and 7.85, Eq. 7.71 can be evaluated analytically to give the ensemble average pulse (EAP)

$$\langle I(\tau) \rangle \propto \exp\left(-\frac{(\tau - \tau_1)^2}{\alpha^2 + \tau_0^2 + \langle \tau^2 \rangle}\right). \tag{7.90}$$

The total pulse time width is the quadrature sum of the unperturbed width α^2, the travel time wander $\langle \tau^2 \rangle$ and the microray spread, τ_0^2.

In general, comparisons between observations and pulse spread predictions using Eq. 7.89 have not been accurate (Reynolds et al., 1985). In the case of extremely long-range, basin-scale observations, the disagreement is dramatic (Colosi et al., 1999). Therefore, microray-induced pulse spread calculations require a more careful approach that is discussed next.

Using an eigenvalue method discussed in Dashen et al. (1985) and in Appendix B, it can be shown that the microray bandwidth function is given by

$$Q(\Delta\omega) = \prod_n \left(1 - \frac{i\Delta\omega}{\lambda_n}\right)^{-1/2}, \tag{7.91}$$

where the values λ_n are the eigenvalues of the equation following the ray

$$\left(\frac{\partial^2}{\partial x^2} + U''\right)\phi_n = -\lambda_n F(x)\phi_n, \tag{7.92}$$

with $F(x) = \langle \mu^2 \rangle L_p \{m^2\}/c_0$ and boundary conditions $\phi_n(0) = \phi_n(R) = 0$. It turns out that the tails of the EAP for both early and late times are controlled by the smallest eigenvalues λ_n (Dashen et al., 1985).

The eigenvalues are controlled by both the background sound-speed profile, through U'', and the nature of the internal-wave spectrum, through $F(x)$. The example for an axial ray is useful here because both factors are constant, namely

$$\lambda_n = \frac{1}{F}\left[\left(\frac{\pi n}{R}\right)^2 - U''\right], \tag{7.93}$$

with

$$Q(\Delta\omega) = \left(\frac{\sin R \sqrt{i\Delta\omega F}}{R \sqrt{i\Delta\omega F}}\right)^{-1/2}, \tag{7.94}$$

where the expression for Q is only for $U'' = 0$, that is, no sound channel. In this case, all the eigenvalues are positive. Negative eigenvalues can only be generated for positive sound-speed curvature that is the normal case in the ocean. It has been found that the number of negative eigenvalues correspond to the number of caustics that the unperturbed ray has passed through (Dashen et al., 1985; Arnold, 1989). The sign of the eigenvalues is important because the sign determines whether the pulse travel time bias is positive or negative. Positive eigenvalues mean a positive bias, that is a pulse that lags behind the unperturbed pulse. The sound channel is seen to be key in producing negative eigenvalues and thus a negative bias.

Figure 7.8 shows a calculation of the mutual coherence function for the case of constant U'' and F, demonstrating the effects of the ocean waveguide, namely U''

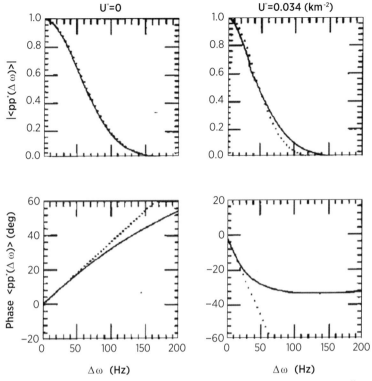

Figure 7.8. Calculation of $\langle pp^*(\Delta\omega)\rangle$ for an axial ray with constant U'' and F. In the left-hand panels, $U'' = 0$, $F = 10^{-2}$ (ms)/km^2, and $\Phi/\omega = 2.8$ ms. For the case with no sound channel Eqs. 7.88 and 7.89 yield $\tau_1 = 1.02$ ms and $\tau_0 = 0.65$ ms. The MCF obtained using the eigenvalue method is shown with a solid curve, while the approximation for small separations, Eq. 7.87, is shown with a dotted line. In the right-hand panel, the same calculation is carried out but $U'' = 0.034$ (km)$^{-2}$, a typical value for the AFAR experiment. In this case, the bias and spread have the values $\tau_1 = -2.61$ ms and $\tau_0 = 2.76$ ms.
Source: Dashen et al. (1985).

zero and nonzero. The parameters are those associated with the AFAR experiment. The cases with and without the waveguide are quite different. With regard to the phase, the waveguide example has more structure, and the initial phase gradients associated with the travel-time bias have different signs. The biases, τ_1, computed from Eq. 7.88 are 1.02 and -2.61 ms for the no waveguide/waveguide cases, respectively. For the no waveguide case the bias is always positive, that is, a longer travel time. This result appears at first to violate the Fermat principle, which would predict an earlier arrival. In fact there is no contradiction, as was pointed out by Codona et al. (1985), who showed that the ray/Fermat result is unweighted with respect to microray intensity, while the path integral as a full field calculation is properly weighted. Physically, high-intensity microrays are created by low sound-speed zones that lead to a longer travel time.

The magnitudes of the MCF are also different, with the no waveguide example showing more coherent bandwidth; the pulse time spreads, τ_0, computed from Eq. 7.89 are 0.65 and 2.76 ms for the two cases. Finally it is seen that the MCF is reasonably modeled using the small $\Delta\omega$ approximation Eq. 7.87. When comparisons are made to observations however, especially for the ensemble average pulse, it has been found that the more accurate eigenvalue approach is needed.

7.5.2 Observations

To date only one experiment has obtained good results for the frequency separation MCF, and this is the AFAR (Reynolds et al., 1985). The San Diego Experiment described by Worcester et al. (1981) also presents results for EAPs at short range, but here the short duration of the experiment leads to large uncertainty. There was also the SLICE89 experiment, which did not examine pulse shape per se but found that the pulse travel-time correlation function showed rapid decorrelation in both depth and time (Duda et al., 1992). Using Monte Carlo simulation, this rapid decorrelation was found to be associated with fracturing of the pulse into multiple peaks, and therefore pulse time spread (Colosi et al., 1994). Finally there is the 3250-km, 75-Hz AET that was analyzed in terms of pulse time spread, but no MCF was estimated (Colosi et al., 1999).

Figure 7.9 shows a comparison between observed and modeled MCFs and EAPs for the AFAR experiment (Reynolds et al., 1985). Here the MCFs and EAPs are calculated using the more accurate eigenvalue method. In this example, the predictions are somewhat sensitive to the smoothing of the background sound-speed profile because the calculation requires U''. Two reasonable smoothings based on internal-wave cut-off scales and the Fresnel zone bracket the observations. The observed EAP shows a clear asymmetry with an obvious precursor at lags earlier than -8 ms, and a hint of a long tail for lags greater than 10 ms. The path integral result for the less smoothed profile reproduces the precursor quite well. This precursor is a result of the waveguide and is controlled by the smallest eigenvalues in Q (Dashen et al., 1985; Reynolds et al., 1985). It is also important to point out that in this example, the EAP width is dominated by the travel-time wander, and so the precursor is seen only for significantly large values of the lag.

A quite different mean pulse was observed during the AET (Figure 5.15). Here the pulse time wander was removed from the EAP by stacking the observed pulses with the peaks aligned. This EAP has a symmetrical central peak with little broadening and a strong plateau-like precursor and tail. Gaussian fits to these average pulses gives an effective pulse time spread τ_0 between 0 and 5.3 ms, while predictions using Eq. 7.89 yield values in the hundreds of ms range.

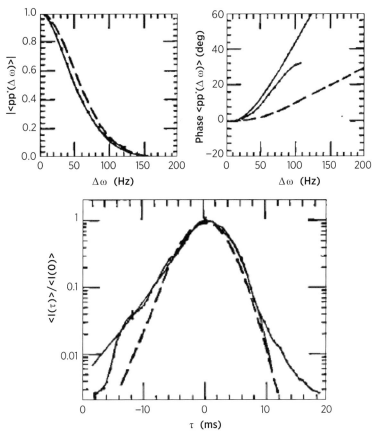

Figure 7.9. Observations and path integral estimates of the MCF (top) and EAP (bottom) for the AFAR experiment transmissions at 3200 Hz. The observations are plotted with a dotted line, and path integral predictions for two different smoothings of the sound-speed profile are shown with a solid and dashed line. For the MCF the slope at zero frequency separation is set to zero, reflecting our lack of knowledge of the travel-time bias. The path integral predictions are computed using the eigenvalue method.
Source: Reynolds et al. (1985).

Using range-independent climatological profiles of sound speed and buoyancy frequency for the AET along with the GM spectrum, a calculation of the EAP using the eigenvalue method predicts significantly less spread (Figure 7.10). The computed bias and spread for time front ID 125 are approximately −2 and 17 ms, respectively. Path integral calculations for other arrivals with larger IDs give similar results, with biases from +10 to −10 ms and spreads from 10 to 20 ms. These computed spreads are still much larger than observed. In addition, the new calculation gives a somewhat symmetrical EAP that does not reproduce the precursor and tail plateau that is seen in the observations. The AET mean pulse remains somewhat of a mystery and is likely influenced by ray chaos effects.

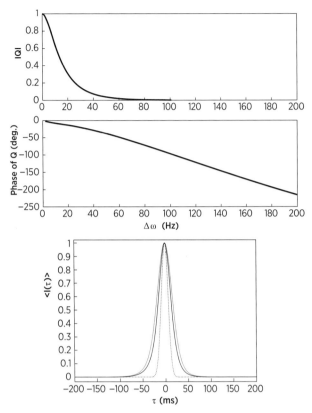

Figure 7.10. Path integral calculation using the eigenvalue method for the micro-ray bandwidth function Q (top panels) and the EAP (bottom) for the 3250-km, 75-Hz AET experiment. The calculation here is done using range-independent climatological profiles, and the parameters in Table 7.2. The wave front ID is +125. Estimates of the EAPs with (gray) and without (black) time wander are shown. The transmitted pulse (dot) has been time shifted to line up with the EAP without time wander. The approximate bias and spread are −2 and 17 ms, respectively.

7.6 Time-Lagged Intensity Covariance

The results from AFAR appear not to be unusual in the respect that travel-time wander is typically much larger than pulse spread. Thus to observe pulse spread an observable is needed that is wander independent or insensitive. One way to deal with this situation with regard to the EAP is to line up peaks (such as was done for the AET), but it is unclear how this rather ad hoc procedure alters the spread estimate from what is predicted by theory. Therefore a wander independent observable is needed and one such observable is the time-lagged intensity covariance:

$$f(\Delta\tau) = \left\langle \int_{-\infty}^{\infty} I(\tau + \Delta\tau)I(\tau)d\tau \right\rangle \Big/ \left\langle \int_{-\infty}^{\infty} I^2(\tau)\, d\tau \right\rangle. \tag{7.95}$$

The intensity covariance function is seen to be a fourth moment, which presents some added difficulties over the simpler second-moment EAP. Using Eq. 7.70, the intensity covariance function can be written in terms of four integrals over frequency and one over time, but the time integral yields a delta function that then leaves the triple integral (Dashen, 1979; Worcester et al., 1981):

$$f(\Delta\tau) = \frac{1}{2\pi} \int_{-\infty}^{\infty} d\omega \int_{-\infty}^{\infty} d\omega' \int_{-\infty}^{\infty} d\Delta\omega \, T_{XX}(\omega, \omega', \Delta\omega)$$

$$\times \langle p(\omega + \Delta\omega/2)p^*(\omega - \Delta\omega/2)p(\omega' - \Delta\omega/2)p^*(\omega' + \Delta\omega/2)\rangle \exp(i\Delta\omega\Delta\tau),$$

(7.96)

where

$$T_{XX}(\Delta\omega, \omega, \omega') = X(\omega + \Delta\omega/2)X^*(\omega - \Delta\omega/2)X(\omega' - \Delta\omega/2)X^*(\omega' + \Delta\omega/2).$$

(7.97)

The fourth moment $\langle p(\omega_1)p^*(\omega_2)p(\omega_3)p^*(\omega_4)\rangle$ has been worked out (Dashen, 1979, see Eq. 8.9) in both the partially saturated and fully saturated regimes. The intensity covariance in both cases is (Dashen, 1979; Worcester et al., 1981)[5]

$$f(\Delta\tau) = \frac{1}{2\pi} \int_{-\infty}^{\infty} d\omega \int_{-\infty}^{\infty} d\omega' \int_{-\infty}^{\infty} d\Delta\omega \, T_{XX}(\omega, \omega', \Delta\omega)$$

$$\times \left[|Q(\Delta\omega)|^2 + |Q(\omega - \omega')|^2 \right] \exp(i\Delta\omega\Delta\tau). \quad (7.98)$$

To gain some physical insight into this complicated equation, some analytic progress can be made by again utilizing Eq. 7.87 and convolving with a Gaussian pulse (X) to get

$$f(\Delta\tau) = \frac{1}{2}\left[\exp\left(-\frac{\Delta\tau^2}{2\alpha^2}\right) + \exp\left(-\frac{\Delta\tau^2}{2\alpha^2 + 4\tau_0^2}\right)\right]. \quad (7.99)$$

The first term represents the covariance of the unperturbed pulse, while the second term includes the microray-induced spreading τ_0^2.

7.6.1 Observations

Figure 7.11 shows a comparison between observed and modeled intensity covariance for the AFAR (Flatté et al., 1987b) and San Diego experiments (Worcester et al., 1981). For AFAR, the spread is rather small, and the estimate

---

[5] There is an expression for $f(\Delta\tau)$ in the literature for the saturated regime that is in error (Spiesberger and Worcester, 1981; see Eq. 4 and A.2). This expression for the intensity covariance function incorrectly depends on the pulse wander, that is, $\tau^2 = \Phi^2/\omega^2$. The calculation for the AFAR (Flatté et al., 1987b) and San Diego (Worcester et al., 1981) experiments are done correctly.

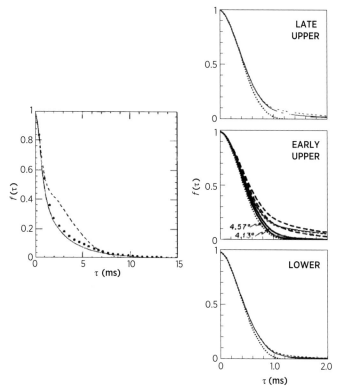

Figure 7.11. Observed and modeled intensity covariance function from the 35-km, 3200-Hz AFAR transmissions (left panel) and the 23-km San Diego Experiment (right panels). For AFAR, the observations are shown with dots, and the theoretical curves use calculations of Q utilizing Eq. 7.91 (solid) and 7.87 (dash). For the San Diego experiment, the intensity covariance for three ray arrivals is displayed. The observations for two different receiver depths are shown with light dash and light solid lines. For the late upper and lower ray paths, the theoretical result is shown with a dotted curve; for the early upper path the dotted curve is the result for the transmitted pulse. For the early upper arrival, theoretical results are shown for two different smoothings of the sound-speed profile (heavy solid and heavy dash).
Sources: Worcester et al. (1981) and Flatté et al. (1987b).

of the intensity covariance using Eq. 7.87 does not accurately describe the observations. A full calculation, estimating Q using Eq. 7.91, is much better. For the San Diego experiment, results are shown for three different ray arrivals. Here the spread is seen to be quite small indeed, and the theory for the early upper arrival using Eq. 7.91 correctly shows that there is little time spread.

7.7 Intensity Coherence and Spectra

The unresolved problem of the disagreement between observed intensity spectra in the Cobb/MATE and AFAR experiments and those predicted by weak fluctuation theory is now taken up. From Chapter 6 it was shown that the observed intensity spectra had much more energy at high frequencies than predicted by weak fluctuation theory. This intensity variability at high frequency is due to microray interference effects, which are indicative of stronger scattering. In this section it will be shown that the path integral method brings theory and observation into relatively good agreement. Similarly good results were obtained for the MATE intensity spectra using moment equations techniques (Macaskill and Ewart, 1996).

The intensity coherence function, like the previously discussed intensity covariance, is a fourth moment. When the coherence is a function of only one coordinate, say Δt or Δz, then the Fourier transform gives the intensity spectrum. Here the result is

$$\langle I(1)I(2) \rangle = \langle p(1)p^*(1)p(2)p^*(2) \rangle, \tag{7.100}$$

where 1 and 2 represent specific observation endpoints in either space or time. When discussing the intensity coherence, it is important to distinguish between the partially and fully saturated regimes. In the saturated regime, the statistics of p are nearly Gaussian. For Gaussian variables, fourth moments can be written in terms of second moments, giving

$$\langle p(1)p^*(2)p(3)p^*(4) \rangle = \langle p(1)p^*(2) \rangle \langle p(3)p^*(4) \rangle + \langle p(1)p^*(4) \rangle \langle p(3)p^*(2) \rangle, \tag{7.101}$$

where (1), (2), (3), and (4) correspond to four different end points. Using the Gaussian rule, the intensity coherence function in the saturated regime has the simple form

$$\langle I(1)I(2) \rangle = 1 + \exp[-D(1,2)], \tag{7.102}$$

where $D(1,2)$ is the familiar phase structure function. The normalization has been chosen so that $\langle I \rangle = 1$, and so for zero separation the result is $\langle I^2 \rangle = 2$, which is consistent with the saturated exponential distribution (Chapter 4). Simple expressions therefore exist for intensity spectra and correlation functions in the asymptotic limits of weak fluctuations and of saturation.

In the partially saturated regime, a more careful calculation taking into account the correlations between microrays gives

$$\langle I(\Delta)I(0) \rangle = 1 + K(\Delta), \tag{7.103}$$

where the calculation of $K(\Delta)$ involves the numerical integration of a second-order differential equation along the unperturbed ray path (Flatté et al., 1987a). For small lags

$$K(\Delta) \simeq \exp[-(\nu'\Delta)^2], \tag{7.104}$$

where ν' is a representative intensity decorrelation scale.

Here the rather technical path integral derivation of $K(\Delta)$ is sketched out so that the reader has some appreciation of the approximations that are essential to the theory. The intensity coherence function is written as a quadruple path integral (Flatté et al., 1987a, 1983):

$$\langle I(\Delta)I(0)\rangle = |N|^4 \int D^4(path)\exp\big(i(\Theta_0(path_1) - \Theta_0(path_2) +$$

$$\Theta_0(path_3) - \Theta_0(path_4)) - \frac{M}{2}\big), \tag{7.105}$$

with

$$M = q_0^2\Big\langle\Big(\int_{path_1}\mu\,dx - \int_{path_2}\mu\,dx + \int_{path_3}\mu\,dx - \int_{path_4}\mu\,dx\Big)^2\Big\rangle. \tag{7.106}$$

Recall that the form of M is a result of the ZMGRV approximation for the sound-speed fluctuations, so that the expectation value can be brought up into the exponent of the path integral. As for previous derivations, there is an expansion about the unperturbed ray, using a new set of variables $v_i(x) = z_i(x) - z_r(x)$, where the i index goes from 1 to 4 representing the four paths. Also as before, the y-deviations of the paths will be ignored. Usually this was done at the end of the calculation, but for simplicity in this more complicated case the y coordinate will be ignored from the start. The quantities $v_i(x)$ constitute four new path variables, but it proves expedient to define four new variables by the transformation (Codona et al., 1986a):

$$2\alpha = v_1 + v_2 + v_3 + v_4,$$

$$2\beta = (v_1 + v_2) - (v_3 + v_4),$$

$$2\gamma = (v_1 + v_4) - (v_2 + v_3),$$

$$2\delta = (v_1 + v_3) - (v_2 + v_4). \tag{7.107}$$

These variables $(\alpha,\beta,\gamma,\delta)$ physically represent the mean vertical position of the four paths, the difference between the mean of paths (1,2) and (3,4), the difference between the mean of (1,4) and (2,3), and finally the difference between the mean of (1,3) and (2,4). If it is assumed, as before, that the correlation function of μ is not a function of the mean vertical location (i.e., α), then M is independent of α. This means that the α path integral can be carried out, and essentially absorbed into the normalization. Also, for the case of intensity coherence, paths 1 and 2 end at the same point, and similarly for paths 3 and 4. The δ variable is then zero (Flatté et al., 1987a).

Thus what remains is only a double-path integral over the variables β and γ, which after a great deal of manipulation is given by (Flatté et al., 1987a)

$$\langle I(\Delta)I(0) \rangle = |N|^2 \int D\beta D\gamma \exp\left(-iq_0 \int_0^R [\dot\beta\dot\gamma - U_0''\beta\gamma]\,dx - \frac{M}{2}\right),$$

(7.108)

with

$$M = \int_0^R [2d(z(x);\gamma,0) + 2d(z(x);\beta,\Delta t) - d(z(x);\beta+\gamma,\Delta t) - d(z(x);\beta-\gamma,\Delta t)]\,dx.$$

(7.109)

Under the ray-tangent approximation, the phase structure function density for vertical and temporal separations is given by

$$d(z(x);\Delta z,\Delta t) = 2q_0^2\langle\mu^2\rangle L_p f(z(x);\Delta z,\Delta t),$$

(7.110)

where $f(z(x);\Delta z,\Delta t) = \{1 - \cos(m\Delta z(x))\cos(\sigma\Delta t)\}$. For small separations in time and depth only there are the approximate relations,

$$f(z(x);0,\Delta t) = \{\sigma^2\}\frac{\Delta t^2}{2},$$

(7.111)

$$f(z(x);\Delta z,0) = \{m^2\}\xi_1^2(x)\frac{\Delta z(R)^2}{2},$$

(7.112)

where $\Delta z(R)$ is the separation at the receiver. The dummy variables β and γ have the boundary conditions $\beta = \gamma = 0$ at $x = 0$ and $\beta = \Delta z(R)$, $\gamma = 0$ at $x = R$.

Equation 7.108 gives important insight into the wave propagation physics. It is seen that the paths contribute pairwise to the intensity fluctuations, where the pairs are (1,2) and (3,4) (i.e., the β coordinate) and (1,4) and (2,3) (i.e., the γ coordinate). Thus evaluating the path integrals in Eq. 7.108 requires a search for regions of path space for which these pairs significantly contribute. These paths are of course those whose values of M are not too large, because each path's contribution is attenuated by the term $\exp(-M/2)$. The value of M will be large if the paths deviate from one another by more than an acoustic coherence length, so these regions for the ray pairs are approximately given by (Flatté et al., 1983)

$$\text{Region A} = \left(|z_1(x) - z_2(x)| < \frac{\hat{L}_z}{\Phi} \text{ and } |z_3(x) - z_4(x)| < \frac{\hat{L}_z}{\Phi}\right),$$

(7.113)

$$\text{Region B} = \left(|z_1(x) - z_4(x)| < \frac{\hat{L}_z}{\Phi} \text{ and } |z_3(x) - z_2(x)| < \frac{\hat{L}_z}{\Phi}\right),$$

(7.114)

where $\hat{L}_z$ is the vertical correlation length of the internal waves, and $\hat{L}_z/\Phi$ is the acoustic vertical coherence scale (Chapter 4, Eq. 4.66). Now that these two regions of path space have been identified, the key issue is the extent to which these regions overlap; that is, what kinds of correlations exist between these pairs of paths?

In the limit that the paths are uncorrelated, the result is

$$M \simeq 2 \int_0^R [d(z(x); \gamma, 0) + d(z(x); \beta, \Delta t)] \, dx. \qquad (7.115)$$

In evaluating the path integral consider the total solution to be the sum of the contributions from regions (A) and (B), and the concern is mostly for small β and γ. In this approximation for M, the parts separate nicely, and so the contribution from region (B) is

$$M_B = 2 \int_0^R d(z(x); \gamma, 0) \, dx. \qquad (7.116)$$

Because this is independent of both $\Delta z(R)$ and Δt and because the answer is normalized for these separations equal to zero, the (B) regions contributes unity to the path integral. For the (A) region the equation is

$$M_A = 2 \int_0^R d(z(x); \beta, \Delta t) \, dx. \qquad (7.117)$$

Substituting into Eq. 7.108 and integrating by parts gives

$$\langle I(\Delta) I(0) \rangle_A = |N|^2 \int D\beta \exp\left(-\int_0^R d(z(x); \beta, \Delta t) \, dx\right)$$

$$\times \int D\gamma \exp\left(iq_0 \int_0^R \gamma[\ddot{\beta} + U_0'' \beta] dx\right). \qquad (7.118)$$

Hence, the double-path integral can be done easily because the γ path integral gives a product of delta functions $\delta(\ddot{\beta} + U_0'' \beta)$, enforcing the ray tube equations along the unperturbed path. Summing the (A) and (B) contributions, the result is

$$\langle I(\Delta) I(0) \rangle = 1 + \exp\left(-D(\Delta z(R), \Delta t)\right). \qquad (7.119)$$

This is precisely the result that was obtained from the Gaussian rule, Eq. 7.101. Thus the trick to finding the solution in the partially saturated regime will be to understand the pair correlations in the complete path integral relation. To accomplish this goal, analytic progress is best achieved by examining the cases of temporal and depth separations separately.

7.7.1 Separations in Time

In the previous section the main contributions to the path integrals came from two regions in which $|\beta| < \hat{L}_z/\Phi$ and $|\gamma| < \hat{L}_z/\Phi$. Thus to simplify the equations, M is approximated for small β and γ. For small γ in region (B) the result is

$$M_B \simeq \int_0^R d(z(x);\gamma,0)\, dx, \tag{7.120}$$

which is seen to be independent of Δt. Again because of the normalization of the coherence, this term contributes unity to the path integral. For small β in region (A) the result is

$$M_A \simeq \int_0^R [d(z(x);\gamma,0) + d(z(x);\beta,\Delta t) - d(z(x);\gamma,\Delta t)]\, dx. \tag{7.121}$$

The β and γ contributions are seen to separate. Hence the double-path integral of Eq. 7.108 in this regime is written as (Flatté et al., 1987a)

$$\langle I(\Delta t)I(0)\rangle = 1 + |N|^2 \int D\gamma \exp\left(-\frac{1}{2}\int_0^R [d(z(x);\gamma,0) - d(z(x);\gamma,\Delta t)]\, dx\right)$$

$$\times \int D\beta \exp\left(iq_0 \int_0^R \beta[\ddot{\gamma} + U_0''\gamma]dx - \frac{1}{2}\int_0^R d(z(x);\beta,\Delta t)\, dx\right), \tag{7.122}$$

where integration by parts has been used to yield the factor $\beta[\ddot{\gamma} + U_0''\gamma]$. To make further analytic progress, the path integrals have to be reduced to a quadratic form so that standard techniques can be applied. This means that the quadratic phase structure function density approximation is used, yielding

$$d(z(x);\gamma,0) - d(z(x);\gamma,\Delta t) = q_0^2\langle\mu^2\rangle L_p\left(\frac{1}{2}\{m^2\}\{\sigma^2\}\gamma^2\Delta t^2 + \{\sigma^2\}\Delta t^2\right), \tag{7.123}$$

$$d(z(x);\beta,\Delta t) = q_0^2\langle\mu^2\rangle L_p\left(\{m^2\}\beta^2 + \{\sigma^2\}\Delta t^2 - \frac{1}{2}\{m^2\}\{\sigma^2\}\beta^2\Delta t^2\right). \tag{7.124}$$

The double quadratic path integral can now be cast in matrix form, giving (Flatté et al., 1987a)

$$K(\Delta t) = N \int D\beta D\gamma \exp\left[-\frac{iq_0}{2}\int_0^R (\gamma,\beta)\right.$$

$$\left.\times\left[\left(\frac{d^2}{dx^2} + U_0''\right)\begin{pmatrix} 0 & 1 \\ 1 & 0 \end{pmatrix} - \frac{2i}{q_0}\begin{pmatrix} H_1 & 0 \\ 0 & H_2 \end{pmatrix}\right]\begin{pmatrix} \gamma \\ \beta \end{pmatrix} dx\right], \tag{7.125}$$

where

$$H_1(x) = q_0^2\langle\mu^2\rangle L_p\{m^2\}, \tag{7.126}$$

$$H_2(x) = \frac{1}{2}q_0^2\langle\mu^2\rangle L_p\{m^2\}\{\sigma^2\}\Delta t^2. \tag{7.127}$$

Here N is determined by the requirement that $K(\Delta t) \to 1$ as $\Delta t \to 0$. The path integral is evaluated by solving a set of coupled ordinary differential equations (Feynman and Hibbs, 1965) for a matrix $\hat{\mathbf{K}}$ given by

$$\left[\left(\frac{d^2}{dx^2} + U_0'' \right) \begin{pmatrix} 0 & 1 \\ 1 & 0 \end{pmatrix} - \frac{2i}{q_0} \begin{pmatrix} H_1 & 0 \\ 0 & H_2 \end{pmatrix} \right] \hat{\mathbf{K}} = 0. \tag{7.128}$$

The initial conditions for these equations are $\hat{\mathbf{K}}(x = 0) = 0$ and $\hat{\mathbf{K}}'(x = 0) = I$ (The identity Matrix), and the solution is

$$K(\Delta t) = \left[\frac{\det \hat{\mathbf{K}}_0(R)}{\det \hat{\mathbf{K}}(R)} \right], \tag{7.129}$$

where $\hat{\mathbf{K}}_0$ is the solution for $\Delta t = 0$.

The solution for $K(\Delta t)$ by these means is clearly arduous, and so an approximate solution for small Δt helps give insight into the problem. Using the same methods described by Dashen et al. (1985) to estimate τ_0 and τ_1, it is found that for small time separations $K(\Delta t) \simeq \exp(-\Delta t^2 / t_I^2)$ where the intensity coherence time t_I is given by (Flatté et al., 1987a)

$$t_I^{-2} = q_0^2 \int_0^R dx \langle \mu^2 \rangle L_p \{m^2\} \{\omega^2\} \int_0^R dx' \langle \mu^2 \rangle L_p \{m^2\} [g(x, x')]^2. \tag{7.130}$$

7.7.2 Separations in Depth

The calculations for depth separations with $\Delta t = 0$ are in fact quite easy. Again, regions of path space are examined for which β and γ are small, thus yielding an equation similar to Eq. 7.122. However, in this case the γ path integral is considerably easier because the structure function parts disappear. Hence the coherence takes the form

$$\langle I(\Delta z(R)) I(0) \rangle = 1 + |N|^2 \int D\beta \exp\left(-\frac{1}{2} \int_0^R d(z(x); \beta, 0) \, dx \right)$$
$$\times \int D\gamma \exp\left(iq_0 \int_0^R \gamma [\ddot{\beta} + U_0'' \beta] \, dx \right), \tag{7.131}$$

where the integration by parts has been done differently than for Eq. 7.122. The γ path integral gives a product of delta functions, enforcing the ray-tube equation along the unperturbed ray, and thus the coherence takes the same form as in the saturated regime, that is,

$$\langle I(\Delta z(R)) I(0) \rangle = 1 + \exp\left[-D(\Delta z) \right]. \tag{7.132}$$

7.7.3 Separations in Frequency

As previously mentioned, the fourth moment $\langle p(\omega_1)p^*(\omega_2)p(\omega_3)p^*(\omega_4)\rangle$ has been worked out in both the partially and fully saturated regimes (Dashen, 1979, Eq. 8.9). The result is that the intensity coherence in both cases is written (Dashen, 1979; Flatté et al., 1987a)

$$\langle I(\Delta\omega)I(0)\rangle = 1 + |Q(\Delta\omega)|^2. \tag{7.133}$$

The intensity coherence function in frequency is closely related to the time-lagged intensity covariance function (Eq. 7.96) and is an important observable because it is independent of the pulse wander. This function is perhaps more useful than the intensity covariance function for evaluating pulse spread and bias from observations, because it does not involve the source amplitude spectrum $X(\omega)$. Observations of this function have only appeared for the San Diego experiment (Worcester et al., 1981).

7.7.4 Observations

Comparisons between observations and theory have been done primarily for time separations, although there are a few results for depth and frequency separations (Flatté, 1983). Here the AFAR and MATE experiments provided the best acoustic and oceanographic data for comparing observations to theory. Figure 7.12 shows the observed and predicted time-lagged intensity coherences for AFAR at 35-km range and for three acoustic frequencies of 410, 1010, and 4670 Hz (Flatté et al., 1987b). Because the AFAR results are on the border between the partially and fully saturated regimes, the prediction is computed from a calculation of $K(\Delta t)$. The right panel of Figure 7.12 shows an expanded view of the 4670 Hz coherence. The fully saturated calculation utilizing $\exp(-D(\Delta t))$ clearly does not accurately predict the coherence. Taking the Fourier transform of the coherence function, Figure 7.13 shows the theory/observation comparison in the frequency domain; the agreement is excellent.

For the MATE experiment, Figure 7.14 compares the observed intensity spectra and those computed using the moment equation method (Macaskill and Ewart, 1996), which is formally equivalent to the path integral solution when the ray tangent or Markov approximation is made (Codona et al., 1986b; Henyey and Macaskill, 1996). The agreement is also exceptional although the MATE ocean model used in the moment equation calculation included both Garrett-Munk type internal waves and fine structure or spice (Levine and Irish, 1981; Levine et al., 1986). In the AFAR path integral calculation only Garrett-Munk internal waves were needed.

At longer range and lower frequency, the 220-Hz, 900-km Bermuda experiment quantified the temporal variability of intensity for three time-resolved acoustic

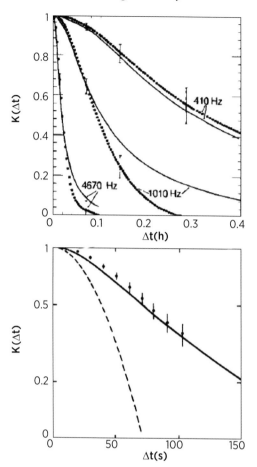

Figure 7.12. Intensity coherence for three acoustic frequencies from AFAR at 35-km range. The dotted curves show the observations, and the solid curves are the theoretical predictions based on $K(\Delta t)$. The bottom panel shows an expanded view of the 4670-Hz results; the solid curve gives the $K(\Delta t)$ prediction while the dashed curve is the full saturation result, that is, $\exp(-D(\Delta t))$.
Source: Flatté et al. (1987b).

paths (Spiesberger and Worcester, 1981). This experiment is predicted to be in the fully saturated regime. Figure 7.15 compares theory and observation for the intensity frequency spectrum; the results are not nearly as good as for the AFAR or MATE.

Finally results are shown for the intensity coherence as a function of frequency separation for the 3.2-kHz, 35-km AFAR (Flatté et al., 1987b) and the 2.2-kHz, 23-km range San Diego experiment (Worcester et al., 1981). Figure 7.16 shows observed and predicted (Eq. 7.133) coherence functions. In both cases the observed and predicted coherence functions show good agreement.

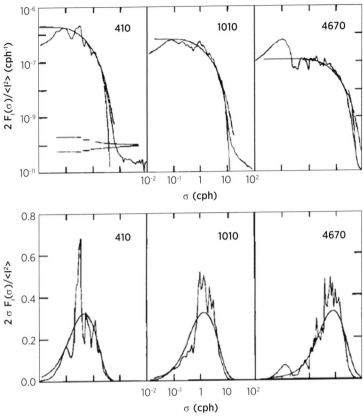

Figure 7.13. Frequency spectra of intensity from AFAR for three acoustic frequencies compared with theoretical predictions based on the Fourier transform of $K(\Delta t)$. The top panel is a log-log display, while the bottom panel displays the spectra in variance preserving form.
Source: Flatté et al. (1987b).

7.8 Intensity Moments: Microray Focusing Parameter

The path integral formalism also allows for the calculation of intensity moment corrections in the fully saturated regime. As was shown in Chapter 4, the PDF of intensity, in the asymptotic limit of long range, is exponential with moments $\langle I^n \rangle = n!$, where it is assumed that $\langle I \rangle = 1$. Corrections to these moments have been worked out by Flatté et al. (1983) such that

$$\langle I^n \rangle = n![1 + n(n-1)\gamma_f/2], \qquad (7.134)$$

where the scintillation index is $1 + 2\gamma_f$. The microray focusing parameter, γ_f, can be computed via path integral methods, which are reviewed here. While Eq. 7.134 gives the corrections for all moments, the path integral results are expected to be most accurate for $n = 2$.

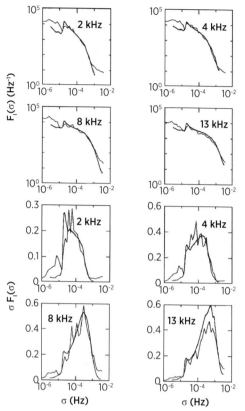

Figure 7.14. Frequency spectra of intensity from MATE for four acoustic frequencies compared with theoretical predictions based on moment equation theory. The top panels show a log-log display, while the bottom panels display the spectra in variance preserving form.
Source: Macaskill and Ewart (1996).

The mean square intensity was shown in the previous section to reduce to a double-path integral over ray pairs, that is, (Flatté et al., 1987b),

$$\langle I^2 \rangle = |N|^2 \int D\beta D\gamma \exp\left(-iq_0 \int_0^R [\dot{\beta}\dot{\gamma} - U_0''\beta\gamma]\, dx - \frac{M}{2}\right), \quad (7.135)$$

with

$$M = \int_0^R [2d(z(x);\gamma) + 2d(z(x);\beta) - d(z(x);\beta+\gamma) - d(z(x);\beta-\gamma)]\, dx. \quad (7.136)$$

Here the dependence on Δt has been dropped and both β and γ must therefore vanish at the end points.[6] As previously shown, contributions are summed from

---

[6] One might be tempted to carry out a calculation similar to what was done for the depth lagged intensity covariance but because β vanishes at the endpoints the result is simply $\langle I^2 \rangle = 2$ (e.g., the full saturation result). Thus we have to take the calculation to higher order.

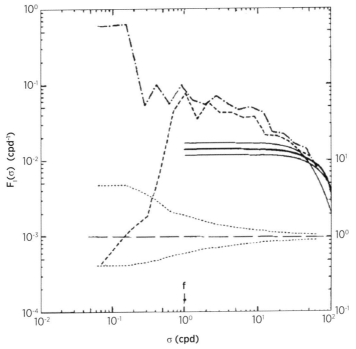

Figure 7.15. Frequency spectra of intensity at 220-Hz and 900-km range from the late arrival of the Bermuda experiment (dashed and dash dot) for two different time periods. Theoretical predictions based on the Fourier transform of $\exp(-D(\Delta t))$ are shown for internal-wave energy levels of 1/2, 1.0, and 2.0 the standard GM level. The 95% confidence intervals are drawn relative to unity. *Source*: Spiesberger and Worcester (1981).

the two regions that contribute to the path integral, that is, β and γ small. For these two cases, which are called (A) and (B), the result is

$$M_{A0} = \int_0^R d(z(x);\beta)\, dx, \quad M_{B0} = \int_0^R d(z(x);\gamma)\, dx. \qquad (7.137)$$

Substituting into the double-path integral, however, yields the uncorrected result

$$\langle I^2 \rangle = |N|^2 \int D\beta D\gamma \left[\exp\left(-iq_0 \int_0^R [\dot\beta\dot\gamma - U_0''\beta\gamma]\, dx - \frac{M_{A0}}{2}\right) \right.$$
$$\left. + \exp\left(-iq_0 \int_0^R [\dot\beta\dot\gamma - U_0''\beta\gamma]dx - \frac{M_{B0}}{2}\right) \right] = 2. \qquad (7.138)$$

The corrections to saturated statistics, therefore, come from the differences $M - M_{A0}$ and $M - M_{B0}$. This has been handled by expanding the path integral in

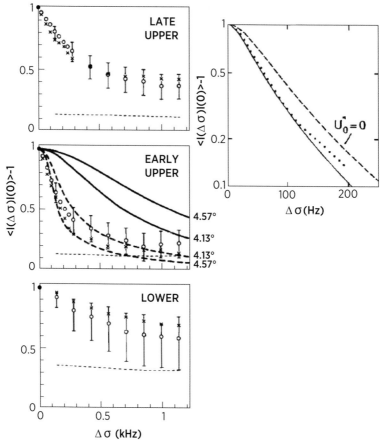

Figure 7.16. Left three panels Intensity coherence as a function of frequency separation for the 2.2-kHz San Diego experiment. Results are displayed for two different receivers (X, O) and three different arrivals. Path integral theory curves using Eq. 7.133 are given for the early upper arrival for two smoothings of the sound-speed profile (heavy solid and heavy dashed). Points above the horizontal dash lines differ significantly from zero. Right panel Intensity coherence as a function of frequency separation for the 3200-Hz, 35-km AFAR transmissions (dots). Path integral predictions are for $U'' = 0$ (dashed) and U'' given by the observed profiles (solid).
Source: Worcester et al. (1981).

Eq. 7.135 in a power series of these differences, that is,

$$|N|^2 \int D\beta D\gamma \exp\left(-iq_0 \int_0^R [\dot\beta\dot\gamma - U_0''\beta\gamma]\, dx - \frac{M}{2}\right)$$

$$\simeq \sum_{r=1}^{\infty} \int D\beta D\gamma \exp\left(-iq_0 \int_0^R [\dot\beta\dot\gamma - U_0''\beta\gamma]\, dx - \frac{M_{A0}}{2}\right) \frac{[(M_{A0} - M)/2]^r}{r!}$$

$$+ \sum_{r=1}^{\infty} \int D\beta D\gamma \exp\left(-iq_0 \int_0^R [\dot\beta\dot\gamma - U_0''\beta\gamma]\, dx - \frac{M_{B0}}{2}\right) \frac{[(M_{B0} - M)/2]^r}{r!}.$$

$$(7.139)$$

It has been found that this series is not convergent because the small contributions from the overlapping regions of (A) and (B) have not been treated exactly. However, the first few terms appear to have some practical value when one is relatively close to saturation (Flatté et al., 1983).

The first few terms have been worked out for the Garrett-Munk internal-wave spectrum, giving (Flatté, 1983; Flatté et al., 1983)

$$\gamma_f = 2q_0^2 \int_0^R dx \langle \mu^2 \rangle L_p \{P(j,x)\}, \tag{7.140}$$

where

$$\{P(j,x)\} = M_j \sum_{j=1}^J \frac{P(j,x)}{j(j^2 + j_*^2)}, \tag{7.141}$$

and

$$P(j,x) = \left(1 - \cos\left[\frac{m^2(j,z(x))g(x,x)}{q_0}\right]\right) \exp\left[-2q_0^2 \int_0^R dx' \langle \mu^2 \rangle L_p\right.$$

$$\left. \times M_j \sum_{k=1}^J \frac{1 - \cos[m(j,z(x))m(k,z(x'))g(x,x')/q_0]}{k(k^2 + k_*^2)}\right]. \tag{7.142}$$

Here the vertical wavenumber is given by the WKB approximation $m(j,z(x)) = \pi j N(z(x))/N_0 B$, and the normalizing factor is

$$M_j^{-1} = \sum_j \frac{1}{j(j^2 + j_*^2)}. \tag{7.143}$$

Various approximations to γ_f are discussed in detail in Flatté et al. (1983) and Flatté et al. (1987b). One approach is as follows. The function $P(j,x)$ can be approximately expressed in terms of two functions of x, called α and β:

$$P(j,x) = (1 - \cos\beta j^2)\exp(-\alpha j^2), \tag{7.144}$$

where

$$\alpha(x) = M_j \left(\frac{\pi N(z(x))}{N_0 B}\right)^2 \int_0^R dx' \langle \mu^2 \rangle L_p \left(\frac{\pi N(z(x'))}{N_0 B}\right)^2 [g(x,x')]^2, \tag{7.145}$$

$$\beta(x) = \left(\frac{\pi N(z(x))}{N_0 B}\right)^2 \frac{g(x,x)}{q_0}. \tag{7.146}$$

For small $\alpha < 10^{-3}$ an asymptotic expansion can be done to obtain

$$\{P(j,x)\} = \frac{1}{4} \frac{\beta^2}{\alpha}. \tag{7.147}$$

Table 7.4. *Observed and modeled scintillation index for several experiments*

Experiment	SI (observed)	SI (modeled)	Notes
MATE (2 kHz, lower path)	0.59	0.54	ME
MATE (4 kHz, lower path)	1.15	1.0	ME
MATE (8 kHz, lower path)	1.34	1.5	ME
MATE (13 kHz, lower path)	1.40	1.4	ME
MATE (2 kHz, upper path)	0.63	N/A	
MATE (4 kHz, upper path)	0.89	N/A	
MATE (8 kHz, upper path)	1.06	N/A	
MATE (13 kHz, upper path)	0.93	N/A	
AFAR (410 Hz)	1.24	1.30	PI
AFAR (1010 Hz)	1.21	1.22	PI
AFAR (4670 Hz)	1.47	1.13	PI
NPAL/SOSUS (75 Hz; Kauai-N)	1.53	1.47	PI
NPAL/SOSUS (75 Hz; Kauai-O)	1.69	1.32	PI

In the rightmost column ME stands for Moment Equation solution and PI stands for Path Integral solution. The MATE and AFAR values are from Ewart and Reynolds (1993) and Flatté et al. (1987b), respectively, and the NPAL results are from Andrew et al. (2005).

7.8.1 Observations

Table 7.4 gives a summary of the comparisons between observations and predictions of scintillation index. Given the large number of experiments conducted over the last several decades, there have been scant few comparisons of these intensity moments.

The most detailed comparison between predicted and observed intensity moments where there is extensive environmental information on internal waves and other processes has been for the AFAR observations (Flatté et al. 1987) and the MATE observations (Macaskill and Ewart, 1996). For the 35-km AFAR path, the observed variances of intensity were 1.24, 1.21, and 1.47 for the acoustic frequencies of 410, 1010, and 4670 Hz. Theoretical computations of γ_f for these three frequencies using Eq. 7.134 gives intensity variance predictions of 1.3, 1.22, and 1.13. As frequency increases the signal is expected to get closer and closer to saturation. The agreement between observed and predicted variances is rather good for the two lower frequencies. The cause of the disagreement at 4670 Hz is unknown.

For the MATE lower path observations, there is excellent agreement between observed and modeled scintillation index (Table 7.4), but again the good agreement involved modeling both internal waves and spicy thermohaline structure (Ewart and Reynolds, 1993; Macaskill and Ewart, 1996). In Table 7.4 the MATE upper path scintillation index is also shown. The moment equation method failed to provide accurate results for the upper path due to the Markov approximation. Path integral calculations without the Markov approximation have not been carried out for this path.

Another comparison between path integral calculations and observations was done by Andrew et al. (2005) for 2500-km range, 75-Hz propagation in the central North Pacific. The ray arrivals were predicted to be in the fully saturated regime with *SI* values for two paths from the Kauai source of 1.53 and 1.69. The path integral prediction gives corresponding values of 1.47 and 1.32.

As with other theoretical expressions using the path integral, the calculation of γ_f has been found to be rather sensitive to the detailed shape of the sound-speed profile. The new methods of computing ray tube equations using the Helmholtz equation may significantly reduce these sensitivities that were seen using the parabolic equation ray tube functions (Chapter 6).

7.9 Summary

The method of path integrals completes our theoretical development of the notion of microray interference and acoustic fluctuations in both phase and amplitude along specific ray paths. Importantly, for phase, the basic results of ray theory are still valid even into the fully saturated regime. Indeed, the phase structure function was shown to play a critical role in many path integral formulae.

For amplitude fluctuations, on the other hand, critical diffractive corrections are required to handle all regimes from unsaturated to fully saturated propagation. The treatment of diffractive corrections, however, requires some approximations, namely expansion about the deterministic ray, truncation of that expansion at second order, and while not absolutely necessary from a computational standpoint, the ray tangent or Markov approximation is made.

One important limitation in path integral results as well as weak fluctuation theory is that the receiver cannot be too close to a caustic. In this case the ray tube Green's functions do not properly describe near by rays. The breakdown of this approach at long range is inevitably tied to ray instability/chaos, but we are only just beginning to understand the limits of the formalism for the various acoustical observables that can be considered. More studies, in which comparisons are made between observations and theory, are urgently needed.

Appendix A Path Integrals as Products of Delta Functions

The focus here is on path integrals of the form

$$\int D(\alpha)\exp\left[iq_0\int_0^R F(x)\alpha(x)\,dx\right].\tag{A.1}$$

To evaluate this path integral, the path $\alpha(x)$ is expanded using an orthonormal set (usually a sine expansion). Writing $\alpha(x) = \sum_{n=1}^{\infty} a_n \phi_n(x)$ the total path integral can be viewed as an integral over all possible a_n. Thus, using Eq. 7.13 the path integral is written as

$$\int_{-\infty}^{\infty} da_1 .... \int_{-\infty}^{\infty} da_n \exp\left[iq_0 \sum_{n=1}^{\infty} \int_0^R F(x) a_n \phi_n(x)\, dx \right]. \qquad (\text{A.2})$$

But this is just a product of terms of the form

$$\int_{-\infty}^{\infty} da_n \exp\left[iq_0 a_n \int_0^R F(x) \phi_n(x)\, dx \right] = 2\pi\delta(F(x)). \qquad (\text{A.3})$$

Appendix B Solution of the Gaussian Path Integral for Frequency Correlations

Here it is shown how to evaluate the path integral in Eq. 7.85, by expanding the u-path in an orthonomal set of functions. Because of the operator M in the path integral, it proves expedient to choose the orthonormal set using the eigenvalue equation

$$L\phi_n = \lambda_n F(x)\phi_n, \qquad (\text{B.1})$$

where $F(x) = \langle \mu^2 \rangle L_p \{m^2\}/c_0$, and the boundary conditions are $\phi_n(0) = \phi_n(R) = 0$ (i.e., $\delta z = 0$ at the end points). The orthonormality relation is

$$\int_0^R F(x)\phi_n(x)\phi_m(x)\, dx = \delta_{nm}. \qquad (\text{B.2})$$

In the path integral exponent the result is

$$\int_0^R uMu\, dx = \int_0^R uM \sum_{n=1}^{\infty} a_n \phi_n\, dx,$$

$$= \int_0^R \sum_{m=1}^{\infty} \sum_{n=1}^{\infty} a_m a_n\, F\phi_m \phi_n (\lambda_n - i\Delta\omega)\, dx,$$

$$= \sum_{n=1}^{\infty} a_n^2 (\lambda_n - i\Delta\omega)\, dx, \qquad (\text{B.3})$$

where the last line follows from the orthogonality relation. The path integral in this eigenmode expansion is again viewed as an infinite product of integrals over all possible a_n, that is,

$$\int D(u) = \int_{-\infty}^{\infty} da_1 \int_{-\infty}^{\infty} da_2 ..... \int_{-\infty}^{\infty} da_n. \qquad (\text{B.4})$$

Written this way the path integral is given by

$$Q(\Delta\omega) = \prod_{n=1}^{\infty} \int_{-\infty}^{\infty} da_n \exp\left[-\frac{i\bar{q}}{2\Delta q}(\lambda_n - i\Delta\omega)a_n^2\right] = \prod_{n=1}^{\infty} \frac{C}{(\lambda_n - i\Delta\omega)^{1/2}}, \qquad (B.5)$$

where C is a constant that will be determined by the normalization condition, that is, $Q = 1$ for $\Delta\omega = 0$. It is then useful to write

$$Q(\Delta\omega) = \prod_{n=1}^{\infty} \left(\lambda_n - i\Delta\omega\right)^{-1/2} = \prod_{n=1}^{\infty} \lambda_n^{-1/2} \prod_{n=1}^{\infty} \left(1 - \frac{i\Delta\omega}{\lambda_n}\right)^{-1/2}. \qquad (B.6)$$

The first product does not depend on $\Delta\omega$, so it can be subsumed into the normalization, giving the final result:

$$Q(\Delta\omega) = \prod_{n=1}^{\infty} \left(1 - \frac{i\Delta\omega}{\lambda_n}\right)^{-1/2}. \qquad (B.7)$$

Appendix C GM Spectral Averages for Path Integral Equations

The spectral average of the squared internal-wave frequency under the perpendicular wavenumber constraint is

$$\{\sigma^2\} = M_\sigma \int_{\sigma_L}^{N} \frac{\sigma_L^2}{\sigma} \left(\frac{\sigma^2 - f^2}{\sigma^2 - \sigma_L^2}\right)^{1/2} d\sigma,$$

$$= \frac{M_\sigma \sigma_L^2}{2}\left[\ln\left(\frac{2N^2 - \sigma_L^2 - f^2 + 2(N^2 - \sigma_L^2)^{1/2}(N^2 - f^2)^{1/2}}{\sigma_L^2 - f^2}\right)\right.$$

$$\left. + \frac{f}{\sigma_L}\ln\left(\frac{N^2(\sigma_L^2 + f^2) - 2f^2\sigma_L^2 - 2f\sigma_L(N^2 - \sigma_L^2)^{1/2}(N^2 - f^2)^{1/2}}{N^2(\sigma_L^2 - f^2)}\right)\right], \qquad (C.1)$$

where the normalization factor is

$$M_\sigma^{-1} = \int_{\sigma_L}^{N} \frac{\sigma_L^2}{\sigma^3} \left(\frac{\sigma^2 - f^2}{\sigma^2 - \sigma_L^2}\right)^{1/2} d\sigma,$$

$$= \frac{1}{2}\left[\left(1 - \frac{\sigma_L^2}{N^2}\right)^{1/2}\left(1 - \frac{f^2}{N^2}\right)^{1/2} + \left(\frac{\sigma_L}{f} - \frac{f}{\sigma_L}\right)\right.$$

$$\left. \ln\left(\frac{(f/\sigma_L)(1 - \sigma_L^2/N^2)^{1/2} + (1 - f^2/N^2)^{1/2}}{(1 - f^2/\sigma_L^2)^{1/2}}\right)\right]. \qquad (C.2)$$

With the perpendicular wavenumber constraint, the dispersion relation gives $l^2 = (\pi j/N_0 B)^2(\sigma^2 - \sigma_L^2)$, so the spectral average of the horizontal wavenumber becomes

$$\{l^2\} = M_\sigma \left(\frac{\pi}{N_0 B}\right)^2 \frac{\langle j \rangle}{\langle j^{-1} \rangle} \int_{\sigma_L}^{N} \frac{\sigma_L^2}{\sigma^3} (\sigma^2 - f^2)^{1/2}(\sigma^2 - \sigma_L^2)^{1/2} \, d\sigma, \qquad \text{(C.3)}$$

$$= \frac{M_\sigma}{2} \left(\frac{\pi \sigma_L}{N_0 B}\right)^2 \frac{\langle j \rangle}{\langle j^{-1} \rangle} \left[\frac{f^2 + \sigma_L^2}{f \sigma_L} \ln\left(\frac{N^2(\sigma_L^2 + f^2) - 2f\sigma_L(a + f\sigma_L)}{N^2(\sigma_L^2 - f^2)}\right) \right.$$

$$\left. + \ln\left(\frac{2(a + N^2) - f^2 - \sigma_L^2}{\sigma_L^2 - f^2}\right) - \frac{a}{N^2} \right], \qquad \text{(C.4)}$$

where $a = (N^4 - N^2(f^2 + \sigma_L^2) + f^2\sigma_L^2)^{1/2}$.

8

Mode Transport Theory

8.1 Introduction

There are several situations in which the methods of the previous chapters cannot be applied or fail outright. In particular, there are cases in which an unperturbed path cannot be identified. This occurs in shadow zone regions, in the finale of long-range deep-water transmissions, and in shallow-water propagation (Figure 1.9). Single-frequency transmissions in which several unperturbed paths add together coherently are a special case of this. There is also the case of long-range propagation for which ray chaos effects can become important and cumulative errors from the ray-tangent approximation become nonnegligible. Finally, there is the issue of attenuation, which is not treated in the previous methodologies and is critical for shallow-water applications.

Figure 8.1 shows the first observations that revealed the limitations of path integral theories for describing acoustic fluctuations. The data are from the 1989, 1000-km, 250-Hz SLICE89 experiment, where scattering into the deep shadow zone near the transmission finale is evident (also see Figures 1.6 and 1.7 showing ensonification of shadow zones for the early arriving time fronts). The SLICE89 observations also revealed that internal-wave-induced scattering leads to a loss of the coherent time front branch pattern in the finale. This branch pattern is needed to identify unperturbed ray paths.

This leads to the method of modal transport equations that have been known for some time outside of the ocean acoustics community. The development of this body of work appears in the physics and chemistry literature (see Van Kampen, 1981, and references therein). Transport equations have been used to describe diverse stochastic phenomena, such as waves in plasmas, Anderson localization, random walks and oscillations, many body problems, and spectral line broadening and narrowing. The key connection to the acoustics problem comes from the fact that the one-way coupled mode equation in Chapter 2 (Eq. 2.128) is in the same form as the Schrödinger equation in the "interaction" representation.

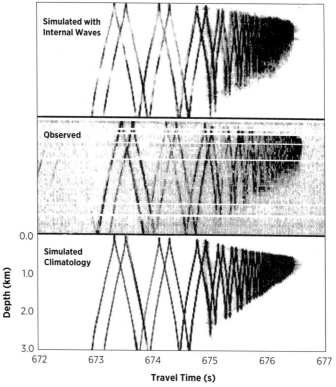

Figure 8.1. Observed and simulated time fronts for the 1000-km, 250-Hz SLICE89 experiment (Figure 7.3). The upper panel shows a parabolic equation simulation with a realization of random sound-speed perturbations obeying the Garrett-Munk internal-wave spectrum at half the reference energy, while the lower panel shows the same simulation without the internal-wave perturbations. The observations and simulation with internal waves shows significant depth broadening in the finale, which is indicative of mode coupling. *Source*: Colosi et al. (1994).

Introduction of the transport theory method to the ocean acoustics community is due to Dozier and Tappert (1978a), whose "Master Equations" for mode energy were far ahead of their time. After a long period of neglect, development of the mode transport equation approach in the last decade has seen remarkable success. Comparisons to Monte Carlo simulations in both deep and shallow-water environments (Dozier and Tappert, 1978b; Colosi and Morozov, 2009; Colosi et al., 2013a, 2011) show that the method is extremely accurate. Comparisons to deep-water observations are also quite good (Chandrayadula et al., 2013). This method does not suffer from inaccuracies as the acoustic angle increases, because the expansion in terms of the unperturbed modes has the high-angle ray geometry built in. The limitations are primarily associated with the weak coupling approximation, which suggests that the method will break down at higher frequencies; however, no such breakdown has been identified for ocean cases. It is also the case that the calculations can become overly burdensome when high frequencies or broadband, impulsive signals are considered though work in this direction has just begun. A ray-based description of mode coupling and of the transport theory results to be discussed in this chapter has recently been provided by Virovlyansky (2014).

8.2 Solutions to the Coupled Mode Equation

This analysis starts with the coupled mode equations in Chapter 2, where the acoustic pressure is written in terms of a sum over unperturbed modes $\phi_n(z)$,

$$p(r,z;\omega) = \sum_n \frac{a_n(r)\,\phi_n(z)}{\sqrt{k_n r}}, \tag{8.1}$$

and the mode amplitudes obey the one-way coupled equations in the weak multiple forward scattering approximation:

$$\frac{d\hat{a}_n}{dr} = -i \sum_{m=1}^{N} \rho_{mn}(r) e^{i l_{mn} r} \hat{a}_m(r). \tag{8.2}$$

Here $\hat{a}_n(r) = a_n(r) e^{-i l_n r}$ is the demodulated mode amplitude, and the complex modal wavenumber is given by $l_n = k_n + i\alpha_n$ and $l_{mn} = l_m - l_n$. The symmetric coupling matrix involves the projection of the fluctuating vertical sound-speed structure, $\mu = \delta c/c_0$, onto the modes m and n. It was shown in Chapter 2 that the coupling matrix is given by

$$\rho_{mn}(r) = \frac{q_0^2}{\sqrt{k_n k_m}} \int_0^\infty \frac{\phi_n(z)\phi_m(z)}{\rho_0(z)} \mu(r,z)\, dz, \tag{8.3}$$

where $q_0 = \omega/c_0$ is a typical acoustic wavenumber and $\rho_0(z)$ is the background density profile. Equation 8.2 is termed a linear stochastic differential equation with multiplicative noise, because the right-hand side is linear and because the stochastic term ρ_{mn} multiplies $\hat{a}_m$. As noted earlier, Eq. 8.2 is of the same form as the Schrödinger equation in the interaction representation and so it is expedient to utilize the well-developed tools of quantum mechanical time-dependent perturbation theory for analysis (Merzbacher, 1961; Sakurai, 1985; Shankar et al., 1994, just to mention a few). Here of course it is understood that the acoustical range coordinate corresponds to the quantum mechanical time coordinate so the approach could more aptly be named range-dependent perturbation theory.

For the present purposes it is useful to write the coupled mode differential equations (Eq. 8.2) as an integral equation:

$$\hat{a}_n(r) = \hat{a}_n(0) - i \sum_{m=1}^{N} \int_0^r dr' \rho_{mn}(r') e^{il_{mn}r'} \hat{a}_m(r').$$ (8.4)

Because the coupling matrix is small, Eq. 8.4 is amenable to solution by iteration in repeated powers of ρ_{mn}. The resulting series is often termed the Dyson series (Sakurai, 1985). The zeroth-order solution is simply the initial condition, $\hat{a}_n(0)$, and subsequent orders p are expressible in terms of the previous order solution $p-1$ such that

$$\hat{a}_{n_p}(r) = -i \sum_{m=1}^{N} \int_0^r dr' \rho_{mn}(r') e^{il_{mn}r'} \hat{a}_{m_{p-1}}(r').$$ (8.5)

Leaving aside questions of convergence, the first few terms of the Dyson series are

$$\hat{a}_n(r) = \hat{a}_n(0) - i \sum_{m=1}^{N} \hat{a}_m(0) \int_0^r dr' \tilde{\rho}_{mn}(r')$$

$$+ (-i)^2 \sum_{m=1}^{N} \sum_{q=1}^{N} \hat{a}_q(0) \int_0^r dr' \int_0^{r'} dr'' \tilde{\rho}_{mn}(r') \tilde{\rho}_{qm}(r'') + \ldots,$$ (8.6)

where $\tilde{\rho}_{mn}(r) = \rho_{mn}(r) e^{il_{mn}r}$ is a short-hand notation.[1] It is important to point out that while ρ_{mn} is a Hermitian matrix, $\tilde{\rho}_{mn}$ is not due to attenuation, which leads to nonconservation of energy. Physically the Dyson series can be thought of as a multiple scattering series (Colosi, 2008), that is, the zeroth-order term corresponds to no mode coupling, the first-order term has one coupling event, the second-order term has two coupling events, and so on.

From inspection of Eq. 8.6 and of higher order terms in the Dyson series, it is seen that the operators $\tilde{\rho}$ occur in a special order in which those at the largest range are on the left, and the ranges progressively decrease moving to the right. This special ordering of the matrices allows use of a standard technique from time-dependent perturbation theory, so that the solution to Eq. 8.4 is written in terms of a range-ordered product operator (Merzbacher, 1961; Baym, 1973; Codona et al., 1986b; Creamer, 1996).[2] To utilize this approach it is convenient to

---

[1] The Dyson series was first introduced to the ocean acoustic literature by Dozier and Tappert (1978a). Equation 8.6 has also been used to describe mode coupling caused by nonlinear shallow-water internal waves (Colosi, 2008).

[2] Those interested in studying the quantum mechanics texts should be mindful that they almost exclusively treat the case in which the system is initially in a single eigenstate of the unperturbed Hamiltonian. This corresponds to the ocean case in which all the initial energy is in a single mode.

write Eq. 8.2 in matrix notation

$$\frac{d\hat{\mathbf{a}}}{dr} = -i\tilde{\boldsymbol{\rho}}\hat{\mathbf{a}}(r), \tag{8.7}$$

where the vector of N mode amplitudes a_n is given by $\mathbf{a} = e^{i\mathbf{L}r}\hat{\mathbf{a}}$, $\mathbf{L}$ is a diagonal $N \times N$ matrix of the l_n complex mode wavenumbers, and $\tilde{\rho} = e^{-i\mathbf{L}r}\rho e^{+i\mathbf{L}r}$ is the $N \times N$ transformed coupling matrix.[3] For the three-dimensional problem in which azimuthal coupling is treated using Eq. 2.133 then $\mathbf{L}$ would include the Laplacian operator as well. Given the form of Eq. 8.7 it is not surprising that the solution is in the form of an exponential. The details of the solution are given in Baym (1973), with the result that

$$\hat{\mathbf{a}}(r) = \mathcal{R}\left[\exp\left(-i\int_0^r dr'\,\tilde{\rho}(r')\right)\right]\hat{\mathbf{a}}(0), \tag{8.8}$$

where the symbol $\mathcal{R}$ represents the range-ordering operator. The action of $\mathcal{R}$ is best understood by Taylor expanding the exponential, where the second-order term is given by

$$\frac{1}{2}\mathcal{R}\left[\int_0^r dr'\,\tilde{\rho}(r')\right]^2 = \int_0^r dr'\int_0^{r'} dr''\,\tilde{\rho}(r')\tilde{\rho}(r''), \tag{8.9}$$

which is exactly the second-order term in Eq. 8.6. Thus, the range-ordering operator enforces a structuring of the coupling matrices not in the order that they appear, but in such a way that larger values of range occur on the left with progressively smaller values moving to the right. This ordering is important because of the multiple scattering interpretation in which the matrices must be evaluated at the proper intermediate ranges.

Equation 8.8 appears to be a rather complicated expression, but it will be found that this form is quite useful when expectation values are considered. Because $\tilde{\rho}$ can be assumed to have a Gaussian PDF, due to the Gaussianity of δc, expectation values of Eq. 8.8 and other similar equations can be brought up into the argument of the exponential as was done in the path integral expressions.

Although two different notations for the coupled mode theory have been introduced, in the subsequent development the nonmatrix formulation of the problem will be primarily used (i.e., Eq. 8.2). This approach is taken because in the nonmatrix form it is clearer how various parameters enter the problem. The matrix form, Eq. 8.7, will be most useful in the formal derivation of the transport theories.

---

[3] Use of a lowercase symbol for the coupling matrix ρ is a break from standard notation. This is necessary to maintain continuity with the literature.

8.2.1 Adiabatic Solution

An important approximate solution to Eq. 8.2 is the adiabatic solution in which mode coupling is ignored, but some degree of mode phase fluctuation can be estimated. One might think this extreme approximation has no practical value, but this is not the case. In the adiabatic approximation, the sound-speed perturbations are assumed small enough and/or the frequency low enough that the coupling matrix is nearly diagonal, that is, $\rho_{mn} = \rho_{nn}\delta_{mn}$. In this case the solution to Eq. 8.2 is

$$\hat{a}_n(r) = \hat{a}_n(0)\exp\left[-i\int_0^r \rho_{nn}(r')\,dr'\right]. \qquad (8.10)$$

The modes here are uncoupled, and $a_n(r)$ varies only in phase, not amplitude.[4] Physically the phase fluctuations occur because $\rho_{nn}(r)$ acts like a local horizontal wavenumber perturbation, altering the mode phase speed. The accuracy of the adiabatic solution depends on the condition that the coupling matrix is much smaller than the unperturbed horizontal wavenumber; that is $|\rho_{nn}(r)| << k_n$. This condition is best satisfied for small δc and/or low frequency. This solution is useful because of its simple physical interpretation, and its relevance is usually best determined by comparison to coupled mode results.

In long-range deep-water propagation, adiabatic theory fails in many cases to accurately describe observed acoustic fluctuations, even at frequencies as low as 28 Hz (Colosi and Flatté, 1996; Worcester et al., 2000). One exception to this characterization of deep-water propagation is the accurate prediction of modal time coherences using adiabatic theory (Colosi et al., 2013a). In the Arctic where scattering from internal waves is weak, adiabatic theory has also successfully described 20-Hz trans-Arctic acoustic transmissions (Michalevsky et al., 1999), and it is expected that the theory will be useful at higher frequencies. Adiabatic theory has also had some success in describing low- and mid-frequency shallow-water propagation through the stochastic internal-wave field (Colosi et al., 2011; Raghukumar and Colosi, 2014). The utility of the adiabatic approximation will be further discussed in Sections 8.5 and 8.6.

8.3 Coupling Matrix Correlation Function

In the subsequent analysis of sound propagation through the stochastic internal-wave field, an important statistical function is the range correlation function of the coupling matrices, that is, $\langle \rho_{mn}(r)\rho_{qp}^*(r - \delta r)\rangle$, where δr is the range lag. Other correlations of the coupling matrices are important as well, including those

---

[4] If Eq. 2.126 is solved using the WKB method (see Jensen, et al. 1993) one finds that there are phase and small amplitude fluctuations in a_n. In the limit of small ρ_{nn} the WKB solution reduces to Eq. 8.10.

associated with time, horizontal position, and frequency, but they will be discussed in sections to follow when acoustic coherence is addressed.

Writing $\mu(z,r)$ as a linear superposition of internal waves with mode number j and x component of the horizontal wavenumber k, the coupling matrix becomes (Colosi and Morozov, 2009)

$$\rho_{mn}(r) = \frac{q_0^2}{c_0 \sqrt{k_m k_n}} \int_0^\infty dz \left(\frac{\partial c(z)}{\partial z}\right)_p \frac{\phi_n(z)\phi_m(z)}{\rho_0(z)} \zeta(z,r),$$

$$\zeta(z,r) = \zeta_0 \left(\frac{2N_0}{N(z)}\right)^{1/2} \sum_{j=1}^J h_j \psi_j(z) \int_{-\infty}^\infty dk \, b_j(k) \exp(ikr) \tag{8.11}$$

where $\psi_j(z) = \sin[\pi j \hat{z}(z)]$ are the WKB internal-wave mode shapes with $\hat{z}(z)$ the WKB stretched vertical coordinate (see Chapter 3).[5] The terms h_j and $b_j(k)$ are complex Gaussian random amplitudes whose variance is given by the unit normalized GM spectrum, so that $\langle \zeta^2 \rangle(z) = \zeta_0^2 N_0 / N(z)$. Multiplying by $\rho_{qp}^*(r - \delta r)$ and taking the expectation, the result is

$$\Delta_{mn,qp}(\delta r) = \langle \rho_{mn}(r)\rho_{qp}^*(r - \delta r)\rangle = \sum_{j=1}^J H(j)G_{mn}(j)G_{qp}(j) \int_{-\infty}^\infty dk F_1(k) \, e^{-ik\delta r},$$

$$\tag{8.12}$$

where

$$G_{mn}(j) = \frac{\zeta_0 q_0^2}{c_0} \sqrt{\frac{2}{k_n k_m}} \int_0^\infty dz \left(\frac{\partial c(z)}{\partial z}\right)_p \left(\frac{N_0}{N(z)}\right)^{1/2} \sin[\pi j \hat{z}(z)] \frac{\phi_n(z)\phi_m(z)}{\rho_0(z)}.$$

$$\tag{8.13}$$

The unit normalized GM spectrum is given by

$$H(j) = \frac{1}{M_j} \frac{1}{j^2 + j_*^2}, \tag{8.14}$$

$$F_1(k) = \frac{2}{\pi^2} \left[\frac{\hat{\kappa}_j}{k^2 + \hat{\kappa}_j^2} + \frac{1}{2} \frac{k^2}{(k^2 + \hat{\kappa}_j^2)^{3/2}} \log\left(\frac{(k^2 + \hat{\kappa}_j^2)^{1/2} + \hat{\kappa}_j}{(k^2 + \hat{\kappa}_j^2)^{1/2} - \hat{\kappa}_j}\right) \right], \tag{8.15}$$

with $\hat{\kappa}_j = \pi f j / N_0 B$. This result for the range correlation function can be simplified using an approximate form for the GM spectrum given in Chapter 3 (Appendix A). There it was shown that a Lorentzian function,

$$F_1(k) \simeq \frac{1}{\pi} \frac{\tilde{\kappa}_j}{k^2 + \tilde{\kappa}_j^2}, \tag{8.16}$$

---

[5] The WKB modes with no turning point corrections depend only on mode number j and not the total horizontal wavenumber κ_h. This leads to important simplification of Eq. 8.11. As mentioned in Chapter 3, use of the WKB results is internally consistent with the use of the GM spectrum.

is a close approximation to Eq. 8.15, with $\tilde{\kappa}_j = \sqrt{2}\hat{\kappa}_j$ giving a slightly better Lorentzian fit to the GM spectrum than $\hat{\kappa}_j$. The Lorentzian shape for the wavenumber spectrum means that the frequency spectrum has a pure power law form of σ^{-2}. Using the Lorentzian spectrum then gives an exponential correlation function for each mode, yielding the simple result:

$$\Delta_{mn,qp}(\delta r) = \sum_{j=1}^{J} H(j) G_{mn}(j) G_{qp}(j)\, e^{-\tilde{\kappa}_j |\delta r|}. \tag{8.17}$$

8.3.1 Scattering Matrix

An important quantity related to the correlation function is the $N^2 \times N^2$ scattering matrix:

$$I_{mn,qp} = \int_0^\infty d\delta r\, \Delta_{mn,qp}(\delta r)\, e^{il_{pq}\delta r}. \tag{8.18}$$

Some useful symmetry properties of Eq. 8.18 are

$$I_{mn,qp} = I_{nm,qp}, \quad I_{mn,pq} = I_{mn,qp}^*. \tag{8.19}$$

In the absence of attenuation, the real part of this function is closely related to the wavenumber spectrum of the coupling matrix, that is,

$$\frac{1}{\pi}\mathrm{Re}(I_{mn,qp}) = \langle \hat{\rho}_{mn}\hat{\rho}_{qp}^*(k_{pq}) \rangle = \frac{1}{2\pi} \int_{-\infty}^{\infty} d\delta r\, \Delta_{mn,qp}(\delta r)\, e^{ik_{pq}\delta r}. \tag{8.20}$$

In most situations, even in shallow water, modal attenuation is sufficiently weak that the imaginary part of l_{pq} in Eq. 8.18 can be ignored to great analytical and computational simplification (Creamer, 1996; Colosi et al., 2011). Because the real part of the scattering matrix is associated with the internal-wave wavenumber spectrum for $k = k_{qp}$, it is apparent that there are resonance conditions for mode coupling that pick out specific internal-wave wavenumbers. It will also be found that adiabatic effects are controlled by scattering matrices in which the indices are the same, namely

$$I_{nn,pp} = \int_0^\infty d\delta r\, \Delta_{nn,pp}(\delta r) = \Delta_{nn,pp}(0)\, L_H(n,p), \tag{8.21}$$

where $L_H(n,p)$ is the horizontal correlation length between the adiabatic terms ρ_{nn} and ρ_{pp}.

Analytic forms for the scattering matrix are available for both the GM spectrum and the Lorentzian approximation. The case for the Lorentzian spectrum is straightforward. For complex wavenumbers, l_n, and making use of Eq. 8.17, the scattering matrix for the Lorentzian spectrum is

$$I_{mn,qp} = \sum_{j=1}^{J} H(j)\, G_{mn}(j) G_{qp}(j) \left(\frac{\tilde{\kappa}_j + \alpha_{pq} + i k_{pq}}{(\tilde{\kappa}_j + \alpha_{pq})^2 + k_{pq}^2} \right) \qquad |\alpha_{pq}| < \tilde{\kappa}_j. \quad (8.22)$$

The GM calculation involves some backtracking and the additional well-substantiated approximation that the imaginary part of the wavenumber l_{pq} in Eq. 8.18 can be neglected. Writing the correlation function integrals in polar coordinate form, the scattering matrix becomes (Abramowitz and Stegun, 1964; Colosi et al., 2013a)

$$I_{mn,qp} = \sum_{j=1}^{J} H(j) G_{mn}(j) G_{qp}(j) \int_0^\infty d\kappa_h F_1(\kappa_h) \int_0^\infty d\delta r\, e^{i k_{pq} \delta r} \int_{-\pi}^{\pi} \frac{d\vartheta}{2\pi} \cos(\kappa_h \delta r \cos\vartheta),$$

$$= \sum_{j=1}^{J} H(j)\, G_{mn}(j) G_{qp}(j) \int_0^\infty d\kappa_h F_1(\kappa_h) \begin{cases} \dfrac{1}{\sqrt{\kappa_h^2 - k_{pq}^2}}, & 0 \le |k_{pq}| < \kappa_h \\[2ex] \dfrac{i\,\mathrm{sign}(k_{pq})}{\sqrt{k_{pq}^2 - \kappa_h^2}}, & 0 \le \kappa_h < |k_{pq}| \end{cases},$$

$$(8.23)$$

where the unit normalized internal-wave spectrum in terms of the magnitude of the horizontal wavenumber, κ_h, is given by (see Chapter 3)

$$F_1(\kappa_h) = \frac{4}{\pi} \frac{\kappa_h^2 \hat{\kappa}_j}{(\kappa_h^2 + \hat{\kappa}_j^2)^2}. \quad (8.24)$$

The resonance condition in the scattering matrix is again evident due to the integrable singularity at $\kappa_h = |k_{pq}|$. An analytic solution to the wavenumber integral in Eq. 8.23 using the GM spectrum is given in Appendix A.

For the adiabatic case where the indices are pairwise matching, the GM scattering matrices have the simple form

$$I_{nn,pp} = \frac{2}{\pi} \sum_{j=1}^{J} H(j) \frac{G_{nn}(j) G_{pp}(j)}{\hat{\kappa}_j}, \quad (8.25)$$

with a similar expression for the Lorentz spectrum (Eq. 8.22). Note here that the adiabatic scattering matrix depends on $H(j)/j$ and is therefore sensitive to large vertical scale internal waves.

With this background, the statistical moments of the acoustic field can be analyzed.

8.4 Mean Pressure

The mode expression for the acoustic pressure (Eq. 8.1) can be used to examine any moment of the acoustical field, although the higher order moments can become quite complicated because of the multiple sums involved. Here moments up to four will be considered, thereby obtaining expressions for mean intensity, intensity variance, and coherence.

The simplest moment is the mean pressure or coherent field which is given by

$$\langle p(r,z;\omega)\rangle = \sum_n \langle a_n(r)\rangle \, \frac{\phi_n(z)}{\sqrt{k_n r}}. \tag{8.26}$$

As shown in Chapter 7, the mean field is expected to decay exponentially with range with a rate dictated by the growth of the phase variance. The mean field is therefore generally not terribly interesting, because it rapidly decays to small levels. Be that as it may, in the mode approach the evaluation of the mean field is reduced to the examination of the range decay of the demodulated mode amplitudes $\langle \hat{a}_n(r)\rangle$. It is useful to start with the adiabatic approach to gain some insight.

8.4.1 Adiabatic Theory

In the adiabatic approximation there are only phase fluctuations (Eq. 8.10), and therefore the mean mode amplitude is given by

$$\langle \hat{a}_n(R)\rangle = \hat{a}_n(0)\left\langle \exp\left(-i\int_0^R \rho_{nn}(r)dr\right)\right\rangle = \hat{a}_n(0)\exp\left(-\frac{1}{2}\left\langle \left(\int_0^R \rho_{nn}(r)\,dr\right)^2\right\rangle\right), \tag{8.27}$$

where the last line follows from the assumption that the coupling matrix ρ_{nn} is a ZMGRV. The quantity in angular brackets is the adiabatic phase variance of the nth mode, so that $\langle \hat{a}_n(R)\rangle = \hat{a}_n(0)\exp(-\Phi_n^2(R)/2)$. In mean and difference coordinates, the phase variance is

$$\Phi_n^2(R) = \int_0^R d\bar{r} \int_{\bar{r}-R}^{\bar{r}} d\delta r \, \Delta_{nn,nn}(\delta r) \simeq R\int_{-\infty}^{\infty} d\delta r \, \Delta_{nn,nn}(\delta r) = 2RI_{nn,nn}, \tag{8.28}$$

assuming that the range R is much larger than the correlation length of $\Delta_{nn,nn}$, so that the limits on the relative coordinate integral can go to $\pm\infty$. The mean mode amplitude and thus the mean pressure decay exponentially with range, as shown in the path integral analysis in Chapter 7. In the next section, it will be shown that mode coupling hastens the exponential range decay of $\langle \hat{a}_n\rangle$ due to coupling-induced phase variation.

8.4.2 Transport Theory

Here the mean mode amplitude $\langle \hat{a}_n\rangle$ evolution is examined using the Dyson series and the range-ordered exponential solutions. A heuristic derivation using the Dyson series is presented first.

Dyson Series Approach

Using Eq. 8.6, the conceptual challenge is to compute the change in mode amplitude at range r for an initial field $\hat{a}(0)$, where the range r is loosely defined such that $r \gg r_c$, with r_c being the correlation length of $\langle \rho_{mn}(r) \rho_{qm}(r') \rangle$. Taking the expectation value of Eq. 8.6 and making the reasonable assumption that the initial mode amplitudes $\hat{a}_m(0)$ are uncorrelated with the subsequent coupling matrices, the result is

$$\langle \hat{a}_n(r) \rangle = \langle \hat{a}_n(0) \rangle - \sum_{m=1}^{N} \sum_{q=1}^{N} \langle \hat{a}_q(0) \rangle \int_0^r dr' \int_0^{r'} dr'' \langle \rho_{mn}(r') \rho_{qm}(r'') \rangle \, e^{i(l_{qm}r'' + l_{mn}r')},$$

(8.29)

where the first-order term is zero because $\langle \rho_{mn} \rangle = 0$. The correlation function $\langle \rho_{mn}(r') \rho_{qm}(r'') \rangle$ is only a function of the difference coordinate $\delta r = r' - r''$, assuming that ρ_{mn} is a stationary random process. Thus the r'' integral can be changed to a δr integral. Furthermore, assuming that $r \gg r_c$, the limits on the δr integral can be taken to infinity to obtain

$$\langle \hat{a}_n(r) \rangle = \langle \hat{a}_n(0) \rangle - \sum_{m=1}^{N} \sum_{q=1}^{N} \langle \hat{a}_q(0) \rangle I_{mn,qm} \int_0^r dr' \, e^{i l_{qn} r'}.$$

(8.30)

Taking the derivative of Eq. 8.30 with respect to r gives,

$$\frac{d\langle \hat{a}_n(r) \rangle}{dr} = -\sum_{q=1}^{N} \langle \hat{a}_q(0) \rangle e^{i l_{qn} r} \sum_{m=1}^{N} I_{mn,qm} \simeq -\sum_{q=1}^{N} \langle \hat{a}_q(r) \rangle e^{i l_{qn} r} \sum_{m=1}^{N} I_{mn,qm},$$

(8.31)

where the last step follows because $\langle \hat{a}_n(0) \rangle \simeq \langle \hat{a}_n(r) \rangle$ (the first correction is at second order) and that the incident field $\langle \hat{a}_n(0) \rangle$ should evolve with range. The derivation of this transport equation involves the Markov approximation (Van Kampen, 1981; Creamer, 1996; Henyey and Ewart, 2006), because it is assumed $r \gg r_c$, but in the end r is taken to be infinitesimally small. In this approximation, it is often asserted that the random medium is delta-correlated. Converting back to the standard mode amplitude a_n then

$$\frac{d\langle a_n(r) \rangle}{dr} - i l_n \langle a_n(r) \rangle = -\sum_{q=1}^{N} \langle a_q(r) \rangle \Gamma_{qn},$$

(8.32)

where $\Gamma_{qn} = \sum_{m=1}^{N} I_{mn,qm}$. The conceptual derivation just described can be obtained more rigorously using operator methods (see Van Kampen (1981) and Creamer (1996)) as discussed next.

Range-Ordered Exponential Approach

The starting place for the more formal derivation of the mean mode amplitude equation is the solution to the coupled mode equations using the range ordering operator (Eq. 8.8). Because ρ_{mn} can be considered a ZMGRV due to the Gaussianity of μ, the expectation value of Eq. 8.8 yields the relation

$$\langle \hat{\mathbf{a}}(r) \rangle = \mathcal{R} \left[\exp \left[-\frac{1}{2} \left\langle \left(\int_0^r dr' \tilde{\rho}(r') \right)^2 \right\rangle \right] \right] \langle \hat{\mathbf{a}}(0) \rangle. \tag{8.33}$$

Note that the Gaussian rule, $\langle e^{i\alpha} \rangle = e^{-\langle \alpha^2 \rangle/2}$, for α a ZMGRV also applies to the case of the range-ordered exponential. While Eq. 8.33 represents a formal solution to the problem, evaluation of the expression is extremely difficult, and thus it is useful to seek something simpler using the fact the the correlation length of the coupling matrices r_c is small. Following the work of Van Kampen (1981) and Creamer (1996), the range derivative of Eq. 8.33 yields

$$\frac{d}{dr} \langle \hat{\mathbf{a}}(r) \rangle = -\mathcal{R} \left[\left\langle \tilde{\rho}(r) \int_0^r dr' \tilde{\rho}(r') \right\rangle \exp \left[-\frac{1}{2} \left\langle \left(\int_0^r dr' \tilde{\rho}(r') \right)^2 \right\rangle \right] \right] \langle \hat{\mathbf{a}}(0) \rangle. \tag{8.34}$$

In this expression the action of the range-ordering operator is somewhat tricky. The first factor of $\tilde{\rho}$ on the left is at the range r and therefore is properly range-ordered, but the second factor of $\tilde{\rho}$ in the integral needs to be ordered properly with the other matrices in the exponential. In the Markov approximation, it is assumed that the first term in Eq. 8.34 is already properly ordered and therefore the range-ordering operator can be pulled through this term and only works on the exponential. In this approximation, the last term in Eq. 8.34 is simply Eq. 8.33, and the final result is (Creamer, 1996)

$$\frac{d}{dr} \langle \hat{\mathbf{a}}(r) \rangle = - \left\langle \tilde{\rho}(r) \int_0^r dr' \tilde{\rho}(r') \right\rangle \langle \hat{\mathbf{a}}(r) \rangle. \tag{8.35}$$

Expanding the right-hand side of this equation gives

$$- \sum_{m=1}^{N} \sum_{q=1}^{N} \hat{a}_q(r) \int_0^r dr' \langle \rho_{mn}(r) \rho_{qm}(r') \rangle e^{il_{mn}r} e^{il_{qm}r'}, \tag{8.36}$$

which can be manipulated to obtain the right-hand side in Eq. 8.31.

8.4.3 *Solution for Mean Pressure*

Equation 8.32 can be solved by eigenvector analysis because the coefficients Γ_{qn} are independent of range. Writing Eq. 8.32 in matrix form gives $d\mathbf{a}(r)/dr = -\mathbf{\Xi}\mathbf{a}(r)$, where $\mathbf{a}(r)$ is the vector of the mean mode amplitudes and $\mathbf{\Xi}$ is the $N \times N$ matrix

$\Gamma_{qn} - il_n\delta_{qn}$. The solution of Eq. 8.32 is

$$\mathbf{a}(r) = e^{-\boldsymbol{\Xi}r}\mathbf{a}(0) = \mathbf{P}e^{-\mathbf{D}r}\mathbf{P}^{\mathrm{T}}\mathbf{a}(\mathbf{0}), \tag{8.37}$$

where $\mathbf{P}^{\mathrm{T}}\boldsymbol{\Xi}\mathbf{P} = \mathbf{D}$ is the diagonal matrix of the eigenvalues of $\boldsymbol{\Xi}$, and $\mathbf{P}$ is a matrix whose columns are the eigenvectors of $\boldsymbol{\Xi}$.

To understand the relationship between the Dyson and adiabatic solutions, it is useful to consider a simple two-mode example. In this case, the solution of Eq. 8.32 can be written as

$$\langle a_1(r)\rangle = c_1 P_{1+}e^{-\lambda_+ r} + c_2 P_{1-}e^{-\lambda_- r}, \qquad \langle a_2(r)\rangle = c_1 P_{2+}e^{-\lambda_+ r} + c_2 P_{2-}e^{-\lambda_- r}, \tag{8.38}$$

where c_1 and c_2 are integration constants to be determined from the initial conditions and (P_{1+}, P_{2+}) and (P_{1-}, P_{2-}) are the eigenvectors. The eigenvalues λ_+ and λ_- come from the characteristic equation giving

$$\lambda_+ \simeq -il_2 - \Gamma_{22}, \qquad \lambda_- \simeq -il_1 - \Gamma_{11}, \tag{8.39}$$

where the approximate eigenvalues follow from the fact that $l_1, l_2 \gg \Gamma_{mn}$. The eigenvalues have important physical significance becuase they dictate the exponential rate at which the mean mode amplitudes decay, with the smallest eigenvalue setting the overall decay range. To first order, the coherent field decays at the rate Γ_{11} and Γ_{22}, relative to the unperturbed field. Explicitly writing the eigenvalues in terms of the scattering matrices, the result is

$$\lambda_- = -il_1 + \left[I_{11,11} + I_{12,12}\right], \quad \lambda_+ = -il_2 + \left[I_{22,22} + I_{12,12}^*\right]. \tag{8.40}$$

The range variation of $\langle \hat{a}_1 \rangle$ and $\langle \hat{a}_2 \rangle$ is controlled by the terms in the square brackets of Eq. 8.40. Within the square bracket, the first term is simply the adiabatic term, and the second is due to coupling. The coupling term has real and imaginary parts, and thus the coupling introduces both an exponential decay (real part is positive definite) and oscillatory variation with range. The additional exponential decay over the adiabatic component can be understood in terms of the extra phase randomization that coupling provides. The oscillatory part is the result of mode coupling back and forth between the two modes.

8.4.4 Connection between Transport and Adiabatic Theory

Lastly, it is useful to look at the case of adiabatic propagation in the transport theory. The adiabatic approximation to Eq. 8.31 is obtained by setting $q = m = n$, yielding

$$\langle \hat{a}_n(r)\rangle = \langle \hat{a}_n(0)\rangle \exp(-RI_{nn,nn}), \tag{8.41}$$

which is the result obtained in Section 8.4.1. Therefore, the transport theory has the proper adiabatic treatment of the mode fluctuations, that is, the transport theory approximations (namely the Markov approximation) have not corrupted the adiabatic limit. This will be the case for all other moments treated using transport theory.

8.5 Mean Intensity

A more interesting observable is now examined, namely mean intensity. Mean intensity is of great interest because of the discovery that internal-wave-induced scattering can lead to significant mean ensonification of acoustic shadow zones for long-range deep-water propagation. The mean intensity is also of intense interest in shallow water, where both stochastic and nonlinear internal waves cause large changes in intensity relative to the unperturbed field.

The second moment, mean intensity, is given by

$$\langle |p(r,z)|^2 \rangle = \sum_{n=1}^{N} \sum_{p=1}^{N} \frac{\langle a_n(r)a_p^*(r) \rangle}{r} \frac{\phi_n(z)\phi_p(z)}{\sqrt{k_n k_p}}, \tag{8.42}$$

where $\langle a_n(r)a_p^*(r) \rangle$ is the $N \times N$ cross-mode coherence matrix. As before, it is useful to work with the demodulated mode amplitudes, and so $\langle \hat{a}_n(r)\hat{a}_p^*(r) \rangle$ is examined. Writing $\gamma_{np} = \hat{a}_n\hat{a}_p^*$ and using Eq. 8.2, the evolution equation for γ_{np} is

$$\frac{d\gamma_{np}}{dr} = \hat{a}_p^* \frac{d\hat{a}_n}{dr} + \hat{a}_n \frac{d\hat{a}_p^*}{dr} = -i \sum_{m=1}^{N} \left(\tilde{\rho}_{mn}(r)\gamma_{mp} - \tilde{\rho}_{mp}^*(r)\gamma_{nm} \right), \tag{8.43}$$

which is of the same form as Eq. 8.2. In matrix form, Eq. 8.43 is

$$\frac{d\boldsymbol{\gamma}(r)}{dr} = -i \, \tilde{\boldsymbol{\Upsilon}}\boldsymbol{\gamma}(r), \tag{8.44}$$

where

$$\tilde{\Upsilon}_{mn,qp} = \delta_{mq}\tilde{\rho}_{np} - \delta_{np}\tilde{\rho}_{mq}^*. \tag{8.45}$$

Because of the common form of the equations, γ_{np} can also be analyzed using the Dyson series and the range-ordered exponential. As before, it is useful to first examine the adiabatic approximation to gain some insight.

8.5.1 Adiabatic Theory

In the adiabatic approximation, the cross-mode coherence decay is driven by the differential phase variations between modes. Using either Eq. 8.10 or Eq. 8.43, the result is

$$\langle \gamma_{np}(R) \rangle = \langle \gamma_{np}(0) \rangle \exp\left(-\frac{1}{2}\left\langle\left(\int_0^R \rho_{nn}(r)\,dr - \int_0^R \rho_{pp}(r)\,dr\right)^2\right\rangle\right), \quad (8.46)$$

where again the assumption has been made that ρ_{nn} are ZMGRVs. The expectation value in the exponent is the phase structure function, which played an important role in coherence in path integral theory. Writing the structure function $D(n,p)$ using the GM scattering matrices (Eq. 8.25), the result is

$$D(n,p) = 2R\left(I_{nn,nn} + I_{pp,pp} - 2I_{nn,pp}\right) = \frac{4R}{\pi}\sum_{j=1}^J \frac{H(j)}{\hat{k}_j}\left(G_{nn}(j) - G_{pp}(j)\right)^2, \quad (8.47)$$

where the double range integrals in Eq. 8.46 have been simplified by assuming that the range is much larger than a correlation length of the n and p coupling matrices. The decorrelation of the modes is driven primarily by the different depth projections of the internal-wave structure onto the mode eigenfunctions, which enters through the G terms. But of course in some cases mode coupling is expected to modify the decorrelation of modes. The structure function depends on $H(j)/j$ and is therefore influenced most strongly by large vertical scale internal waves.

In the adiabatic approximation, the mode energy and cross-mode coherence matrix are given by the simple analytic expressions (Colosi et al., 2011):

$$\langle |a_n|^2 \rangle(r) = \langle |a_n|^2 \rangle(0)\ e^{-2\alpha_n r},$$
$$\langle a_n a_p^* \rangle(r) = \langle a_n a_p^* \rangle(0)\ e^{i(l_n - l_p^*)r}\ e^{-(I_{nn,nn} - 2I_{nn,pp} + I_{pp,pp})r}. \quad (8.48)$$

These results have been particularly useful in shallow-water problems in which the acoustic frequency is below 1 kHz. The loss of cross-mode coherence in shallow water has critical implications for the acoustic interaction with the intense nonlinear shallow-water internal-wave field (Colosi et al., 2011). Loss of cross-mode coherence means that the scattering by these nonlinear waves is much less coherent, and it also means that the acoustic field memory of a nonlinear wave scattering event diminishes with range.

8.5.2 Transport Theory

As in the case of the transport theory for the mean mode amplitude, in this section two derivations of the transport equation for the cross-mode coherence matrix are presented, that is: one based on the Dyson series and the other based on the range-ordered exponential.

Dyson Series Approach

Equation 8.43 describes the range evolution of the cross-mode coherence, and has the same form as the evolution equation for the mode amplitude $\hat{a}_n$, which was solved using the Dyson series (Eq. 8.6). The successive terms in the Dyson series for Eq. 8.43 up to order 2 are

$$[\gamma_{np}(r)]_0 = \gamma_{np}(0),$$

$$[\gamma_{np}(r)]_1 = -i \sum_{j=1}^{N} \int_0^r dr' \left(\rho_{jn}(r') e^{il_{jn}r'} [\gamma_{jp}]_0(r') - \rho_{jp}(r') e^{-il_{jp}^* r'} [\gamma_{nj}]_0(r') \right),$$

$$[\gamma_{np}(r)]_2 = -i \sum_{m=1}^{N} \int_0^r dr'' \left(\rho_{mn}(r'') e^{il_{mn}r''} [\gamma_{mp}]_1(r'') - \rho_{mp}(r'') e^{-il_{mp}^* r''} [\gamma_{nm}]_1(r'') \right),$$

where $[]_n$ represents the nth order approximation. The procedure for finding the evolution equation here is precisely the same as the case for the mean mode amplitude. Applying the expectation value, and making the assumption that the initial coherences are uncorrelated with the subsequent coupling matrices, the result is

$$\langle \gamma_{np}(r) \rangle = \langle \gamma_{np}(0) \rangle - \sum_{m=1}^{N} \sum_{j=1}^{N} \int_0^r dr' \int_0^{r'} dr''$$

$$\left(\langle \gamma_{jp}(0) \rangle \, \Delta_{mn,jm}(\delta r) \, e^{i(l_{mn}r' + l_{jm}r'')} - \langle \gamma_{mj}(0) \rangle \, \Delta_{mn,jp}(\delta r) e^{i(l_{mn}r' - l_{jp}^* r'')} \right.$$

$$\left. - \langle \gamma_{jm}(0) \rangle \, \Delta_{mp,jn}(\delta r) e^{-i(l_{mp}^* r' - l_{jn}r'')} + \langle \gamma_{nj}(0) \rangle \, \Delta_{mp,jm}(\delta r) e^{-i(l_{mp}^* r' + l_{jm}^* r'')} \right),$$

$$(8.49)$$

where $\delta r = r' - r''$. As before, the first-order terms do not contribute because the coupling matrix has a zero mean. Changing the r'' integration variable to δr, assuming $r \gg r_c$ so that the δr integration limits can go to infinity, and differentiating with respect to r, as was done in Section 8.4.2, the result is

$$\frac{d\langle \gamma_{np}(r) \rangle}{dr} = -\sum_{m=1}^{N} \sum_{j=1}^{N} \left(\langle \gamma_{jp}(r) \rangle \, I_{mn,jm} \, e^{il_{jn}r} - \langle \gamma_{mj}(r) \rangle \, I_{mn,jp}^* \, e^{i(l_{mn}r - l_{jp}^* r)} \right.$$

$$\left. - \langle \gamma_{jm}(r) \rangle \, I_{mp,jn} \, e^{-i(l_{mp}^* r - l_{jn}r)} + \langle \gamma_{nj}(r) \rangle \, I_{mp,jm}^* \, e^{-il_{jp}^* r} \right). \qquad (8.50)$$

Here the incident coherences $\langle \gamma_{np}(0) \rangle$ have been replaced with $\langle \gamma_{np}(r) \rangle$ in Eq. 8.50 for the reasons discussed in Section 8.4.2. The right-hand side of this equation has factors that depend on r, making numerical or further theoretical progress difficult. Again, the r dependence is removed by undoing the transformation $\hat{a}_n = a_n e^{-il_n r}$;

the result is

$$\frac{d\langle a_n a_p^*\rangle(r)}{dr} - i(l_n - l_p^*)\langle a_n a_p^*\rangle = -\sum_{m=1}^{N}\sum_{j=1}^{N} \Big(\langle a_j a_p^*\rangle\, I_{mn,jm} - \langle a_m a_j^*\rangle\, I_{mn,jp}^*$$

$$-\langle a_j a_m^*\rangle\, I_{mp,jn} + \langle a_n a_j^*\rangle\, I_{mp,jm}^*\Big). \qquad (8.51)$$

This is the fundamental equation that will be used in much of the analysis of mode fluctuations, and it will be slightly modified to handle cross-mode coherences with lags not just in mode number, but in time, transverse direction, and frequency.

Range-ordered Exponential Approach

Equation 8.50 has been obtained by Creamer (1996) in the ocean acoustics context using the operator methods that were outlined in Section 8.4.2. Because the mode coherence equation (Eq. 8.44) is of the same form as the coupled mode equation (Eq. 8.2), the solution can be written in terms of a range-ordered exponential, that is,

$$\boldsymbol{\gamma}(r) = \mathcal{R}\left[\exp\left(-i\int_0^r dr'\, \tilde{\boldsymbol{\Upsilon}}(r')\right)\right]\boldsymbol{\gamma}(0). \qquad (8.52)$$

Therefore, the same methodology applied to the derivation of the mean mode amplitude (Section 8.4.2) can be applied to the coherence matrix, yielding the matrix transport equation:

$$\frac{d}{dr}\langle\boldsymbol{\gamma}(r)\rangle = -\left\langle\tilde{\boldsymbol{\Upsilon}}(r)\int_0^r dr'\, \tilde{\boldsymbol{\Upsilon}}(r')\right\rangle\langle\boldsymbol{\gamma}(r)\rangle. \qquad (8.53)$$

This equation is identical to Eq. 8.50.

8.5.3 Mode Energy

Mode energy, that is, the diagonal of the cross-mode coherence matrix, is an acoustic observable that has received much attention (Colosi and Flatté, 1996; Creamer, 1996; Wage et al., 2005) since the seminal papers by Dozier and Tappert (1978a). For the case of no attenuation, equipartitioning of modal energy is understood to be a modal manifestation of full saturation. Thus, a fundamental question in ocean acoustic wave propagation is the nature of the evolution of the cross-mode coherence matrix to the state in which the diagonal terms are all equal and the off-diagonal terms are zero.

Early studies worked under the premises that cross-mode coherences decay rapidly in range and that mode energy equilibration proceeds more slowly (Dozier and Tappert, 1978a). Recent work, however, has revealed that the two processes evolve over similar range scales (Colosi and Morozov, 2009). Furthermore, it is now known that the cross-mode coherences have little influence on the mode energy evolution (Colosi and Morozov, 2009). To see this, consider Eq. 8.51, which gives the evolution of the mode energy:

$$\frac{d\langle |a_n|^2 \rangle}{dr} = -2\alpha_n \langle |a_n|^2 \rangle - \sum_{m=1}^{N} (\langle |a_n|^2 \rangle\, g_{mn} - \langle |a_m|^2 \rangle f_{mn})$$

$$- \sum_{m=1}^{N} \sum_{j=1, j\neq n}^{N} 2\mathrm{Re}(\langle a_j a_n^* \rangle\, I_{mn,jm}) + \sum_{m=1}^{N} \sum_{j=1, j\neq m}^{N} 2\mathrm{Re}(\langle a_j a_m^* \rangle\, I_{mn,jn}),$$

$$(8.54)$$

where the diagonal contributions (single-sum terms) have been separated out from the cross-mode coherence contributions (double sums). In this equation there are two important factors:

$$f_{mn} = 2 \int_0^\infty d\delta r\, \Delta_{mn,mn}(\delta r)\, \cos(k_{mn}\delta r)\, e^{+\alpha_{mn}\delta r}, \qquad (8.55)$$

$$g_{mn} = 2 \int_0^\infty d\delta r\, \Delta_{mn,mn}(\delta r)\, \cos(k_{mn}\delta r)\, e^{-\alpha_{mn}\delta r}. \qquad (8.56)$$

Because the matrices f_{mn} and g_{mn} are positive definite quantities, the single-sum terms associated with mode energies have a significant contribution to Eq. 8.54. However, the terms $\mathrm{Re}(\langle a_j a_n^* \rangle\, I_{mn,jm})$ and $\mathrm{Re}(\langle a_j a_m^* \rangle\, I_{mn,jn})$ are highly oscillatory functions of the indexes and are not likely to contribute significantly to the mode evolution. This has been verified by direct numerical evaluation of the cross-mode coherence matrix equations and Monte Carlo numerical simulation (Colosi and Morozov, 2009; Colosi et al., 2011) for both deep and shallow-water environments (also see Section 8.5.6).

As previously mentioned, a good assumption is that the mode attenuation is weak enough that the variation of the exponentials in f_{mn} and g_{mn} is small over the short correlation length of the coupling matrix; in this approximation, $g_{mn} = f_{mn}$ and Eq. 8.54 becomes

$$\frac{\partial \langle |a_n|^2 \rangle}{\partial r} + 2\alpha_n \langle |a_n|^2 \rangle = \sum_{m=1}^{N} f_{mn} \left(\langle |a_m|^2 \rangle - \langle |a_n|^2 \rangle \right), \qquad (8.57)$$

where

$$f_{mn} = 2 \int_0^\infty d\delta r\, \Delta_{mn,mn}(\delta r)\, \cos(k_{mn}\delta r) = 2\pi \langle |\hat{\rho}_{mn}(k = k_{mn})|^2 \rangle = 2\mathrm{Re}(I_{mn,mn}).$$

$$(8.58)$$

Here $\langle |\hat{\rho}_{mn}(k)|^2 \rangle$ is the wavenumber spectrum of the coupling matrix between modes n and m. Equation 8.57 was obtained by Creamer (1996) and with $\alpha_n =$

0, Dozier and Tappert's "master equations" are recovered (Dozier and Tappert, 1978a). Physically, this equation indicates that the strength of interaction between modes n and m is given by the difference in energy between the modes with the weighting factor f_{mn} that is often termed the transition matrix. The transition matrix, which is in units of m^{-1}, gives the effective mode coupling rate between modes n and m. That is, one would expect significant coupling between modes n and m after a range $1/f_{mn}$. Because ocean internal waves cause small-angle scattering, f_{mn} decreases rapidly as mode number separation increases. Lastly, f_{mn} scales approximately as ω^2, so the strength of the coupling increases significantly with acoustic frequency.

An interesting limit is the case in which only nearest neighbor coupling is considered and so Eq. 8.57 can be written as

$$\frac{\partial \langle |a_n|^2 \rangle}{\partial r} + 2\alpha_n \langle |a_n|^2 \rangle = D_n \left[\langle |a_{n-1}|^2 \rangle - 2 \langle |a_n|^2 \rangle + \langle |a_{n+1}|^2 \rangle \right], \tag{8.59}$$

where $D_n = f_{n-1,n}$ and it is assumed $f_{n-1,n} \simeq f_{n+1,n}$. Equation 8.59 has the form of a diffusion equation, where the term in square brackets is the discrete representation of the second derivative and D_n is the diffusion constant.

For no attenuation, $\alpha_n = 0$, the mode energy equation (Eq. 8.57)) conserves the total energy, that is,

$$\frac{d}{dr} \sum_n \langle |\hat{a}_n|^2 \rangle (r) = 0. \tag{8.60}$$

Asymptotic Behavior: Equipartion of Energy
The asymptotic behavior of the mode energies is of fundamental interest because the approach to equipartitioning of energy is strongly connected to the approach to saturation. At long range, where the cross-mode coherences have decayed to small values, the mean intensity has the simple form:

$$\langle I \rangle = \langle |p(r,z)|^2 \rangle = \sum_{n=1}^{N} \frac{\langle |a_n|^2 \rangle (r)}{rk_n} \phi_n^2(z). \tag{8.61}$$

To examine the asymptotic behavior, it is useful to consider Eq. 8.57 with no attenuation. There is a stable equilibrium toward which the system will evolve as the range increases. For $d\langle |\hat{a}_n|^2 \rangle (r)/dr \to 0$, a state of equipartitioning of modal energy is obtained, that is, $\langle |\hat{a}_1|^2 \rangle (r) = \langle |\hat{a}_2|^2 \rangle (r)..... = \langle |\hat{a}_N|^2 \rangle (r)$.

This result and the effects of attenuation can be better understood by looking at the mode energy equation in matrix form, that is,

$$\frac{d\gamma_E}{dr} = -\mathbf{\Pi}\gamma_E, \tag{8.62}$$

where $\gamma_E = \langle|a_n|^2\rangle$ is a vector of mode energies, and the $N \times N$ matrix Π is given by

$$\Pi_{nm} = \delta_{nm}\left(2\alpha_n + \sum_{k \neq n}^{N} f_{nk}\right) - (1 - \delta_{nm})f_{nm}. \tag{8.63}$$

Because Π is a constant matrix, the solution of Eq. 8.62 is of the same form as Eq. 8.37. Here Π is diagonalized by writing

$$\Pi = \mathbf{P}\mathbf{D}_{\Pi}\mathbf{P}^T, \tag{8.64}$$

where $\mathbf{P}$ is an orthogonal matrix of eigenvectors, and the diagonal matrix $\mathbf{D}_{\Pi} = \text{diag}[\lambda_1, \lambda_2, ..., \lambda_N]$ holds the eigenvalues. According to Gerschgorin's theorem (Isaacson and Keller, 1966; Dozier and Tappert, 1978a), $0 \leq \lambda_1 \leq \lambda_2 \leq ...\lambda_N \leq \max[|\sum_{k \neq n} f_{kn} + \alpha_n|]$. Writing out the solution explicitly,

$$\langle|a_n|^2\rangle(r) = \sum_{k=1}^{N} P_{nk}\left[\sum_{m=1}^{N} P_{mk}\langle|a_m|^2\rangle(0)\right]e^{-\lambda_k r}. \tag{8.65}$$

At long range the solution is dominated by the smallest eigenvalue, that is,

$$\langle|a_n|^2\rangle(r) \simeq P_{n1}\left[\sum_{m=1}^{N} P_{m1}\langle|a_m|^2\rangle(0)\right]e^{-\lambda_1 r}. \tag{8.66}$$

As a result, the limiting shape of the modal energy distribution is dictated by the first eigenvector components, namely

$$\lim_{r\to\infty} \frac{\langle|a_n|^2\rangle(r)}{\langle|a_m|^2\rangle(r)} \simeq \frac{P_{n1}}{P_{m1}}. \tag{8.67}$$

For the case of no attenuation, $\alpha_n = 0$, Π is a singular matrix because its columns sum to zero. This means one of the eigenvalues will be zero, corresponding to the case $d\gamma_E/dr = 0$, that is, equipartion of energy among the modes. In this case, the equipartion range, R_{eq}, is defined in terms of the smallest nonzero eigenvalue, giving (Dozier and Tappert, 1978a)

$$R_{eq} = \frac{1}{\lambda_2}. \tag{8.68}$$

8.5.4 Two-Mode Example

A two-mode example is useful to understand the relationship between coupling and adiabatic effects, and to see how various internal-wave parameters enter the problem. Using Eq. 8.51 and the approximation that attenuation is weak enough

to be ignored in the scattering matrices, several terms simplify to yield the three
evolution equations:

$$\frac{d\langle|a_1|^2\rangle}{dr} + 2\alpha_1\langle|a_1|^2\rangle = 2\mathrm{Re}(I_{12,12})(\langle|a_2|^2\rangle - \langle|a_1|^2\rangle)$$

$$+ 2(I_{21,11} - I_{21,22})\mathrm{Re}(\langle a_1 a_2^*\rangle), \qquad (8.69)$$

$$\frac{d\langle|a_2|^2\rangle}{dr} + 2\alpha_2\langle|a_2|^2\rangle = 2\mathrm{Re}(I_{12,12})(\langle|a_1|^2\rangle - \langle|a_2|^2\rangle)$$

$$+ 2(I_{21,22} - I_{21,11})\mathrm{Re}(\langle a_1 a_2^*\rangle), \qquad (8.70)$$

$$\frac{d\langle a_1 a_2^*\rangle}{dr} + i(l_2^* - l_1)\langle a_1 a_2^*\rangle = -(I_{11,11} + I_{22,22} - 2I_{11,22} + 2I_{12,12})\langle a_1 a_2^*\rangle$$

$$+ (I_{11,21} - I_{22,21})\langle|a_1|^2\rangle + (I_{22,21} - I_{11,21})\langle|a_2|^2\rangle. \qquad (8.71)$$

These equations are still a bit complicated, and further simplification is needed.
In Eqs. 8.69 and 8.70, the effects of the cross-mode coherence should be small
because $\mathrm{Re}(\langle a_1 a_2^*\rangle))$ is oscillating. Further, the terms $I_{21,11} - I_{21,22}$ and $I_{11,21} - I_{22,21}$
can also be small if the projection of the sound-speed fluctuations on the two-mode
functions is not a strong function of mode number. If this is the case, the last terms
in Eqs. 8.69–8.71 can be ignored, yielding coupled equations for the mode energies
and an uncoupled equation for the coherence:

$$\frac{d\langle|a_1|^2\rangle}{dr} + 2\alpha_1\langle|a_1|^2\rangle = 2\mathrm{Re}(I_{12,12})(\langle|a_2|^2\rangle - \langle|a_1|^2\rangle), \qquad (8.72)$$

$$\frac{d\langle|a_2|^2\rangle}{dr} + 2\alpha_2\langle|a_2|^2\rangle = 2\mathrm{Re}(I_{12,12})(\langle|a_1|^2\rangle - \langle|a_2|^2\rangle), \qquad (8.73)$$

$$\frac{d\langle a_1 a_2^*\rangle}{dr} + i(l_2^* - l_1)\langle a_1 a_2^*\rangle = -(I_{11,11} + I_{22,22} - 2I_{11,22} + 2I_{12,12})\langle a_1 a_2^*\rangle. \qquad (8.74)$$

The transfer of energy between modes 1 and 2 is controlled by the transition matrix
$f_{12} = 2\mathrm{Re}(I_{12,12})$ (i.e., the same result as the mode energy equation). The decay
of the coherence, however, is controlled by more terms. The first three terms on
the right-hand side of the equation are the adiabatic contribution associated with
the phase structure function. The fourth term $2I_{12,12}$ is a coupling term that also
controls the modal energy transfer.

Solutions of equations like Eqs. 8.72 and 8.73 are easily obtained by eigenvector
analysis, because the coefficients are independent of range. Here the eigenvalues,
which give the exponential decay rates, are given by

$$\lambda_\pm = \alpha_1 + \alpha_2 + f_{12} \pm \sqrt{f_{12}^2 + (\alpha_1 - \alpha_2)^2}. \qquad (8.75)$$

With this equation the dependence on internal-wave spectral parameters can be understood.

In the absence of attenuation, the eigenvalues are zero and $\lambda_+ = 2f_{12}$, so the modal energy equipartion range is given by $R_{eq} = 1/(2f_{12})$. For the simple Lorentzian internal-wave wavenumber spectrum (i.e., a correlation function given by Eq. 8.17) and no attenuation, the result is

$$f_{12} = 2 \sum_{j=1}^{J} \frac{H(j)}{\tilde{\kappa}_j} G_{12}^2(j) \frac{1}{1 + (k_{12}/\tilde{\kappa}_j)^2}. \tag{8.76}$$

The mode coupling rate is therefore dictated by $k_{12}/\tilde{\kappa}_j$, and a similar result holds for the GM spectrum, where the important term is $k_{12}/\hat{\kappa}_j$. These terms are a function of acoustic frequency and the background sound-speed profile. In both cases the coupling is most strongly influenced by the low internal-wave modes because of the factor $H(j)/\tilde{\kappa}_j$.

With small attenuation such that $f_{12} \gg \alpha_1, \alpha_2$, the eigenvalues are slightly modified, giving $\lambda_+ = 2f_{12} + \alpha_1 + \alpha_2$ and $\lambda_- = \alpha_1 + \alpha_2$. There is a coupling-induced rate to equipartion superimposed on the slow overall attenuation decay of the mode amplitudes. This might be the case for high acoustic frequency or for deep-water propagation where α is small. If the attenuation is larger, such that $f_{12} \ll \alpha_1, \alpha_2$, the approach to equipartion is dramatically changed. In this case the eigenvalues are $f_{12} + 2\alpha_1$ and $f_{12} + 2\alpha_2$, and the attenuation dominates the exponential decay of the modes; this is the case in shallow water.

For the cross-mode coherence, the solution of Eq. 8.74 gives

$$\langle a_1 a_2^* \rangle(R) = \langle a_1 a_2^* \rangle(0) \, e^{i(l_1 - l_2^*)R} \, e^{-(I_{11,11} - 2I_{11,22} + I_{22,22})R} \, e^{-2I_{12,12}R}. \tag{8.77}$$

For the coherence, attenuation enters most significantly through the first exponential term and results in a slow decay of each of the initial mode amplitudes. The remaining terms are interpreted as phase randomization terms caused by adiabatic effects and mode coupling. As previously mentioned, the second exponential term is the adiabatic contribution (see Eq. 8.47) resulting from the adiabatic phase structure function. The last exponential term comes from mode coupling and has both real and imaginary parts. Calculations by Colosi and Morozov (2009) for a deep-water environment show that coupling terms like $e^{-2I_{12,12}R}$ have a significant impact on the loss of cross-mode coherence, but for shallow-water cases that have been examined (Colosi et al., 2011; Raghukumar and Colosi, 2014), the adiabatic term is the dominant one.

8.5.5 Hybrid Theory

In some cases of interest, particularly shallow-water propagation, mode coupling significantly alters mode energy distributions, but modal phase variations have a strong adiabatic component (Raghukumar and Colosi, 2014). In this case one can consider a hybrid theory for cross-mode coherence in which the mode energies, $\langle |a_n|^2 \rangle$, evolve according to the mode energy equation (Eq. 8.57), but the phase randomization comes from adiabatic effects. Such an equation is given by

$$\langle a_n a_p^* \rangle(r) = \langle |a_n|^2 \rangle^{1/2}(r) \langle |a_p|^2 \rangle^{1/2}(r) \, \text{sign}[a_n(0)] \text{sign}[a_p(0)]$$
$$e^{i(k_n - k_p)r} e^{-(I_{nn,nn} - 2I_{nn,pp} + I_{pp,pp})r}, \tag{8.78}$$

where the sign functions account for the initial phase of the modes for a point source. Attenuation is built into the first two terms because they are solved for using the mode energy equation (Eq. 8.57) with $\alpha_n \neq 0$. This approach only requires the calculation of an $N \times N$ scattering matrix, a considerable reduction in computational complexity compared to the $N^2 \times N^2$ problem for the full transport theory.

8.5.6 Monte Carlo Validation

Transport theory for the cross-mode coherence matrix (Eq. 8.51) has been validated using Monte Carlo numerical simulations for both deep and shallow-water environments, and the hybrid theory (Eq. 8.78) has been validated in the same way for kilohertz shallow-water propagation. The agreement between theory and simulation in virtually all cases is exceptional, as described below.

Deep Water and Mid-Latitude

Mode statistics for a low-frequency (100 Hz) acoustical beam, in deep water, like that shown in Figure 8.2 have been examined (Colosi and Morozov, 2009). In this case attenuation is taken to be zero. For the beam shown here, the initial condition yields a narrow excitation of modes centered on mode 20. Because of the dispersive nature of the ocean waveguide, the unperturbed beam does not stay perfectly collimated in range but slowly disperses. When internal-wave-induced sound-speed fluctuations are added to the calculation, the beam broadens and disintegrates much more rapidly. The gradual break-up of the beam is a visual indication of mode phase randomization and loss of cross-mode coherence due to coupling and adiabatic effects. In addition, the depth spreading of the beam demonstrates mode coupling and energy transfer into higher and lower modes that were not excited at the source.

The statistics of the cross-mode coherence matrix of these beams is shown in Figure 8.3, and the transport theory is seen to agree quite well with the Monte Carlo simulation results for both mode energy and cross-mode coherence. The mean mode energy evolution with range also closely follows the curve predicted using the mode energy equation (Eq. 8.57), thus demonstrating that mode energy

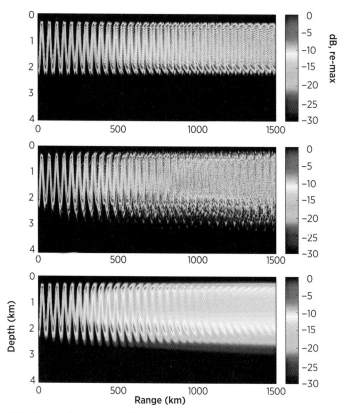

Figure 8.2. Acoustic beams from 100-Hz numerical simulations using a Munk canonical profile and an exponential buoyancy frequency profile. The top panel displays the unperturbed acoustical beam, while the middle panel shows a realization of the beam propagation through a random realization of internal-wave-induced sound-speed perturbations obeying the Garrett-Munk spectrum. The bottom panel shows the mean intensity averaged over 500 realizations of the internal-wave field. Cylindrical spreading is not included in the mean intensity.
Source: Colosi and Morozov (2009).

is insensitive to the cross-mode coherence terms in the transport theory. Mode energy evolution for a point source and for frequencies from 75 to 250 Hz also shows excellent agreement between Monte Carlo simulation and transport theory (Colosi et al., 2013a).

Next, the ensemble average of the intensity of the simulations gives the mean intensity shown in the bottom panel of Figure 8.2. By 1500-km range, the acoustic energy is smoothly distributed between 0.2 and 3.0 km, with no indication of the interference pattern seen in the unperturbed case; this again is a result of the loss of cross-mode coherence with the mean intensity roughly given by Eq. 8.61. Comparisons between the Monte Carlo calculation and transport theory for mean intensity are shown in Figure 8.4. Again the agreement between

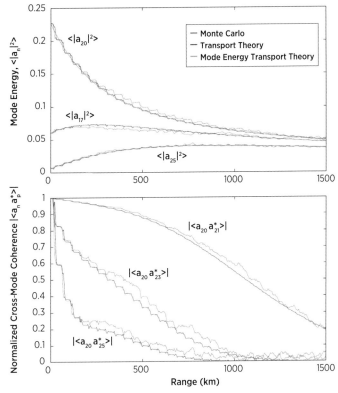

Figure 8.3. Range evolution of mean mode energy (top) and cross-mode coherence (bottom) for various modes from the simulations shown in Figure 8.2 and predictions from transport theory. Monte Carlo numerical simulation results are shown in red, while transport theory results (Eq. 8.51) are shown in blue. For mode energy, predictions using the mode energy equation (Eq. 8.57) are plotted in green.
Source: Colosi and Morozov (2009).

Monte Carlo simulation and transport theory is excellent, and the mean intensity differs significantly from the unperturbed intensity. The transport theory accurately estimates the ensonification of the beam shadow zone (lower panel). Path integral methods cannot handle this situation becuase there is no unperturbed ray.

Shallow Water and Continental Shelf

In shallow-water cases attenuation effects are critical. The fundamental issue is the relative rates of mode coupling given by the f_{mn} matrix and the rate of attenuation given by α_n, that is, one would like to know if the modes will couple before they experience significant attenuation or the other way around.

This situation has been examined by Colosi et al. (2011) and Raghukumar and Colosi (2014)[6] for acoustic frequencies between 200 Hz and 1 kHz for a

---

[6] Also see earlier numerical studies by Tielburger et al. (1997).

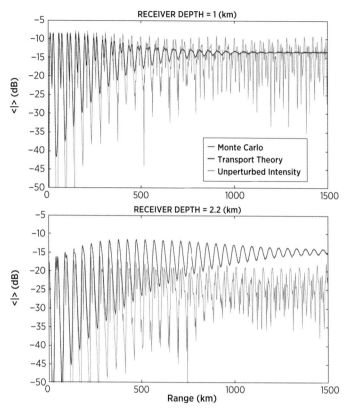

Figure 8.4. Range evolution of mean intensity for the acoustical beams in Figure 8.2 and predictions from transport theory. The top panel shows mean intensity for a receiver on the sound-channel axis, and the bottom panel shows a receiver depth in the shadow zone of the unperturbed beam. Monte Carlo numerical simulation results are shown in red, while transport theory results (Eq. 8.51) are shown in blue. Unperturbed intensity is shown in green. Cylindrical spreading has been removed.
Source: Colosi and Morozov (2009).

continental shelf environment typical of the mid-Atlantic Bight (see Figure 2.3). In these studies Monte Carlo simulations were carried out for a point source and internal-wave-induced sound-speed perturbations obeying a shallow-water version of the GM model (see Chapter 3). Figure 8.5 compares Monte Carlo calculations, transport theory, and adiabatic theory for cross-mode coherences and mode energies at 200 and 400 Hz. Three critical conclusions are to be made. First, the propagation is strongly adiabatic with little change in mode energy aside from bottom attenuation. Second, the cross-mode coherences decay quite rapidly, which will have important consequences for the mean intensity and loss of the mode interference pattern. Third, the standard adiabatic theory predicts the loss of cross mode coherence to occur too rapidly: this is from an edge effect (Colosi et al., 2011; Raghukumar and Colosi, 2014), which is due to the approximation

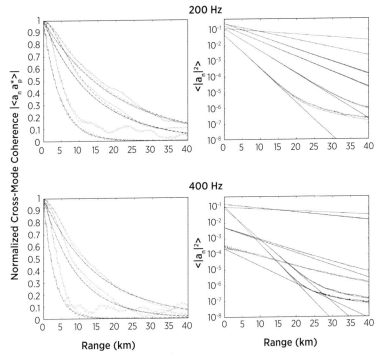

Figure 8.5. Range evolution of cross-mode coherence (left) and mean mode energy (right) for a shallow-water environment similar to the summer time mid-Atlantic Bight (SW06). Top/bottom panels respectively show results for 200 and 400 Hz. Cross-mode coherences are shown for mode pairs (1,2), (1,3), and (2,3). Monte Carlo numerical simulation results are shown in red, while transport theory results (Eq. 8.51) are shown in blue. For cross-mode coherence, adiabatic results are shown in a blue dash, and a modified adiabatic calculation taking account of the edge effect is shown in a red dash. For mode energy, adiabatic curves are shown in black.
Source: Colosi et al. (2011).

that the range must be larger than the correlation length of the internal waves. This allowed certain integrals over the correlation function to be brought out to $\pm\infty$. When the range is too short, this approximation leads to overcounting, and the transport theory cross-mode coherences decay too quickly. A simple correction to the adiabatic phase structure function (Eq. 8.47) has been provided by Raghukumar and Colosi (2014), giving the result for the GM spectrum:

$$D(n, p) = \frac{4}{\pi} \sum_{j=1}^{J} \frac{H(j)}{\hat{\kappa}_j} (G_{nn} - G_{pp})^2 \left(R + \frac{e^{-\hat{\kappa}_j R} - 1}{\hat{\kappa}_j} \right), \qquad (8.79)$$

which in the limit of large R is the standard result. Using this corrected structure function brings the Monte Carlo simulation and adiabatic theory into good agreement (Figure 8.5).

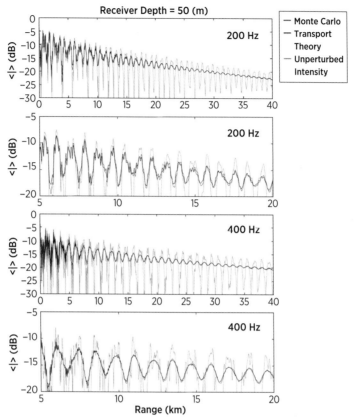

Figure 8.6. Range evolution of mean intensity for a shallow-water environment similar to the summer time mid-Atlantic Bight (SW06). Top/bottom two panels show results for 200 and 400 Hz. Monte Carlo numerical simulation results are shown in red, while transport theory results (Eq. 8.51) are shown in blue. Unperturbed intensity is shown in green. Cylindrical spreading has been removed. The second and fourth panels show an expanded view between 5- and 20-km range.
Source: Colosi et al. (2011).

Next, the mean intensity is shown in Figure 8.6. Transport theory again gives excellent predictions of mean intensity, and the rapid loss of cross-mode coherence removes the modal interference pattern evident in the unperturbed calculation. Similar results are obtained for 1-kHz propagation, where the hybrid model works exceptionally well (Raghukumar and Colosi, 2014).

8.5.7 Observations

To date, comparisons between observations and transport theory for cross-mode coherence and mode energy have been primarily carried out for deep-water, long-range experiments. Because of the difficulty in deploying dense, large-aperture vertical arrays, these observations have focused on low-frequency, low-order modes trapped near the sound-channel axis. In shallow water, essentially no attention has been paid to the observable of the cross-mode coherence matrix. The focus has been primarily on mode coupling events caused by nonlinear internal-wave packets.

Deep Water

Low-frequency multimegameter acoustic transmission data in the North Pacific Ocean obtained by the Acoustic Thermometry of Ocean Climate (ATOC) program has been analyzed in terms of normal modes by Wage et al. (2003, 2005) (see Figure 8.7). In these analyses a 1400-m vertical aperture array was used to isolate modes 1 through 10 at ranges of 3500 and 5200 km, and frequencies of 28, 75, and 84 Hz. For these observations, the source was close to the sound-channel axis.

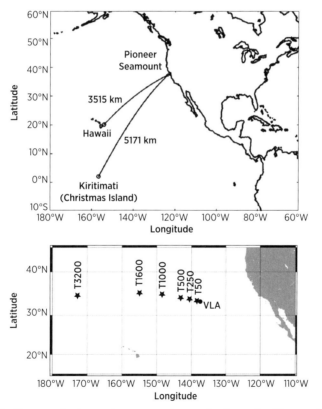

Figure 8.7. Top panel: Acoustic propagation paths from the Pioneer Seamount source to two wide aperture vertical receiving arrays for the ATOC program. Lower panel: The LOAPEX transmit stations (T50, T250, etc.) along the geodesic and the moored wide aperture vertical receiving array (VLA). The number following the "T" roughly corresponds to the transmission range in kilometers. *Sources*: Wage et al. (2005) and Chandrayadula et al. (2013).

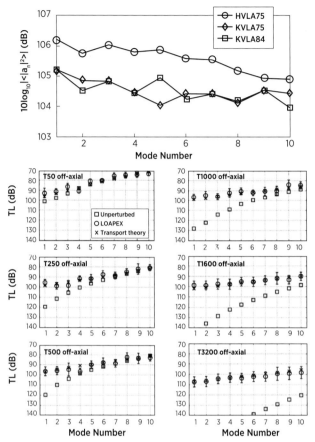

Figure 8.8. Top panel: Mean 75- and 84-Hz mode energy measured for two different multimegameter acoustic paths from an acoustic source on Pioneer Seamount (Figure 8.7). KVLA denotes the Pioneer to Kiritimati path and HVLA denotes the Pioneer to Hawaii path. Six bottom panels: Range evolution of 75-Hz mean mode transmission loss for the off-axis source depth in LOAPEX (Figure 8.7). The unperturbed estimates give the transmission loss if there were no mode coupling.
Sources: Wage et al. (2005) and Chandrayadula et al. (2013).

In another study named the Long-range Ocean Acoustic Propagation Experiment (LOAPEX), a similar receiving array was utilized and 75-Hz normal mode statistics for modes 1 through 10 at ranges of 50, 250, 500, 1000, 2300, and 3200 km were studied (Chandrayadula et al., 2013) . The LOAPEX analysis was unique in having multiple ranges, as well as on- and off-axis source depths corresponding to different initial mode energy distributions.

The acoustic observable of mode energy $\langle |a_n|^2 \rangle$ is considered first (Figure 8.8). For the ATOC observations at the extremely long ranges of 3500 and 5200 km, the first 10 modes have the same energy within about 1 dB. For a near-axial source

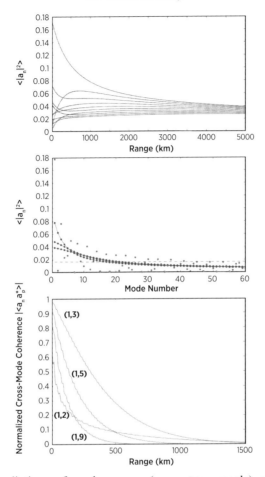

Figure 8.9. Predictions of mode energy (upper two panels) and cross-mode coherence (lower panel) as a function of range for 75-Hz long-range propagation through GM internal waves and a Munk canonical sound-speed profile (Figure 2.1). A point source located on the sound-channel axis ($z = 1000$ m) is modeled, and a total of $N = 60$ modes are used in the calculation. The first upper panel shows the range evolution of mode energy for modes 1 through 10. The middle panel shows the mode energy for all 60 modes at ranges of zero (dots unconnected), 1000, 3000, and 5000 km (dots connected). The connected dot curves separate at low mode number, with the 1000-km points higher and 5000-km points the lowest. A dashed curve shows the equipartition energy. The lower panel shows cross-mode coherences for mode combinations of $(n, p) =$ $(1, 2), (1, 3), (1, 5),$ and $(1, 9)$; these combinations are qualitatively similar to other combinations.

depth and for the low-order modes, odd modes have large initial energy while even modes have smaller initial energies. Thus large differences in mode energy must be equilibrated to give the observed equipartitioning of low mode energy. Figure 8.9 shows a transport theory calculation for a Munk canonical sound-speed profile, an exponential buoyancy frequency profile, and the GM spectrum with standard parameters (a canonical mid-latitude environment). A rapid equilibration of mode energies occurs within the first 1000 km of propagation through the random internal-wave field.[7] A more impressive comparison of mode energies is for the LOAPEX off-axis transmissions that are also shown in Figure 8.8. Here the low-order modes are weakly excited at the source, but significant mode coupling leads to in-filling of the mode population such that the modes are within 10 dB of each other by the 3200-km range. Transport theory does an excellent job at predicting the observed evolution of these mode energies (Chandrayadula et al., 2013).

For cross-mode coherences at a fixed frequency $\langle a_n a_p^* \rangle$, it is convenient to use the normalized coherence given by

$$\frac{|\langle a_n a_p^* \rangle|(r)}{\sqrt{\langle |a_n|^2 \rangle(r) \langle |a_p|^2 \rangle(r)}}. \tag{8.80}$$

ATOC observations show that the first 10 modes have coherences statistically consistent with zero. This result is also seen in the transport theory calculations done by Chandrayadula et al. (2013) out to 3200-km range for LOAPEX, and coherences estimated from the canonical mid-latitude environment discussed above (Figure 8.9). Importantly, the LOAPEX observations show how the low mode coherences evolve at shorter ranges (Figure 8.10). For the on-axis source, the cross-mode coherences are seen to have decayed substantially by 500-km range, and even more so for the off-axis source. Here, however, the transport theory seems to underestimate the cross-mode coherence, for reasons that are not fully understood.

Shallow Water

In spite of the strong focus on normal modes in shallow-water research over the last 20 years, reports of mean mode energy or cross-mode coherence are essentially absent from the literature. It is hoped that this book inspires some research in this direction. There is, however, one result comparing observed and transport theory transmission loss for the SW06 experiment. In this case a towed source emitted 1200 Hz tones that were recorded on a vertical line array. The observed transmission loss is plotted in Figure 8.11 with transport theory estimates of mean transmission loss with error bars. The error bars are computed from transport theory estimates of intensity variance that are discussed later in the chapter. The single realization of transmission loss is seen to fit between the errors of the model over most of the propagation ranges.

---

[7] This result is also consistent with the transport theory calculations done by Chandrayadula et al. (2013) out to 3200-km range for the LOAPEX.

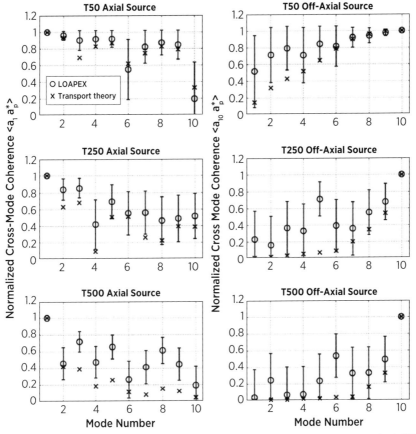

Figure 8.10. Evolution of cross-mode coherence to 500-km range for LOAPEX for the on-axis source (left panels) and the off-axis source (right panels). In the on-axis case, coherences relative to mode 1 are plotted, while for the off-axis case coherences are relative to mode 10. Cross-mode coherences for modes 1–10 are statistically consistent with zero for LOAPEX ranges greater than 500 km.
Source: Chandrayadula et al. (2013).

8.6 Mutual Coherence Functions: Space and Time Separations

Another important second moment, which is somewhat more complicated than the mean intensity, is the mutual coherence function (MCF) at fixed frequency. This should be contrasted to the coherence functions computed in Chapter 7 using path integral theory, which gave coherences along fixed acoustic paths, a more broadband concept. Theoretical development and Monte Carlo simulation testing of transport theory of MCFs were carried out by Colosi et al. (2013a) for a deep-water mid-latitude environment. The coherence function for separations in depth (z), transverse position (y), frequency (ω), and time (t) as a function of

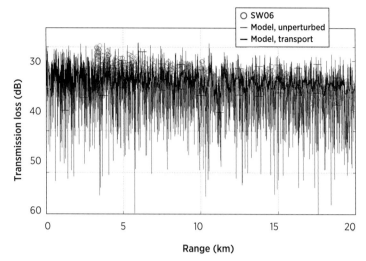

Figure 8.11. A single observation of 1200-Hz transmission loss from SW06 (red dots) and estimates from transport theory (black with error bars computed from fourth moment transport theory). The unperturbed transmission loss is plotted in blue. The 200-dB (re 1 μPa at 1 m) source was towed at a depth of 50 m, and the receiver sat on the ocean bottom at a depth of 82 m. Cylindrical spreading has been removed.

propagation range (r) is written as

$$\langle p(r,z_1,y_1,\omega_1,t_1)p^*(r,z_2,y_2,\omega_2,t_2)\rangle = \sum_{n=1}^{N}\sum_{p=1}^{N}\frac{\langle a_{n1}a_{p2}^*(r)\rangle}{r}\frac{\phi_{n1}(z_1)\phi_{p2}(z_2)}{\sqrt{k_{n1}k_{p2}}}. \quad (8.81)$$

The coherence depends critically on the mode amplitude coherence $\langle a_{n1}a_{p2}^*\rangle$ between points nominally represented by (1) and (2). For the case in which there are no frequency and depth separations, a useful summary coherence is the depth average given by

$$\int_0^\infty \frac{\langle p(r,z,t_1,y_1)p^*(r,z,t_2,y_2)\rangle}{\rho_0(z)}\, dz = \sum_{n=1}^{N}\frac{\langle a_n a_n^*(r,\Delta t,\Delta y)\rangle}{k_n r}, \quad (8.82)$$

where $\Delta t = |t_1 - t_2|$ and $\Delta y = |y_1 - y_2|$. This result follows from the orthogonality of the modes.

The transport equation for cross-mode coherence derived in Section 8.5 can be trivially modified to give the evolution equation for the mode amplitude coherence function with the aforementioned lags, such that

$$\frac{d\langle a_{n_1}a_{p_2}^*\rangle}{dr} - i(l_{n1} - l_{p2}^*)\langle a_{n_1}a_{p_2}^*\rangle = \sum_{m=1}^{N}\sum_{q=1}^{N}\langle a_{m_1}a_{q_2}^*\rangle I_{mn1,qp2}^* + \langle a_{q_1}a_{m_2}^*\rangle I_{mp2,qn1}$$

$$- \langle a_{n_1}a_{q_2}^*\rangle I_{mp2,qm2}^* - \langle a_{q_1}a_{p_2}^*\rangle I_{mn1,qm1}. \quad (8.83)$$

Here the scattering matrices involving the two separate points are given by

$$I_{mn1,qp2} = \int_0^\infty d\delta r \, \Delta_{mn1,qp2}(\delta r, \Delta t, \Delta y) \, e^{ik_{pq2}\delta r}, \tag{8.84}$$

where $\Delta_{mn1,qp2}(\delta r, \Delta t, \Delta y) = \langle \rho_{mn1}\rho_{qp2}(\delta r, \Delta t, \Delta y)\rangle$ is the correlation function of the coupling matrices for range separations (δr) and other frequency, horizontal position, and/or time separations. The mode wavenumber difference is $k_{qp2} = k_{q2} - k_{p2}$ and involves only the second point. An expression similar to Eq. 8.84 gives $I_{mp2,qn1}$. In Eq. (8.84) it is assumed as before that attenuation is weak over a correlation length of $\Delta_{mp1,qp2}$, and thus the imaginary part of the wavenumber difference in the exponential is ignored.

Because new lag parameters have been introduced into the coupling matrix correlation function, it is useful to express this function in polar coordinate form giving

$$\Delta_{mn1,qp2}(\delta r, \Delta t, \Delta y) = \sum_{j=1}^J H(j)G_{mn1}(j)G_{qp2}(j) \int_0^\infty d\kappa_h \, F_1(\kappa_h) \, \cos(\sigma_j(\kappa_h)\Delta t)$$

$$\times \int_{-\pi}^{\pi} \frac{d\vartheta}{2\pi} \cos(\kappa_h \delta r \cos\vartheta)\cos(\kappa_h \Delta y \sin\vartheta). \tag{8.85}$$

Here $\sigma_j(k)$ is the internal-wave dispersion relation, and ϑ is the azimuthal angle. The G functions are still given by Eq. 8.13, but they are computed at the two separate frequencies.

Adiabatic/Hybrid Theory

It has been found that adiabatic theory is useful for understanding and predicting coherence in both deep-water (Colosi et al., 2013a) and shallow-water (Raghukumar and Colosi, 2014) environments. As seen previously, the adiabatic results from transport theory are the same as those derived directly from the mode equations in the adiabatic approximation. The adiabatic approximation to Eq. 8.83 is obtained by considering only those scattering matrices with pairwise identical indices to be nonzero. The result is

$$\langle a_{n_1}a_{p_2}^*\rangle(R, \Delta t, \Delta y) = \langle a_{n_1}a_{p_2}^*\rangle(0) \, e^{i(l_{n1}-l_{p2}^*)R} \, e^{-(I_{nn1,nn1}+I_{pp2,pp2}-2I_{nn1,pp2}(\Delta t,\Delta y))R}. \tag{8.86}$$

Modifying the result to accommodate mode coupling in the form of energy exchanges among the modes (but not coupling induced phase changes), the hybrid theory result is

$$\langle a_{n_1}a_{p_2}^*\rangle(R, \Delta t, \Delta y) = \langle |a_{n1}|^2\rangle^{1/2}(r)\langle |a_{p2}|^2\rangle^{1/2}(r)\text{sign}[a_{n_1}(0)]\text{sign}[a_{p_2}(0)]$$

$$e^{i(k_{n1}-k_{p2})R} \, e^{-(I_{nn1,nn1}+I_{pp2,pp2}-2I_{nn1,pp2}(\Delta t,\Delta y))R}. \tag{8.87}$$

Here the mode energy factors are obtained by solving the mode energy equation (Eq. 8.57) with $\alpha_n \neq 0$. More details of the adiabatic solutions will be given below when specific space and time lags are considered.

At this point it is useful to treat the different coherence functions separately, and the discussion of frequency separations is deferred until Section 8.7.

8.6.1 Depth Separations

The vertical/depth MCF is obtained quite easily from Eq. 8.81, because the depth dependence has been factored out into the unperturbed mode functions. The coherence function at fixed frequency for receiver depths z_1 and z_2 is therefore given by

$$\langle p(r,z_1)p^*(r,z_2)\rangle = \sum_{n=1}^{N}\sum_{p=1}^{N} \frac{\langle a_n a_p^*(r)\rangle}{r} \frac{\phi_n(z_1)\phi_p(z_2)}{\sqrt{k_n k_p}}. \qquad (8.88)$$

No vertical stationarity assumption has been built into this equation, such as was done for the path integral where the coherence function was defined locally in depth near the unperturbed ray. In Eq. 8.88, the depth structure of the modes gives the coherence function for any receiver pair. Estimates of the vertical coherence function are obtained by solving the same transport theory for the cross-mode coherence $\langle a_n a_p^*(r)\rangle$, as was done in the section on mean intensity. The full coherence function is then obtained by the double sum (Eq. 8.88). Unlike the path integral formalism, a natural coherence length does not follow from the modal coherence equation. Nonetheless one can define the coherence length z_0 as the half e-folding distance of the normalized coherence function, that is,

$$\frac{\langle pp^*(r,\Delta z=z_0)\rangle}{\langle |p|^2(r,z_{ref})\rangle} = \exp(-0.5), \qquad (8.89)$$

where z_{ref} is an appropriate reference depth, usually the mean depth.

A few interesting limiting cases are worth mentioning. First, when the modes have decorrelated, the coherence function has the simple form:

$$\langle p(r,z_1)p^*(r,z_2)\rangle = \sum_{n=1}^{N} \frac{\langle |a_n|^2(r)\rangle}{r} \frac{\phi_n(z_1)\phi_n(z_2)}{k_n}. \qquad (8.90)$$

If attenuation is weak, such as in deep-water low-frequency propagation, then the mode energies $\langle |a_n|^2(r)\rangle$ at long ranges will vary only weakly with range. In this case, the normalized coherence function is essentially independent of range and will not scale as $1/\sqrt{R}$ as was the case in the path integral formulation. For shallow-water cases, the range dependence of the depth coherence is strongly influenced by the mode dependence of the attenuation.

Another interesting case is the adiabatic limit and the hybrid theory limit for which the cross-mode coherence is given by the adiabatic relation (Eq. 8.48), and there are simple relations for the scattering matrices. In this limit,

$$\langle a_n a_p^* \rangle(r) = \langle a_n a_p^* \rangle(0) \, e^{i(l_n - l_p^*)r} \, \exp\left[-\frac{1}{2}\frac{r}{R_{np}}\right], \tag{8.91}$$

and the cross-mode coherence decays exponentially with range. The characteristic mode decorrelation range using the GM spectrum is

$$R_{np}^{-1} = \frac{4N_0 B}{\pi^2 f} \sum_{j=1}^{J} \frac{H(j)}{j}(G_{nn}(j) - G_{pp}(j))^2. \tag{8.92}$$

The $H(j)/j$ weighting in the sum means that cross-mode decorrelation is driven primarily by the low mode internal waves. Note also that the decorrelation range depends on the squared difference of the $G(j)$ functions, which will be strongly influenced by the depth structure of $\langle \mu^2 \rangle$. Because the G functions scale linearly with acoustic frequency, R_{np} will scale as one over frequency squared.

Numerical Validation and Results

It has been seen from previously described Monte Carlo simulations that transport theory for the cross-mode coherence matrix (Eq. 8.51) is quite accurate for both shallow and deep-water environments. Thus, estimates of vertical coherence are expected to be excellent as well. Figure 8.12 shows simulated and transport theory calculations of normalized depth coherence functions for a deep-water environment typical of the Philippine Sea (Colosi et al., 2013a). The agreement is good, and it should be noted that the depth coherence function is not necessarily a monotonically decreasing function of lag, nor is it necessarily symmetric. These features are functions of the vertical mode structure and the cross-mode coherence.

The issue of the range and frequency scaling of the depth coherence is fundamental. If a typical vertical coherence length is defined as the length at which the coherence decays to $e^{-1/2}$ of the zero lag, and this value is an average over positive and negative lags, then the scaling of this coherence length as a function of range and frequency can be investigated (Figure 8.12). For this example, the depth coherence scales as frequency to the minus one power, which is the expected result (Colosi et al., 2013a). This scaling is the same as that obtained from ray/path integral theory. The range scaling, however, is not the expected one over square root range, but is much slower, that is, range to the minus one fifth power. This result is due to the loss of cross-mode coherence, which leads to a coherence function of the form of Eq. 8.90. For this deep-water, small-attenuation case the depth coherence will change quite slowly with range once the cross-mode coherences have decayed.

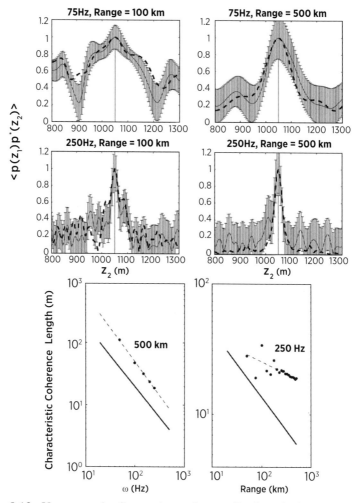

Figure 8.12. Upper panels: Comparison of normalized vertical coherence functions from Monte Carlo simulations (solid with error bars) and transport theory (dashed). The reference depth z_1 is 1050 m. Ranges of 100 and 500 km and frequencies of 75 and 250 Hz are shown. The environment modeled is one typical of the western Philippine Sea. Lower panels: Frequency and range scaling of the characteristic vertical coherence length defined as the $e^{-1/2}$ point. The thick lines in the frequency/range plots show scalings of ω^{-1} and $R^{-1/2}$, respectively. *Source*: Colosi et al. (2013a).

With regard to adiabaticity, the normalized depth coherence function is poorly modeled using the adiabatic approach for a mid-latitude deep-water ocean environment (Colosi et al., 2013a). There is some cause for optimism, however, that the hybrid model could do significantly better, because the deep-water case involves a good deal of mode energy redistribution from coupling. Given the success of the hybrid model in shallow water, it is expected that this approach will work well for shallow-water vertical coherence.

8.6.2 Temporal Separations

The temporal MCF has perhaps garnered the most attention in both the shallow and deep-water acoustics literature. This function has the form

$$\langle p(r,z,t_1)p^*(r,z,t_2)\rangle = \sum_{n=1}^{N}\sum_{p=1}^{N}\frac{\langle a_n a_p^*(r,\Delta t)\rangle}{r}\frac{\phi_n(z)\phi_p(z)}{\sqrt{k_n k_p}},\qquad(8.93)$$

where $\Delta t = |t_1 - t_2|$. Because the dispersion relation depends only on the magnitude of the horizontal wavenumber, κ_h, the scattering matrices involving the time separation are most easily obtained using Eq. 8.85. The integrals over angle and horizontal separation are done analytically to obtain

$$I_{mn1,qp2}(\Delta t) = \sum_{j=1}^{J} H(j)G_{mn}(j)G_{qp}(j)$$

$$\times \int_0^{\kappa_{max}(j)} d\kappa_h\, F_1(\kappa_h)\,\cos(\sigma_j(\kappa_h)\Delta t)\begin{cases}\frac{1}{\sqrt{\kappa_h^2-k_{pq}^2}}, & 0\le |k_{pq}| < \kappa_h \\[2ex] \frac{i\,\text{sign}(k_{pq})}{\sqrt{k_{pq}^2-\kappa_h^2}}, & 0\le \kappa_h < |k_{pq}|\end{cases}.$$

$$(8.94)$$

Here it will be noticed that a maximum horizontal wavenumber, κ_{max}, has been imposed. This is necessary because of singularities associated with the form of the temporal coherence function using the WKB dispersion relation. The value of κ_{max} is obtained by restricting the WKB dispersion relation to have a maximum frequency $\sigma_{max} = N_{max}$. In this case, the maximum wavenumber for mode j is given by

$$\kappa_{max}(j) \simeq \pi j\frac{N_{max}}{N_0 B}.\qquad(8.95)$$

Numerical evaluation of Eq. 8.83 using the scattering matrices from Eq. 8.94 can be done to give the complete coherence function. As previously mentioned, the mode formalism does not provide a clean formula for the coherence scale, so the coherence time t_0 must be defined implicitly as the half e-folding time of the normalized coherence function, that is,

$$\frac{\langle pp^*(r,z,\Delta t = t_0)\rangle}{\langle |p|^2(r,z)\rangle} = \exp(-0.5).\qquad(8.96)$$

As in the path integral chapter, it is helpful to examine the coherence function for small values of lag (Flatté et al., 1979). The cosine in Eq. 8.94 can then be

Taylor expanded and the WKB dispersion relation used to obtain

$$I_{mn1,qp2}(\Delta t) \simeq I_{mn,qp} - \frac{f^2 \Delta t^2}{2} \sum_{j=1}^{J} H(j) G_{mn}(j) G_{qp}(j)$$

$$\times \int_0^{\kappa_{max}} d\kappa_h \, F_1(\kappa_h) \left(1 + \frac{\kappa_h^2}{\hat{\kappa}_j^2}\right) \begin{cases} \frac{1}{\sqrt{\kappa_h^2 - k_{pq}^2}}, & 0 \leq |k_{pq}| < \kappa_h \\ \frac{i \, \text{sign}(k_{pq})}{\sqrt{k_{pq}^2 - \kappa_h^2}}, & 0 \leq \kappa_h < |k_{pq}| \end{cases}.$$

(8.97)

Appendices A and B provide analytic expressions for the wavenumber integrals required to calculate both terms in Eq. 8.97 for the GM spectrum and noninfinite κ_{max}. Coherence results using the full scattering matrix (Eq. 8.94) and the quadratic approximation (Eq. 8.97) have been compared for a mid-latitude environment. The quadratic approximation is found to work best for longer ranges and higher frequencies (Colosi et al., 2013a, see figure 7). A reasonable criteria for the quadratic approximation to be accurate is that the modal phase variance must be greater than 1, that is, $\Phi_n^2 \gg 1$. In the quadratic approximation the coherence decays too rapidly, but if factors of 2 are not important the quadratic approximation is a reasonable choice and confers a significant numerical efficiency.

Temporal coherence in the adiabatic approximation, both with and without the quadratic lag approximation, is of great relevance. The adiabatic scattering matrices are obtained by setting $m = n$ and $q = p$, which allows the wavenumber integrals in the quadratic lag approximation to be done analytically using the GM spectrum (see Appendix B). The scattering matrices are then given by the simple result:

$$I_{nn1,pp2}(\Delta t) = I_{nn,pp} - \frac{f^2 \Delta t^2}{\pi} \sum_{j=1}^{J} H(j) \frac{G_{nn}(j) G_{pp}(j)}{\hat{\kappa}_j} \ln\left[\frac{\hat{\kappa}_j^2 + \kappa_{max}^2}{\hat{\kappa}_j^2}\right]. \quad (8.98)$$

The scattering matrix diverges logarithmically if the maximum wavenumber is allowed to go to infinity, which is why the cut-off is necessary. In the adiabatic approximation, the mode equations decouple. In this quadratic approximation, the result is

$$\langle a_{n_1} a_{p_2}^* \rangle (R, \Delta t) = \langle a_{n_1} a_{p_2}^* \rangle (0) \, e^{i(l_n - l_p^*)R} \, \exp\left[-\frac{1}{2}\frac{R}{R_{np}}\right] \exp\left[-\frac{1}{2}\frac{\Delta t^2}{t_{np}^2}\right], \quad (8.99)$$

where the characteristic adiabatic coherence time t_{np} using the GM spectrum is

$$t_{np}^{-2} \simeq R \frac{8f N_0 B}{\pi^2} \ln\left(\frac{N_{max}}{f}\right) \sum_{j=1}^{J} \frac{H(j)}{j} \, G_{nn}(j) G_{pp}(j). \quad (8.100)$$

Equation 8.100 has a form that is similar to the axial ray coherence time estimate from path integrals (Eq. 7.56). Like the mode decorrelation range R_{np} defined in the previous section, the characteristic coherence time, t_{np}, involves the internal wave mode weighting $H(j)/j$, which means that the coherence time is affected most strongly by low-mode internal waves. This internal-wave mode weighting is of the same form seen for the path integral time coherence. The logarithmic divergence as $\kappa_{max} \to \infty$ also means that this observable is sensitive to small horizontal scale internal waves. Lastly the coherence time t_{np} scales as one over the square root of range (a result that clearly comes about because of the quadratic lag approximation), and it scales as one over frequency because G scales linearly with frequency.

Numerical Validation and Results

Transport theory for $\langle a_n a_p^*(r, \Delta t) \rangle$ has been well validated for deep-water, mid-latitude environments, and given the strong adiabaticity in shallow water, the theory is expected to do well there too. Figure 8.13 shows comparisons between 75- and 250-Hz Monte Carlo simulations and transport theory for depth averaged time coherence (Eq. 8.82) in a deep-water environment typical of the Philippine Sea. The comparisons are quite favorable for the two frequencies and at short and long ranges (Colosi et al., 2013a). Here no quadratic or adiabatic approximation has been made in the transport theory. Similarly good results are found for nondepth averaged coherences.

While the adiabatic approximation gives some insight into temporal coherence scaling with range and frequency, an analysis of the full transport theory is necessary and has been carried out by Colosi et al. (2013a). As with the depth coherences, one can define a characteristic time coherence that is the $e^{-1/2}$ decay point of the depth-integrated temporal coherence function computed from transport theory. These coherence times are plotted as a function of acoustic frequency and range in Figure 8.13. The transport theory result is quite close to that expected from ray/path integral theory, and that is time coherence scaling as ω^{-1} and $R^{-1/2}$.

Lastly, there is the important issue of adiabaticity, which has also been evaluated by Colosi et al. (2013a) for deep-water, mid-latitude conditions. Quite unexpectedly, the adiabatic time coherence approximation is excellent (Colosi et al., 2013a, see figure 6). Given the strong adiabatic conditions encountered in shallow-water environments, it is expected that time coherences in this case will also be well approximated by adiabatic theory.

8.6.3 Transverse Separations

The horizontal coherence function, like the time coherence, has garnered much attention for both deep and shallow-water environments. The horizontal coherence

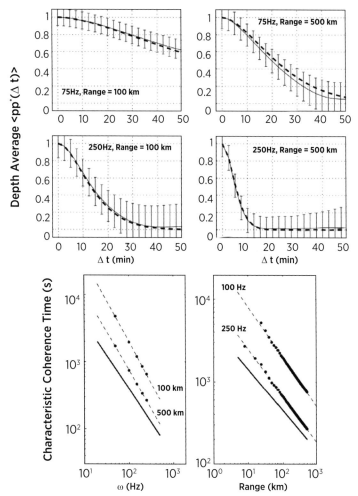

Figure 8.13. Upper panels: Comparison of normalized temporal coherence functions from Monte Carlo simulations (solid with error bars) and transport theory (dashed). Ranges of 100 and 500 km and frequencies of 75 and 250 Hz are shown. The environment modeled is one typical of the western Philippine Sea. Lower panels: Frequency and range scaling of the characteristic vertical coherence length defined as the $e^{-1/2}$ point. The thick lines in the frequency/range plots show scalings of ω^{-1} and $R^{-1/2}$, respectively.
Source: Colosi et al. (2013a).

function has the form

$$\langle p(r,z,y_1)p^*(r,z,y_2)\rangle = \sum_{n=1}^{N}\sum_{p=1}^{N} \frac{\langle a_n a_p^*(r,\Delta y(r))\rangle}{r} \frac{\phi_n(z)\phi_p(z)}{\sqrt{k_n k_p}}, \tag{8.101}$$

where the path-dependent horizontal separation is $\Delta y_p(r) = (r/R)\Delta y$ and $\Delta y = |y_1(R) - y_2(R)|$ is the y-separation at the receiver. Solution of the transport equation for horizontal separations is complicated because the scattering matrix is no longer independent of the range coordinate. The scattering matrix in this case is

$$
I_{mn1,qp2}(\delta r, \Delta y_p(r)) = \sum_{j=1}^{J} H(j) G_{mn}(j) G_{qp}(j) \int_0^{\kappa_{max}} d\kappa_h \, F_1(\kappa_h)
$$

$$
\times \int_0^\infty d\delta r \, e^{ik_{pq}\delta r} \int_{-\pi}^\pi \frac{d\vartheta}{2\pi} \cos(\kappa_h \delta r \cos \vartheta) \cos(\kappa_h \Delta y_p(r) \sin \vartheta).
$$

$$(8.102)$$

To make this problem tractable (even numerically), two of the three integrals in Eq. 8.102 must be done analytically, but this has proven to be a difficult task. However, some analytic progress can be made in the quadratic lag approximation. Expanding the cosine to second order, the integration over angle (Abramowitz and Stegun, 1964, see 9.1.20) and then over δr (Abramowitz and Stegun, 1964, see 11.4.35–36) can be done to obtain

$$
I_{mn1,qp2}(\Delta y) = I_{mn,qp} - \frac{\Delta y_p^2(r)}{2} \sum_{j=1}^{J} H(j) G_{mn}(j) G_{qp}(j)
$$

$$
\times \int_0^{\kappa_{max}} d\kappa_h F_1(\kappa_h) \kappa_h
\begin{cases}
\sqrt{1 - k_{pq}^2/\kappa_h^2} + i k_{pq}/\kappa_h, & 0 \le |k_{pq}| < \kappa_h \\
i \, \text{sign}(k_{pq}) \dfrac{\kappa_h}{|k_{pq}| + \sqrt{k_{pq}^2 - \kappa_h^2}}, & 0 \le \kappa_h \le |k_{pq}|
\end{cases}.
$$

$$(8.103)$$

The wavenumber integral still requires numerical evaluation. The coherence length y_0 is defined as the half e-folding distance of the normalized coherence function, that is,

$$
\frac{\langle pp^*(r, z, \Delta y = y_0) \rangle}{\langle |p|^2(r, z) \rangle} = \exp(-0.5).
$$

$$(8.104)$$

Again, the adiabatic limit is of fundamental interest, and in this case the wavenumber integral can be done analytically to obtain

$$
I_{nn1,pp2}(\Delta y_p(r)) = I_{mn,qp} - \frac{\Delta y^2}{\pi} \frac{r^2}{R^2} \sum_{j=1}^{J} H(j) \, G_{nn}(j) G_{pp}(j)
$$

$$
\times \hat{\kappa}_j \left[\ln\left(\frac{\hat{\kappa}_j^2 + \kappa_{max}^2}{\hat{\kappa}_j^2} \right) - \frac{\kappa_{max}^2}{\hat{\kappa}_j^2 + \kappa_{max}^2} \right].
$$

$$(8.105)$$

There is again a logarithmic divergence if the maximum wavenumber is allowed to go to infinity, indicating some sensitivity to small scales. The adiabatic solution to the transport equation in the quadratic lag approximation is thus

$$\langle a_{n_1} a_{p_2}^* \rangle (R, \Delta y) = \langle a_{n_1} a_{p_2}^* \rangle (0) \; e^{i(l_n - l_p^*)R} \; \exp\left[-\frac{1}{2} \frac{R}{R_{np}}\right] \exp\left[-\frac{1}{2} \frac{\Delta y^2}{y_{np}^2}\right], \quad (8.106)$$

with the characteristic adiabatic horizontal coherence length using the GM spectrum given by

$$y_{np}^{-2} \simeq R \frac{8f}{3N_0 B} \left[\ln\left(\frac{N_{max}}{f}\right) - \frac{1}{2}\right] \sum_{j=1}^{J} H(j) \, j \, G_{nn}(j) G_{pp}(j). \quad (8.107)$$

Equation 8.107 has a form that is similar to the axial ray coherence length estimate from path integrals (Eq. 7.58), a result that was also seen for the modal time coherence. Here y_{np} has a different internal wave mode number weighting than t_{np} and R_{np} and depends on $H(j)j$. The adiabatic horizontal coherence is more sensitive to the higher internal-wave modes. The sensitivity to high mode internal waves was also found in the path integral results for transverse coherence. As for time coherence, the adiabatic horizontal coherence length scales as one over the square root of range, a consequence of the quadratic lag approximation. Numerical evaluation similar to that shown in Figures 8.12 and 8.13 demonstrates that transverse coherence scales closely as ω^{-1} (Colosi et al., 2013a, see figure 10), similar to the results for depth and time separations and that predicted by ray/path integral theory.

Monte Carlo numerical simulations to test the transverse coherence expressions developed here are rather computationally intensive and have not been carried out. The accuracy of the adiabatic approximation has been addressed, however. It is found that the adiabatic approximation for the transverse MCF is not as bad as that for the depth MCF, but not as good as that for the time MCF (Colosi et al., 2013a, see figure 9).

8.6.4 Observations

Observations of temporal and spatial coherence have been of intense interest to ocean acousticians for many decades, because of the large signal gains that can be achievable using coherent processing techniques. More rare, however, are observations of the coherence properties of individual modes, which is the topic here. Thus the present treatment of observations will address both types of coherence observables, that is full-field single-frequency coherence and coherence of individual modes. Tables 8.1 and 8.2 give a summary of the various deep and shallow-water observations and predictions of coherence, some of which are closely related to the coherence discussion in Chapter 7 (see Table 7.1).

Table 8.1. *Coherence times for experiments that resolved individual modes*

Experiment	Range (km)	Mode Number	t_{nn} (obs/pred) (min)
ATOC-KVLA (75 Hz)	5200	1	7.1/7.7
ATOC-HVLA (75 Hz)	3500	1	9.2/9.4
ATOC-HVLA (75 Hz)	3500	5	7.8/6.4
ATOC-HVLA (75 Hz)	3500	10	7.1/6.0
AST-KVLA (28 Hz)	5200	1	21/21
AST-HVLA (28 Hz)	3500	1	28/25
LOAPEX (75 Hz, on axis)	500	1	21/25
LOAPEX (75 Hz, off axis)	500	10	15/20
SWARM(224 Hz)	42	1	90/40
SWARM(400 Hz)	42	1	40/42
SWARM(400 Hz)	32	1	2.0^a/40
SWARM(400 Hz)	32	2	1.8^a/14
SWARM(400 Hz)	32	3	1.7^a/8.1
SWARM(400 Hz)	32	4	1.7^a/6.2

Theoretical predictions for the shallow-water experiments are based on adiabatic formulas, while the deep-water, long-range predictions are based on calculations from Colosi et al. (2013a) and Chandrayadula et al. (2013). Primary references are: ATOC/AST (Wage et al., 2005), LOAPEX (Chandrayadula et al., 2013), and SWARM (Headrick et al., 1999; Rouseff et al., 2002). Several of the experiments are also in Table 7.1.
[a]These SWARM observations are for the short timescale coherence time determined largely by nonlinear internal solitary waves (see Figure 8.16).

Deep Water

Starting with vertical coherence, there are few observational studies for which comparisons have been made. The most detailed analysis comes from Colosi et al. (2005), who examined low-frequency, long-range transmissions in the North Pacific Ocean. Full-field vertical coherences were computed in the finale of acoustic receptions for frequencies of 28–84 Hz, and ranges of 3250–5200 km (see Figure 7.5). The fact that the observations were in the finale is important, because this region is composed of low-order modes. The one-half e-folding coherence length in these data at the higher frequencies of 75 and 84 Hz is in the range 30–50 m, while the coherence length at 28 Hz is in the range 125–150 m. The observations scale approximately as ω^{-1}, and the results for 3250 km and 5200 km are nearly the same. The transport theory results from Colosi et al. (2013a) for low-order modes, gives a 75-Hz vertical coherence length of 75 m at 500-km range. Scaling this result to 4000 km using the $R^{-0.2}$ law (see Figure 8.12) and scaling with frequency, coherence lengths of 75 m at 75 Hz and 225 m at 28 Hz are obtained (see Table 8.2 for further companions). Encouragingly, these rough estimates are within a factor of two of the observations.

Table 8.2. *Full-field coherence times and lengths for experiments that did not resolve ray paths but rather were either a single-frequency or narrowband complex interference pattern of modes*

Experiment	Range (km)	t_0 (obs/pred) (min)	z_0 (obs/pred) (m)	y_0 (obs/pred) (m)	Notes
MIMI (406 Hz)	550	5.0/2.9	NA/14	NA/290	CW
MIMI (406 Hz)	1250	3.3/1.9	NA/12	NA/190	CW
NPAL Billboard (75 Hz)	3900	NA/5.9	NA/50	460/590	NB
NPAL SOSUS O (75 Hz)	2500	NA/7.3	NA/54	410/730	NB
NPAL SOSUS N (75 Hz)	2500	NA/7.3	NA/54	530/730	NB
ATOC-KVLA (75 Hz)	5200	NA/5.1	30/47	NA/510	NB
ATOC-HVLA (75 Hz)	3500	NA/6.2	50/51	NA/620	NB
AST-KVLA (28 Hz)	5200	NA/14	125/130	NA/1400	NB
AST-HVLA (28 Hz)	3500	NA/17	150/140	NA/1700	NB
ASIAEX (300 Hz)	18.9	15.1/14	NA/NA	NA/790	NB
ASIAEX (500 Hz)	18.9	5.9/8.2	NA/NA	NA/470	NB
SW06 (224 Hz)	30	NA/16	NA/NA	66/480	NB
SW06 (400 Hz)	30	NA/9	NA/NA	37/250	NB

The label NA indicates no value available, and CW/NB denotes single-frequency or narrowband observations. Theoretical predictions for the shallow-water experiments are based on adiabatic formulas, while the deep-water, long-range predictions are based on calculations and range/frequency scalings from Colosi et al. (2013a). Primary references are: MIMI (Dyson et al., 1976), NPAL (Andrew et al., 2005; Voronovich et al., 2005), ATOC/AST (Colosi et al., 2005), ASIAEX (Mignerey and Orr, 2004), and SW06 (Duda et al., 2012). Several of the experiments are also in Table 7.1.

Turning to time coherence, there are a number of deep-water observational results that can be examined. Using the observations discussed previously for vertical coherence, Wage et al. (2005) estimated temporal coherences for modes 1 through 10, a few examples of which are shown in Figure 8.14. For the 3500-km path (HVLA) at 75 Hz, the mode 1, 5, and 10 time coherences defined by the half-e-folding time are 10.8, 8.8, and 8.6 min.[8] The transport theory coherence times for these modes as computed by Colosi et al. (2013a) are 20, 17, and 16 min at 500-km range. Extrapolating to 3500-km range using inverse square root scaling leads to predicted values of 7.6, 6.4, and 6.0 min, which are well within a factor of two of the observations. Similarly good agreement is obtained for the 75- and 84-Hz observations at 5200-km range. The 28-Hz mode 1 coherence time (see Figure 8.14 center panel) is estimated to be roughly 30 min. Scaling the transport theory result gives 23 minutes (see Table 8.1).

---

[8] Figure 8.14 plots mean square coherence, and thus a square root is needed to compare to the MCF used in this book.

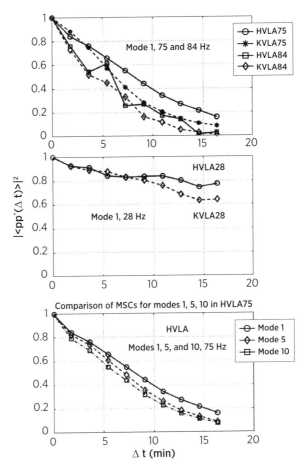

Figure 8.14. Mode mean square coherence (MSC) as a function of time lag for acoustic frequencies of 84, 75, and 28 Hz and for two different multimegameter acoustic paths in the North Pacific Ocean. The ranges for KVLA/HVLA are 5200/3500 km. To compare to the MCF used in the text, a square root is required. *Source*: Wage et al. (2005).

In another modal time coherence analysis, Chandrayadula et al. (2013) estimated 75-Hz modal time coherences at 50, 250, and 500 km for both on- and off-axis source depths and compared the results to transport theory. Results for mode 1 for the on-axis source and mode 10 for the off-axis source are quite good (Figure 8.15).

Lastly, transverse coherence is discussed. Assimilating a large number of observations, Carey (1998) estimated a typical deep-water, mid-latitude transverse coherence length as 100 wavelengths at 500-km range, with variations by factors of 2.[9] Using transport theory, Colosi et al. (2013a) computed 500-km range,

---

[9] This estimate can be range-scaled using the inverse square root.

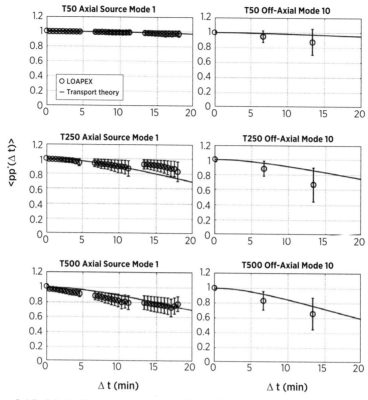

Figure 8.15. Mode time coherences at 75 Hz, for ranges of 50, 250, and 500 km for on-/off-axis source depth (left/right). Observations are plotted with error bars, and transport theory estimates are shown as continuous lines.
Source: Chandrayadula et al. (2013).

transverse coherence lengths of 500 and 1800 m for frequencies of 250 and 75 Hz, that is, 83 and 90 wavelengths respectively, which is quite close to the Carey number. In another study of long-range propagation, Voronovich et al. (2005) estimated 75-Hz transverse coherence at 3900-km range (see Figure 7.4). The observed full-field horizontal coherence length of 500 m can be compared to the 75-Hz range scaled value quoted above giving 680 m, again a favorable comparison.

Shallow Water

Because of bottom attenuation, it is expected that coherence values in shallow water will be a complex function of (1) background profiles and water depth, (2) bottom properties, (3) frequency and range, (4) source depth, and (5) geographic and temporal variations in the internal-wave spectrum. The interplay of these factors has not been studied at all.

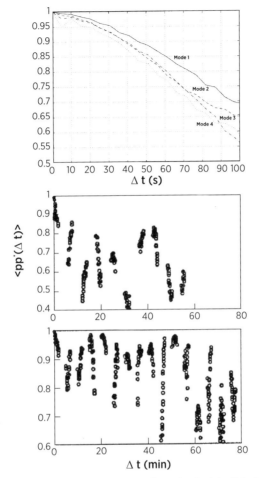

Figure 8.16. Upper panel: Time coherence of modes 1–4 (solid, dashed, dash-dot, and dot) at 32-km range and 400-Hz during the SWARM experiment. Lower two panels: Time coherence of mode 1 at 42-km range and 400 Hz (middle) and 224 Hz (lower) also during SWARM.
Sources: Headrick et al. (1999) and Rouseff et al. (2002).

For shallow water there are observations of temporal coherence from the SWARM Experiment (Headrick et al., 1999; Rouseff et al., 2002), and there are results for transverse coherence for the SW06 Experiment (Duda et al., 2012) (see Figure 7.6). There is also the overview by Carey (1998), in which a typical value of shallow-water transverse coherence is estimated to be 30 wavelengths at 45-km range. Table 8.1 gives examples of observed and predicted coherence times for single modes, while Table 8.2 gives observed and predicted coherence times and lengths for the full field.

For individual modes, the SWARM experiment has provided the most information (Apel et al., 1997). Figure 8.16 shows estimates of temporal coherence for individual modes from SWARM at 224 and 400 Hz. Two important timescales are evident. First, there is the short-time behavior that is dictated by the passage of nonlinear internal solitary waves. Then there is the longer timescale decorrelation caused by stochastic internal waves. Mode 1 has a long coherence time, because the background sound-speed profile traps this mode below the main thermocline. Sound-speed perturbations are greatest in the main thermocline and smaller below. This effect has also been observed in the Barrents Sea polar front experiment (Lynch et al., 1996). The adiabatic predictions (Eq. 8.100) for mode 1 in SWARM give temporal coherences of 40 and 42 minutes at 400 and 224 Hz respectively, which are close to the observations (Table 8.1). The predicted coherence times at the two frequencies are close to one another because of the aforementioned trapping of mode 1. The lower frequency mode extends further up in the water column, thus interacting with stronger sound-speed perturbations, counteracting the expected increase in coherence time relative to 400 Hz. The short-time observations of higher mode coherence time (top panel Figure 8.16 and Table 8.1) are clearly dominated by nonlinear wave effects, and some re-correlation is expected at longer timescales.

The observations of horizontal coherence by Duda et al. (2012) for SW06 show that the coherence length has strong temporal variability and is highly anisotropic. The most time stable coherence values are for across-shelf propagation, yielding coherence lengths of roughly 15 wavelengths (consistent with the Carey result within a factor of 2). Along-shelf propagation paths have coherences that are strongly influenced by the passage of nonlinear wave packets that are predominantly moving perpendicular to the path. The full-field adiabatic transport theory estimates using Eq. 8.107 are much too high (see Table 8.2).

Lastly, there are temporal coherence observations from 18.9 km, and 300 and 500 Hz transmissions in ASIAEX (Mignerey and Orr, 2004, also see discussion in Chapter 7). The full-field adiabatic transport theory estimates using Eq. 8.100 are consistent with the observations and the path integral predictions (see Tables 8.2 and 7.1).

8.7 Mutual Coherence Function: Frequency Separations

Because the statistics of pulse propagation are of interest, the cross-mode coherence function at two different frequencies must be addressed, namely $\langle \hat{a}_{n1} \hat{a}_{p2}^* \rangle$, where (1) and (2) respectively refer to the frequencies ω_1 and ω_2. To date little work has been done on cross-frequency mode statistics using transport theory, and so what is presented here gives mostly the rudimentary beginnings of a treatment of the subject. Thus, while there are potentially many pulse statistics that could be examined, such as was done using the path integral, here only the second moment of mode amplitudes is examined. This allows the treatment of two observables, namely mean pulse intensity,

$$\langle I \rangle(r,z,t) = \int_{-\infty}^{\infty} \int_{-\infty}^{\infty} d\omega_1 \, d\omega_2 \, e^{i(\omega_2-\omega_1)t} \sum_{n=1}^{N} \sum_{p=1}^{N} \frac{\langle a_{n1} a_{p2}^*(r) \rangle}{r} \frac{\phi_{n1}(z) \phi_{p2}(z)}{\sqrt{k_{n1} k_{p2}}},$$

(8.108)

and the mean mode pulse,

$$\langle I_n \rangle(r,z,t) = \int_{-\infty}^{\infty} \int_{-\infty}^{\infty} d\omega_1 \, d\omega_2 \, e^{i(\omega_2-\omega_1)t} \frac{\langle a_{n1} a_{n2}^*(r) \rangle}{r} \frac{\phi_{n1}(z) \phi_{n2}(z)}{\sqrt{k_{n1} k_{n2}}}.$$

(8.109)

The mean pulse shape is interesting observationally because of the ensonification of shadow zones (Dushaw et al., 1999; Van Uffelen et al., 2009), and mode pulses have garnered some attention for their stability properties (Udovydchenkov et al., 2013). The hope is that these results may stimulate more work on other observables.

The treatment of cross-frequency mode coherence differs fundamentally from the narrowband mode physics that has been discussed previously. At fixed frequency, the modes physically interact, exchanging energy and perturbing phases, but when two separate frequencies are considered there is no physical coupling. Coupling between frequencies can only occur when nonlinear effects are considered (see, for example, Lighthill, 1978). The issue that is being addressed in examining cross-frequency correlations is the frequency dependence of the coupling physics.

This being said, the transport equation for the cross-mode cross-frequency coherence is obtained using results already presented for spatial and temporal coherences, that is,

$$\frac{d\langle a_{n_1} a_{p_2}^* \rangle(r)}{dr} - i(l_{n1} - l_{p2}^*)\langle a_{n_1} a_{p_2}^* \rangle = \sum_{m=1}^{N} \sum_{q=1}^{N} \Big(\langle a_{m_1} a_{q_2}^* \rangle I_{mn1,qp2}^* + \langle a_{q_1} a_{m_2}^* \rangle I_{mp2,qn1}$$

$$- \langle a_{n_1} a_{q_2}^* \rangle I_{mp2,qm2}^* - \langle a_{q_1} a_{p_2}^* \rangle I_{mn1,qm1} \Big).$$

(8.110)

Here the coupling matrix correlation function at the two frequencies is given by

$$\Delta_{mn1,qp2}(\delta r) = \langle \rho_{mn1}(r) \rho_{qp2}^*(r - \delta r) \rangle,$$

(8.111)

and the scattering matrix is

$$I_{mn1,qp2} = \int_0^\infty d\delta r \, \Delta_{mn1,qp2}(\delta r) \, e^{ik_{pq2}\delta r}. \tag{8.112}$$

The scattering matrices for this transport equation involve both mixed frequencies (i.e., $I^*_{mn1,qp2}$ and $I_{mp2,qn1}$) and individual frequencies (i.e., $I^*_{mp2,qm2}$ and $I_{mn1,qm1}$). The two acoustic frequencies separate nicely into the G_{mn} functions and the difference wavenumber k_{pq2}, and one can write as before

$$I_{mn1,qp2} = \sum_{j=1}^J H(j) \, G_{mn1}(j) G_{qp2}(j) \int_0^\infty d\kappa_h F_1(\kappa_h) \begin{cases} \dfrac{1}{\sqrt{\kappa_h^2 - k_{pq2}^2}}, & 0 \le |k_{pq2}| < \kappa_h \\[2mm] \dfrac{i \, \text{sign}(k_{pq2})}{\sqrt{k_{pq2}^2 - \kappa_h^2}}, & 0 \le \kappa_h < |k_{pq2}| \end{cases} . \tag{8.113}$$

Therefore all the analytic solutions that have been obtained for the scattering matrices for both the GM and Lorentzian spectra still apply for the two-frequency case. In the adiabatic approximation, the familiar scattering matrices for the GM spectrum are

$$I_{nn1,pp2} = \frac{2}{\pi} \sum_{j=1}^J H(j) \frac{G_{nn1}(j) G_{pp2}(j)}{\hat{k}_j}. \tag{8.114}$$

The difficulty in solving Eq. 8.110 is that $\langle a_{n_1} a^*_{p_2} \rangle(r)$ is not a simple function of frequency or mode number. This means that the mean intensity (Eq. 8.108) and the mean mode pulse (Eq. 8.109) equations are rather cumbersome to evaluate. Thus, a useful place to start is the adiabatic approximation.

8.7.1 Adiabatic Approach

The decoupled adiabatic equations can be solved for the two-frequency case, yielding the result

$$\langle a_{n1} a^*_{p2} \rangle(r) = \langle a_{n1} a^*_{p2} \rangle(0) \, e^{i(l_{n1} - l^*_{p2})r} \, e^{-(I_{nn1,nn1} + I_{pp2,pp2} - 2I_{nn1,pp2})r}. \tag{8.115}$$

As for other adiabatic results, there are no phases associated with the frequency dependence of scattering; there is only the deterministic term $e^{i(l_{n1} - l^*_{p2})r}$. Again, an adiabatic half-e-folding decorrelation range comes easily out of this expression, with the result

$$R^{-1}_{n1,p2} = \frac{4N_0 B}{\pi^2 f} \sum_{j=1}^J \frac{H(j)}{j} [G_{nn1}(j) - G_{pp2}(j)]^2. \tag{8.116}$$

If the frequency separation is sufficiently small, then the G functions will depend nearly linearly on frequency. The decorrelation range can be written as

$$R_{n1,p2}^{-1} \simeq \frac{4N_0 B}{\pi^2 f} \sum_{j=1}^{J} \frac{H(j)}{j} \left[\left(1 - \frac{\Delta\omega}{2\bar{\omega}}\right) G_{nn0}(j) - \left(1 + \frac{\Delta\omega}{2\bar{\omega}}\right) G_{pp0}(j) \right]^2. \qquad (8.117)$$

where $\Delta\omega = \omega_2 - \omega_1$ and $\bar{\omega} = (\omega_1 + \omega_2)/2$. Equation 8.117 can be solved for $\Delta\omega$ as a function of range to give a frequency decorrelation scale. Note that certain mode pairs at different frequencies can be strongly correlated if the terms in the square brackets nearly cancel. This occurs for modes at different frequencies that have similar turning depths or modal group speed. This type of energy is associated with ray-like arrivals (Dzieciuch et al., 2001), that is, modes that constructively interfere to generate wave fronts. Figure 8.17 shows the adiabatic decorrelation range relative to two different modes ($n = 20$ and 60) at 250 Hz. The largest correlation ranges roughly follow contours of constant modal group speed (see Rayleigh formula, Eq. 2.144), that is for frequencies lower/higher than 250 Hz, the modes 20 and 60 show more correlation with mode numbers less/greater than their value.

Another important case that is relevant to the mode pulse is when $p = n$ for which the fluctuations induce a pulse time spread. In this case, Eq. 8.117 can be used to rewrite the cross-mode coherence equation to get

$$\langle a_{n1} a_{n2}^* \rangle (r) \simeq \langle a_{n1} a_{n2}^* \rangle (0) \, e^{i(l_{n1} - l_{n2}^*)r} \, \exp\left[-\frac{1}{2}(\Delta\omega\tau_n)^2 \right], \qquad (8.118)$$

where τ_n, the adiabatic modal Doppler time spread equivalent to Eq. 7.83, is given by (see Eq. 8.28)

$$\tau_n^2 = \frac{1}{\bar{\omega}^2} \frac{4N_0 BR}{\pi^2 f} \sum_{j=1}^{J} \frac{H(j)}{j} G_{nn0}^2(j) = \frac{\Phi_n^2}{\bar{\omega}^2}. \qquad (8.119)$$

The adiabatic effects described here are interpreted physically in terms of the frequency dependence of the phase speed. Next, it is critical to address the effects of mode coupling.

8.7.2 Effects of Coupling

Mode coupling introduces an additional decay in the magnitude of the cross-mode coherence and generates phase fluctuations in the coherence. As previously mentioned, the frequency dependence of Eq. 8.110 is not simple and therefore amenable to perturbation analysis. Hence, the best way to understand the results is to look at some examples, and here it is instructive to draw from both deep and shallow-water environments.

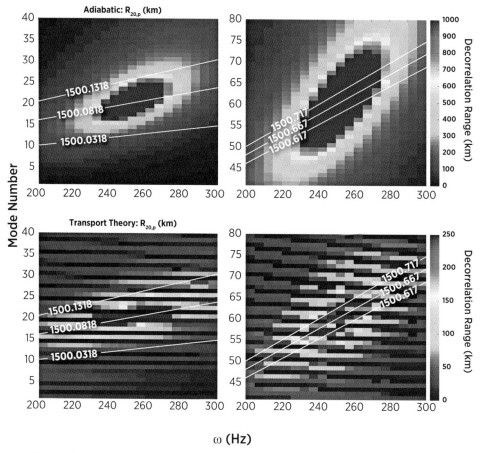

Figure 8.17. Adiabatic decorrelation range, $R_{n1,p2}$, as a function of frequency and mode number relative to mode 20 at 250 Hz (upper left panel) and mode 60 at 250 Hz (upper right panel). Lower panels show the corresponding approximate decorrelation range for the full transport theory for a point source on the sound-channel axis. Contours of modal group speed are also shown. The calculation uses a Munk canonical profile and an exponential buoyancy-frequency profile (Figure 2.1).

Figure 8.17 shows a deep-water example displaying the half e-folding decorrelation range computed both from adiabatic theory and by direct numerical solution of Eq. 8.110. Because the mode energies are changing with range, the half-e-folding range is computed using the normalized cross-mode coherence function, that is,

$$\frac{|\langle a_{n1} a_{p2}^*(r)\rangle|}{\sqrt{\langle |a_{n1}|^2(r)\rangle \langle |a_{p2}|^2(r)\rangle}}. \tag{8.120}$$

In this case, the source is on the sound-channel axis, and Figure 8.18 shows examples of the coherence functions. Unlike the adiabatic case these functions are not pure exponentials, and so the half-e-folding distance is an approximate way of quantifying the decay of the magnitude of the coherence when coupling is important. Comparing the adiabatic and fully coupled cases shown in Figure 8.17, it is apparent that coupling significantly diminishes the decorrelation range and that the pattern of decorrelation for the fully coupled case is a complicated function of mode number and frequency. Highly correlated modes can be adjacent to strongly decorrelated modes because of the initial condition. Figure 8.18 shows an example of the coherence function and mode energy evolution for two strongly correlated modes and another example for two rapidly decorrelating modes. The more strongly correlated modes are seen to have a similar mode coupling evolution while the decorrelated modes have different evolution with range. Even though the maximum decorrelation ranges are of order several hundreds of kilometers, this does not mean that modes cannot combine to form wave fronts such as have been observed in long range experiments including SLICE89 and ATOC. This constructive interference depends on both the amplitudes and phases of the cross-mode cross-frequency coherence matrix.

A shallow-water case for the decorrelation range is shown in Figure 8.19 where the decorrelation range relative to mode 4 at 400 Hz is displayed. Unlike the deep-water case, the adiabatic and full transport theory results are somewhat comparable. This is consistent with what is known regarding adiabaticity in shallow water. Unlike the deep-water cases, the cross-mode frequency coherence decays rapidly across mode number, but less rapidly across frequency. Calculations (not shown) for which the reference mode number is mode 1 at 400 Hz show that mode 1 is nearly perfectly coherent across the 350–450-Hz band, but has little coherence with mode 2.

Analysis of frequency correlations in the mode formalism has barely begun.

8.8 Intensity Variance

Now a fourth moment of the field is considered, that is, the mean square intensity given by

$$\langle I^2 \rangle = \sum_n \sum_p \sum_m \sum_q \frac{\langle a_n a_p^* a_m a_q^* \rangle}{r^2} \frac{\phi_n(z)\phi_p(z)\phi_m(z)\phi_q(z)}{\sqrt{k_n k_p k_m k_q}}, \tag{8.121}$$

where the mode amplitude fourth moment is $\langle a_n a_p^* a_m a_q^* \rangle$. This observable is used primarily to compute the scintillation index, that is, $SI = \langle I^2 \rangle / \langle I \rangle^2 - 1$. The principle work on this observable was carried out by Dozier and Tappert (1978a) and later by Creamer (1996). In what follows, the approach of Creamer (1996) is mostly utilized because of the incorporation of attenuation for shallow-water applications.

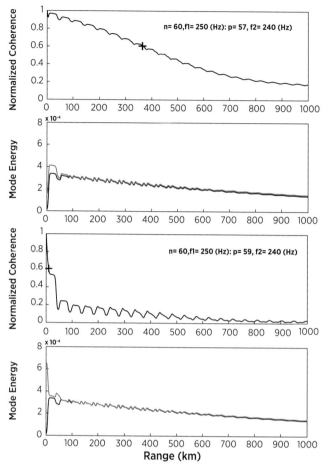

Figure 8.18. Normalized cross-mode cross-frequency coherence $|\langle a_{n1} a^*_{p2}(r)\rangle|/$ $\sqrt{\langle|a_{n1}|^2(r)\rangle\langle|a_{p2}|^2(r)\rangle}$ for the cases in which the coherence is high (upper two panels) and low (lower two panels). The second and fourth panels show the range evolution of mode energy, $\langle|a_{n1}|^2(r)\rangle$ (black), and $\langle|a_{p2}|^2(r)\rangle$ (gray) corresponding to the cross-mode coherences above. The modes being compared are $n = 60$, $f_1 = 250$ Hz, and $p = 57$, $f_1 = 240$ Hz (upper panels), and $n = 60$, $f_1 = 250$ Hz, and $p = 59$, $f_1 = 240$ Hz (lower panels).

As before, using the demodulated mode amplitudes, it is useful to define

$$\psi_{npmq} = \hat{a}_n \hat{a}^*_p \hat{a}_m \hat{a}^*_q, \qquad (8.122)$$

which should be considered a N^4 vector. Taking the range derivative of ψ and using Eq. 8.2, an evolution equation for ψ is obtained that has a familiar form, that is,

$$\frac{d\psi}{dr} = -i\tilde{\Omega}\psi. \qquad (8.123)$$

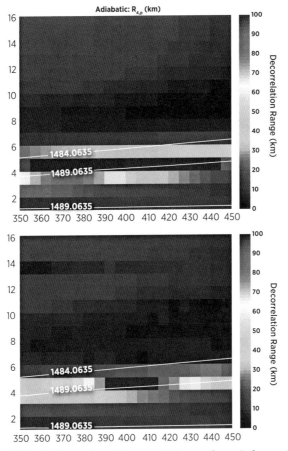

Figure 8.19. Adiabatic (upper) and transport theory (lower) decorrelation range, $R_{n1,p2}$, as a function of frequency and mode number relative to mode 4 at 400 Hz. Contours of modal group speed are also shown. Calculation is for the canonical SW06 shallow-water environment (Figure 2.3).

Here the $N^4 \times N^4$ matrix $\tilde{\mathbf{\Omega}}$ is given by the rather complicated expression:

$$\tilde{\Omega}_{npmq,abcd} = \tilde{\rho}_{pb}\delta_{na}\delta_{mc}\delta_{qd} + \tilde{\rho}_{qd}\delta_{na}\delta_{mc}\delta_{pb} - \tilde{\rho}^*_{na}\delta_{pb}\delta_{mc}\delta_{qd} - \tilde{\rho}^*_{mc}\delta_{na}\delta_{pb}\delta_{qd}. \quad (8.124)$$

Again, because Eq. 8.123 is in the standard form (i.e., Eq. 8.7), the solution can be written in terms of the range-ordered exponential:

$$\psi(r) = \mathcal{R}\left[\exp\left(-i\int_0^r dr'\,\tilde{\mathbf{\Omega}}(r')\right)\right]\psi(0). \quad (8.125)$$

It is useful to first look at the adiabatic solution.

8.8.1 Adiabatic Approach and Hybrid Theory

As has been seen in previous sections, the adiabatic and hybrid approaches work fairly well in shallow water. The situation for deep-water environments is presently unknown. Assuming Gaussian statistics for the adiabatic phases, as was done previously, the adiabatic fourth moment of mode amplitudes can be expressed as

$$\langle a_n a_p^* a_m a_q^* \rangle = \tilde{a}_n \tilde{a}_p^* \tilde{a}_m \tilde{a}_q^* \exp\big[-(I_{nn,nn} + I_{pp,pp} + I_{mm,mm} + I_{qq,qq}$$

$$+ 2[I_{nn,mm} + I_{pp,qq} - I_{nn,pp} - I_{nn,qq} - I_{mm,pp} - I_{mm,qq}])r\big].$$

$$(8.126)$$

In the straight adiabatic approach, the mode amplitudes are given by $\tilde{a}_m(r) = a_m(0)e^{il_n r}$. Using the hybrid approach, the mode amplitudes in the first part of Eq. 8.126 vary with range due to coupling and are of the form

$$\tilde{a}_m(r) = \text{sign}(a_m(0))\langle |a_m|^2 \rangle^{1/2}(r)e^{ik_n r}. \qquad (8.127)$$

Equation 8.126 can be written in more compact form and corrected for the edge effect by defining $\langle a_n a_p^* a_m a_q^* \rangle = \tilde{a}_n \tilde{a}_p^* \tilde{a}_m \tilde{a}_q^* e^{-B_{npmq}(r)}$, where using the GM spectrum the result is

$$B_{npmq}(r) = \frac{4}{\pi} \sum_{j=1}^{J} \frac{H(j)}{\hat{\kappa}_j} (G_{nn} - G_{pp} + G_{mm} - G_{qq})^2 \Big(r + \frac{e^{-\hat{\kappa}_j r} - 1}{\hat{\kappa}_j} \Big). \qquad (8.128)$$

For shallow-water cases typical of the mid-Atlantic continental shelf off the United States, the adiabatic expressions have been utilized by Colosi et al. (2011) to study low-frequency propagation. The hybrid and edge corrections were developed by Raghukumar and Colosi (2014), who considered kilohertz propagation. Figure 8.20 compares scintillation indices computed using adiabatic theory and from Monte Carlo numerical simulations. The adiabatic theory is seen to match the simulation results exceedingly well for both 200 and 400 Hz. Deviations at short range are due to uncorrected edge effects in the adiabatic calculation. The curves oscillate in range due to ducting effects that cause regions of lower/higher intensity to have higher/lower scintillation indices. The *SI* curves decrease with range due to loss of cross-mode coherence and mode stripping. At longer ranges, there are fewer modes to interfere and thus cause intensity fluctuations. Similarly good comparisons to Monte Carlo simulation are found at 1000-Hz frequency using the hybrid theory (Raghukumar and Colosi, 2014). It will be shown later, however, that when attenuation is involved (as it was in these shallow-water examples), the asymptotic behavior of the scintillation index must eventually show exponential growth.

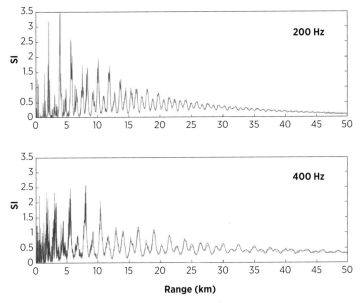

Figure 8.20. Range evolution of scintillation index in the main thermocline at 200 (upper) and 400 (lower) Hz for an environment similar to SW06. Red curves show Monte Carlo simulation results, while blue curves are from adiabatic theory. *Source*: Colosi et al. (2011).

8.8.2 Transport Theory

For the mode amplitude fourth moment, the Dyson series approach becomes too complicated, and so here the compact treatment offered by the range-ordered exponential is followed (Creamer, 1996). Since Eq. 8.125 is in standard form, the transport theory for $\langle \psi \rangle$ is

$$\frac{d}{dr}\langle \psi(r) \rangle = -\left\langle \tilde{\boldsymbol{\Omega}}(r) \int_0^r dr' \tilde{\boldsymbol{\Omega}}(r') \right\rangle \langle \psi(r) \rangle. \tag{8.129}$$

This equation with its $N^4 \times N^4$ matrix is also rather complicated and computationally limiting. Some order of approximation is required, which is discussed in the next section.

8.8.3 Asymptotic Behavior

Here the desire is to examine the fluctuations near saturation. For the asymptotic case in which the modes have attained sufficient phase variation and have largely decorrelated, the primary contributions to the quadruple sum in Eq. 8.121 come from the terms in which all the indices are the same and from those in which $n = p$ and $m = q$ (see also Chapter 4, Eq. 4.15). In all other cases, the large phase fluctuations drive the terms to small values. In this limit, the mean square intensity is

$$\langle I^2 \rangle = \sum_n \sum_p \varphi_{np} \frac{\phi_n^2(z)\phi_p^2(z)}{r^2 k_n k_p}, \qquad (8.130)$$

where the new matrix of interest to be solved for is

$$\varphi_{np} = (2 - \delta_{np})\langle |a_n|^2 |a_p|^2 \rangle. \qquad (8.131)$$

The new matrix is clearly symmetric with $\varphi_{pn} = \varphi_{np}$[10]. Because of symmetry, it proves useful to treat $\boldsymbol{\varphi}$ as an $M = N(N+1)/2$ vector. Setting to zero all terms in Eq. 8.129 that do not correspond to the terms in $\boldsymbol{\varphi}$, the transport equation becomes

$$\frac{d\boldsymbol{\varphi}}{dr} = -\boldsymbol{\Theta}\boldsymbol{\varphi}, \qquad (8.132)$$

where

$$\Theta_{np,mq} = \delta_{nm}\delta_{pq}\left[2(\alpha_n + \alpha_p) + 4(1 - \delta_{np})f_{np} + \sum_{j \neq n,p}(f_{nj} + f_{pj})\right]$$
$$- 2f_{np}(1 - \delta_{np})(\delta_{nm}\delta_{nq} + \delta_{pm}\delta_{pq}) - (1 - \delta_{nm})(1 - \delta_{pm})(f_{nm}\delta_{pq} + f_{pm}\delta_{nq}). \qquad (8.133)$$

Recall $f_{mn} = 2Re(I_{mn,mn})$. Except for the attenuation factors, this result obtained by Creamer (1996) is equivalent to the one obtained by Dozier and Tappert (1978a). Some important results related to the approach to saturation can be obtained if at first attenuation is ignored.

No Attenuation

While $\boldsymbol{\Theta}$ is a symmetric matrix, in the absence of attenuation it is also a singular matrix because its columns sum to zero. This means that one of the eigenvalues of $\boldsymbol{\Theta}$ is zero, corresponding to a stable equilibrium in which

$$\frac{d}{dr} \sum_n \sum_p \varphi_{np} = 0. \qquad (8.134)$$

This result is directly related to the asymptotic stable equilibrium of modal energy in the absence of attenuation (equipartition). Using a unit energy normalization, equipartitioning corresponds to $\langle |a_n|^2 \rangle = 1/N$, where N is the number of modes. Under the same normalization, the M vector $\boldsymbol{\varphi}$ has a stable equilibrium when all its components are given by (Dozier and Tappert, 1978a)

$$\varphi = \frac{2}{N(N+1)}. \qquad (8.135)$$

---

[10] In this approximation for the terms involved, there is no difference between the standard mode amplitudes and the demodulated ones.

Some useful conclusions follow from these results. First, the mode energy correlations diminish to quite small values in the limit of long range and large N, that is,

$$\frac{\langle |a_n|^2 |a_m|^2 \rangle - \langle |a_n|^2 \rangle \langle |a_m|^2 \rangle}{\langle |a_n|^2 \rangle \langle |a_m|^2 \rangle} \simeq -\frac{1}{N+1} \to 0. \tag{8.136}$$

Second, the modal energy scintillation index saturates to 1 according to

$$\frac{\langle |a_n|^4 \rangle - \langle |a_n|^2 \rangle^2}{\langle |a_n|^2 \rangle^2} \simeq \frac{N-1}{N+1} \to 1. \tag{8.137}$$

This is consistent with modal multipathing due to coupling, in which the central limit theorem drives the mode amplitudes to Gaussian statistics. The saturation limit is a function of the number of interfering modes as was described in Chapter 4, Section 1.4. Curiously, the asymptotic modal scintillation index is always less than 1.

Lastly the full-field scintillation index is obtained using Eqs. 8.130 and 8.61 in equilibrium, the result is

$$\langle I^2 \rangle = \sum_n \sum_p \hat{\varphi}_{np} \frac{\phi_n^2(z)\phi_p^2(z)}{r^2 k_n k_p} \simeq \frac{2}{N(N+1)r^2} \left[\sum_n \frac{\phi_n^2(z)}{k_n} \right]^2, \tag{8.138}$$

$$\langle I \rangle = \frac{1}{Nr} \sum_n \frac{\phi_n^2(z)}{k_n}, \tag{8.139}$$

where $\hat{\varphi}_{np}$ is the full $N \times N$ symmetric matrix. The full-field scintillation index then has the same form as the modal scintillation index, that is,

$$SI \simeq \frac{N-1}{N+1}. \tag{8.140}$$

Again, there is the curious result that the asymptotic scintillation index value is always less than 1.

Nonzero Attenuation

Next, the effects of attenuation are considered. While Eq. 8.132 is more complicated than those for the mean mode amplitude or mean mode energy, the form of the equations for the three quantities is the same. The constant matrix $\boldsymbol{\Theta}$ can be diagonalized with the result

$$\boldsymbol{\Theta} = \mathbf{V}\mathbf{D}_\Theta\mathbf{V}^{\mathrm{T}}, \tag{8.141}$$

where as before V is an orthogonal matrix of eigenvectors, and $\mathbf{D}_\Theta$ is a diagonal matrix holding the eigenvalues, that is, $\mathbf{D}_\Theta = \mathrm{diag}[\theta_1, \theta_2, \ldots \theta_M]$. The eigenvalues can be ordered such that $0 \leq \theta_1 \leq \theta_2 \leq ... \leq \theta_M \leq 2\max_{n<m}[|\alpha_m + \alpha_n + 2f_{mn} + \sum_{k \neq n,m}(f_{nk} + f_{mk})|]$. As in the case of the mode energies, the explicit solution can be written as

$$\varphi_{ij}(r) = \sum_{l=1}^{M} \sum_{k \leq l} V_{ij,kl} \left[\sum_{n=1}^{M} \sum_{m \leq n} V_{mn,kl} \varphi_{mn}(0) \right] e^{-\theta_{k+(l-1)N} r}, \tag{8.142}$$

and again the asymptotic behavior is dictated by the smallest eigenvalue, giving

$$\varphi_{ij}(r) \simeq V_{ij,11} \left[\sum_{n=1}^{M} \sum_{m \leq n} V_{mn,11} \varphi_{mn}(0) \right] e^{-\theta_1 r}. \tag{8.143}$$

With this asymptotic result for φ, the scintillation behavior of individual modes, that is, $SI_n = \langle |a_n|^4 \rangle / \langle |a_n|^2 \rangle^2 - 1$, can be examined. Substituting the asymptotic forms for φ_{nn} and $\langle |a_n|^2 \rangle$ (i.e., Eqs. 8.143 and 8.66) into Eq. 8.138, the modal scintillation index is

$$SI_n \simeq B_n e^{\chi r} - 1. \tag{8.144}$$

Here the amplitude factor is given by

$$B_n = \frac{V_{nn,11} \left[\sum_{k=1}^{M} \sum_{m \leq k} V_{mk,11} \varphi_{mk}(0) \right]}{P_{n1}^2 \left[\sum_{m=1}^{N} P_{m1} \langle |a_m|^2 \rangle(0) \right]^2}, \tag{8.145}$$

and the exponential growth rate, independent of mode number, is

$$\chi = 2\lambda_1 - \theta_1. \tag{8.146}$$

The remarkable result of Creamer (1996) is now seen, that is, because SI is a positive definite quantity, the exponential factor χ must also be positive or zero, meaning that the modal scintillations can grow asymptotically without bound. As will be seen, the asymptotic growth of scintillations is due to the differential decay rates of the mode statistics due to attenuation. This situation is markedly different from the expected behavior in strongly scintillating cases (such as in partial saturation), where high-intensity events have a large probability of occurring. In the case where the modal scintillations grow without bound, the full-field scintillation index $SI = \langle I^2 \rangle / \langle I \rangle^2 - 1$ also grows without bound (Creamer, 1996; Tielburger et al., 1997; Tang et al., 2006). The correspondence between modal and full-field scintillation indices was demonstrated in the previous section.

8.8.4 Two-Mode Example

The remarkable result of the exponential growth of the scintillation index is nicely demonstrated with a two-mode example following Creamer (1996). From the mode energy equation (Eq. 8.62), the matrix $\mathbf{\Pi}$ is given by

$$\mathbf{\Pi} = \begin{pmatrix} 2\alpha_1 + f_{12} & -f_{12} \\ -f_{12} & 2\alpha_2 + f_{12} \end{pmatrix}, \tag{8.147}$$

which has the two eigenvalues given by Eq. 8.75. The smallest eigenvalue is $\lambda_1 = \alpha_1 + \alpha_2 + f_{12} - \sqrt{f_{12}^2 + (\alpha_1 - \alpha_2)^2}$.

The 3×3 matrix $\mathbf{\Theta}$ is given by

$$\mathbf{\Theta} = 2 \begin{pmatrix} 2\alpha_1 + f_{12} & -f_{12} & 0 \\ -f_{12} & \alpha_1 + \alpha_2 + 2f_{12} & -f_{12} \\ 0 & -f_{12} & 2\alpha_2 + f_{12} \end{pmatrix}. \tag{8.148}$$

To find the eigenvalues associated with the fourth moment, it is useful to re-write this matrix as $\mathbf{\Theta} = 2(\alpha_1 + \alpha_2 + f_{12})\mathbf{I} + \mathbf{\Theta}'$, where $\mathbf{I}$ is the 3×3 identity matrix and

$$\mathbf{\Theta}' = 2 \begin{pmatrix} \alpha_1 - \alpha_2 & -f_{12} & 0 \\ -f_{12} & f_{12} & -f_{12} \\ 0 & -f_{12} & \alpha_2 - \alpha_1 \end{pmatrix}. \tag{8.149}$$

Writing $d\alpha = |\alpha_2 - \alpha_1|$ it is found that $\chi = -2\sqrt{f_{12}^2 + d\alpha^2} - \theta_1'$, with θ_1' the smallest eigenvalue of $\mathbf{\Theta}'$. Thus, the exponential growth rate only depends on the difference of the modal attenuation, $d\alpha$, and the mode coupling factor f_{12}. Scaling both χ and f_{12} by $d\alpha$, it is apparent that $\chi/d\alpha$ is only a function of f_{12}/α, yielding a universal curve (see Figure 8.21). In all cases, it is found that $\chi > 0$, and thus there will be exponential growth. In the case where $d\alpha = 0$ or there is zero attenuation, it is clear from the matrices $\mathbf{\Pi}$ and $\mathbf{\Theta}$ that $\chi = 0$, and there is no exponential growth. In Figure 8.21 the largest growth rate occurs when differential attenuation and mode coupling rates are roughly equal.

Beyond the simple two-mode case, there are few generalizations that can be made regarding the exponential growth of the scintillation index, and thus calculations for specific experimental situations pitting attenuation against mode coupling must be done (for example, see Tielburger et al., 1997; Tang et al., 2006). The limiting case for the multimode situation, however, is tractable when attenuation is much stronger than mode coupling rates. In this case a perturbation method can be used to compute the correction to the smallest eigenvalues λ_1 and

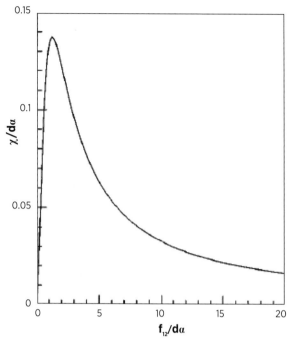

Figure 8.21. For the two-mode example, exponential growth rate of the mode fluctuations, χ, scaled by the differential attenuation $d\alpha$ is plotted versus the mode coupling factor f_{12}, also scaled by the differential attenuation (Creamer, 1996). The maximum growth rate, $f_{12}/d\alpha \simeq 1$, occurs where the mode coupling and differential attenuation are nearly equal.
Source: Creamer (1996).

θ_1 due to small coupling (Creamer, 1996). The result is

$$\lambda_1 \simeq 2\alpha_1 + \sum_{k=2}^{N} f_{1k} + \sum_{n=2}^{N} \frac{f_{1n}^2}{2(\alpha_1 - \alpha_n)}, \tag{8.150}$$

$$\theta_1 \simeq 4\alpha_1 + 2\sum_{k=2}^{N} f_{1k} + \sum_{n=2}^{N} \frac{2f_{1n}^2}{(\alpha_1 - \alpha_n)}, \tag{8.151}$$

which yields the growth rate

$$\chi = 2\lambda_1 - \theta_1 = \sum_{n=2}^{N} \frac{f_{1n}^2}{(\alpha_n - \alpha_1)}. \tag{8.152}$$

In shallow water, attenuation is expected to be an increasing function of mode number, and thus the only way for χ to be small or zero is for mode 1 to have little or no coupling to the other modes, that is, f_{1n} needs to be small.

Based on calculations by Creamer (1996), Tielburger et al. (1997), and Tang et al. (2006), the regime of exponential growth of the scintillation index appears to be difficult to obtain experimentally, primarily because in lossy, shallow-water environments, internal-wave mode coupling is relatively weak (as has been seen). Considering additional sources of randomness, for example, rough surface and bottom effects might create a situation in which the exponential regime could be observed (Tang et al., 2006).

8.8.5 *Observations*

Observations of single-frequency intensity variance are surprisingly rare in the literature. In fact, the MIMI experiment seems to be a singular example of published single-frequency fluctuations. With the advent of broadband sources, the focus has been on single-path intensity statistics, such as were described in the Chapters 5 through 7. Furthermore, the intensity variance of individual modes has also not been treated, which is a curious omission given the importance of the modal representation in shallow water. Therefore, in this section, three types of observations of intensity variance that can be related to the results of transport theory are discussed. They are (1) the MIMI results; (2) narrowband analysis of long-range, deep-water observations in the complex multi-mode interference finale region of the pulse; and (3) analysis of long-range broadband shallow-water observations also in a complex multi-mode interference pattern. In all these cases, the propagation is in the fully saturated regime with scintillation indices close to 1. For this reason, shallow-water, broadband observations behave much like narrowband data, as demonstrated by Colosi et al. (2005) for deep-water observations. In this study, broadband scintillation indices were comparable to, but slightly higher than, the corresponding narrowband estimates, a result that is consistent with the well-known notion that bandwidth slows the approach to saturation. Table 8.3 summarizes the observations to be discussed here. For deep-water cases, the results are taken from MIMI (Dyson et al., 1976) and the North Pacific work of Colosi et al. (2005). For shallow water, the results come from the work of Duda et al. (2004) in the South China Sea and from Fredricks et al. (2005) on the New England continental shelf.

With regard to the deep-water observations, the scintillation indices for the ATOC HVLA and KVLA transmissions are mostly quite close to 1, except for the *SI* value of 1.81 observed at 3500-km range for the 28-Hz transmissions. Hybrid transport theory calculations for the HVLA and KVLA ATOC profiles discussed in Chapter 7 show predictions that are quite a bit less than the observations (Table 8.3). Similarly the modeled scintillation indices are quite low for the MIMI experiment (Table 8.3). While values of *SI* were not reported for MIMI, values of 4.9 and 6.9 dB were given for σ_ι (Dyson et al., 1976), which roughly indicates *SI* values near 1 and slightly above.

Table 8.3. *Observations of scintillation index in both deep and shallow water.*

Experiment	Range (km)	SI (Obs)	SI (Pred)
MIMI (406 Hz)	550	< 1	0.4
MIMI (406 Hz)	1250	> 1	0.6
ATOC-KVLA (75 Hz)	5200	1.15	0.77
ATOC-HVLA (75 Hz)	3500	1.17	0.71
AST-KVLA (28 Hz)	5200	1.13	0. 61
AST-HVLA (28 Hz)	3500	1.81	0.55
Primer-SW (400 Hz)	60	1.08	0.40
Primer-SE (400 Hz)	42	1.05	0.50
ASIAEX-East (400 Hz)	21	2.6	0.75
ASIAEX-South (400 Hz)	31	1.7	0.60

The deep-water examples are from narrowband observations in the complex finale region of the arrival pattern (Colosi et al., 2005). The shallow-water examples come from the analysis by Fredricks et al. (2005) and Duda et al. (2004), where a fixed time in the complex broadband interference pattern is used and has been found to yield result close to a narrowband analysis. Predictions for the deep-water experiments come from calculations using hybrid theory for $\langle I^2 \rangle$ and $\langle I \rangle$ (Eqs. 8.126 and 8.78). Shallow-water experiments derive from the results of Colosi et al. (2011) (see Figure 8.20). For the MIMI experiment, rms intensity values, in dB were reported, but not *SI* values (Dyson et al., 1976); thus one can only estimate that *SI* was greater than or less than 1.

One may call into question the hybrid theory results presented above. However, comparison of the hybrid approach to Monte Carlo simulations (not shown) reveal that the hybrid theory is accurate at the 20% level at 75 Hz and at the 5% level at 28 Hz. Furthermore, the simulations and theory indicate that for the deep-water environment, *SI* approaches saturation from below 1 with no indication of a partially saturated regime. The large observed values of intensity variance are therefore somewhat of a mystery.

Shallow-water observations of scintillation index have been derived from broadband arrival data in which intensity values are collected at a fixed time relative to the on-set of the arrival. Being in the middle of the modal interference pattern and having the propagation regime in full saturation, this broadband data somewhat mimic narrowband data. In the study by Fredricks et al. (2005), 400-Hz scintillation indices were computed for timescales less than and greater than 2 hours. This distinction is important because of the energetic mesoscale and tidal fluctuations in shallow water, so that the rapid timescale fluctuations are most indicative of the stochastic internal-wave field and the nonlinear solitary waves. Table 8.3 shows that the *SI* values for the short timescale are close to full saturation. In the study by Duda et al. (2004), the short and long timescale separation was not accounted for, and their observations show much larger 400-Hz *SI* values.

In both these shallow-water experiments, the acoustic transmissions were over relatively complicated slope and shelf break topography, so comparisons to a simple canonical transport theory model (Colosi et al., 2011) are not entirely reasonable. Comparing these observed *SI* values to the canonical ones shown in Figure 8.20, the model is indeed not accurate (Table 8.3).

8.9 Summary

Because mode transport theory has only recently emerged as a tool for treating sound propagation through the stochastic internal-wave field, there is a paucity of quality comparisons to observations, and efforts to predict and understand observed modal and full-field scintillation indices, as well as other second moments, have only just begun. While much of the theory has been well tested through Monte Carlo simulations, more observational work is critically needed to understand the real utility of the approach. In addition, further theoretical progress is needed to handle a larger array of acoustic observables. Of primary interest are fourth moment quantities, such as intensity correlation functions and spectra, as well as cross-frequency quantities associated with pulse propagation.

Appendix A Integrals Related to Standard Scattering Matrices

The wavenumber integral in Eq. 8.23 for the unit normalized GM spectrum is given by[11]

$$\frac{4}{\pi} \int_0^\infty d\kappa_h \frac{\kappa_h^2 \hat{\kappa}_j}{(\kappa_h^2 + \hat{\kappa}_j^2)^2} \begin{cases} \frac{1}{\sqrt{\kappa_h^2 - k_{pq}^2}}, & 0 \le |k_{pq}| < \kappa_h \\ \frac{i \, \mathrm{sign}(k_{pq})}{\sqrt{k_{pq}^2 - \kappa_h^2}}, & 0 \le \kappa_h < |k_{pq}| \end{cases}$$

$$= \frac{1}{\pi} \left[\frac{2\hat{\kappa}_j}{k_{pq}^2 + \hat{\kappa}_j^2} + \frac{k_{pq}^2}{(k_{pq}^2 + \hat{\kappa}_j^2)^{3/2}} \ln \left(\frac{\sqrt{k_{pq}^2 + \hat{\kappa}_j^2} + \hat{\kappa}_j}{\sqrt{k_{pq}^2 + \hat{\kappa}_j^2} - \hat{\kappa}_j} \right) + i \, \mathrm{sign}(k_{pq}) \frac{k_{pq}^2 \pi}{(k_{pq}^2 + \hat{\kappa}_j^2)^{3/2}} \right].$$

$$(\text{A.1})$$

For the special case associated with adiabatic propagation, that is $k_{pq} = 0$, the result is

$$\frac{4}{\pi} \int_0^\infty d\kappa_h \frac{\kappa_h \hat{\kappa}_j}{(\kappa_h^2 + \hat{\kappa}_j^2)^2} = \frac{2}{\pi \hat{\kappa}_j}. \qquad (\text{A.2})$$

---

[11] The identity $\mathrm{atanh}(x) = (1/2) \ln[(1 + x)/(1 - x)]$ is used.

When a maximum horizontal wavenumber κ_{max} is imposed, then the solution to Eq. 8.23 is given by

$$\frac{4}{\pi}\int_0^{\kappa_{max}} d\kappa_h \frac{\kappa_h^2 \hat{\kappa}_j}{(\kappa_h^2 + \hat{\kappa}_j^2)^2} \begin{cases} \frac{1}{\sqrt{\kappa_h^2 - k_{pq}^2}}, & 0 \le |k_{pq}| < \kappa_h \\ \frac{i\,\text{sign}(k_{pq})}{\sqrt{k_{pq}^2 - \kappa_h^2}}, & 0 \le \kappa_h < |k_{pq}| \end{cases} =$$

$$\frac{1}{\pi}\left[\frac{2\hat{\kappa}_j k_{pq}^2 \cos\theta_{min}}{(k_{pq}^2 + \hat{\kappa}_j^2)(\hat{\kappa}_j^2(1 - \cos 2\theta_{min})/2 + k_{pq}^2)} + \frac{k_{pq}^2}{(k_{pq}^2 + \hat{\kappa}_j^2)^{3/2}}\ln\left(\frac{\sqrt{k_{pq}^2 + \hat{\kappa}_j^2} + \hat{\kappa}_j \cos\theta_{min}}{\sqrt{k_{pq}^2 + \hat{\kappa}_j^2} - \hat{\kappa}_j \cos\theta_{min}}\right)\right.$$

$$\left. +i\,\text{sign}(k_{pq})\frac{k_{pq}^2 \pi}{(k_{pq}^2 + \hat{\kappa}_j^2)^{3/2}}\right], \tag{A.3}$$

where $\sin\theta_{min} = |k_{pq}|/\kappa_{max}$. In the adiabatic case, the result is

$$\frac{4}{\pi}\int_0^{\kappa_{max}} d\kappa_h \frac{\kappa_h \hat{\kappa}_j}{(\kappa_h^2 + \hat{\kappa}_j^2)^2} = \frac{2}{\pi\hat{\kappa}_j}\frac{\kappa_{max}^2}{\kappa_{max}^2 + \hat{\kappa}_j^2}. \tag{A.4}$$

For the unit normalized Lorentzian spectrum, the wavenumber integral required for the scattering matrix is

$$\int_0^\infty d\kappa_h \frac{\kappa_h \tilde{\kappa}_j}{(\kappa_h^2 + \tilde{\kappa}_j^2)^{3/2}} \begin{cases} \frac{1}{\sqrt{\kappa_h^2 - k_{pq}^2}}, & 0 \le |k_{pq}| < \kappa_h \\ \frac{i\,\text{sign}(k_{pq})}{\sqrt{k_{pq}^2 - \kappa_h^2}}, & 0 \le \kappa_h < |k_{pq}| \end{cases} = \frac{\tilde{\kappa}_j + ik_{pq}}{k_{pq}^2 + \tilde{\kappa}_j^2}. \tag{A.5}$$

When a maximum horizontal wavenumber is imposed the integral result becomes

$$\int_0^{\kappa_{max}} d\kappa_h \frac{\kappa_h \tilde{\kappa}_j}{(\kappa_h^2 + \tilde{\kappa}_j^2)^{3/2}} \begin{cases} \frac{1}{\sqrt{\kappa_h^2 - k_{pq}^2}}, & 0 \le |k_{pq}| < \kappa_h \\ \frac{i\,\text{sign}(k_{pq})}{\sqrt{k_{pq}^2 - \kappa_h^2}}, & 0 \le \kappa_h < |k_{pq}| \end{cases} =$$

$$\frac{\tilde{\kappa}_j}{k_{pq}^2 + \tilde{\kappa}_j^2}\frac{\sqrt{2}\cos\theta_{min}}{\sqrt{2 + \tilde{\kappa}_j^2(1 - \cos 2\theta_{min})/k_{pq}^2}} + i\frac{k_{pq}}{k_{pq}^2 + \tilde{\kappa}_j^2}. \tag{A.6}$$

In the adiabatic limit, the integral is

$$\int_0^{\kappa_{max}} d\kappa_h \frac{\tilde{\kappa}_j}{(\kappa_h^2 + \tilde{\kappa}_j^2)^{3/2}} = \frac{1}{\tilde{\kappa}_j}\frac{\kappa_{max}}{\sqrt{\kappa_{max}^2 + \tilde{\kappa}_j^2}}. \tag{A.7}$$

Appendix B Integrals Related to MCFs with Temporal and Transverse Lags

For temporal coherence in the quadratic lag approximation, the wavenumber integral in Eq. 8.97 using the GM spectrum can be solved by making the substitution $x = \kappa_h^2/\hat{\kappa}_j^2$ which gives

$$\int_0^{\kappa_{max}} d\kappa_h \frac{\kappa_h^2}{(\kappa_h^2 + \hat{\kappa}_j^2)} \begin{cases} \frac{1}{\sqrt{\kappa_h^2 - k_{pq}^2}}, & 0 \le |k_{pq}| < \kappa_h \\ \frac{i \, \text{sign}(k_{pq})}{\sqrt{k_{pq}^2 - \kappa_h^2}}, & 0 \le \kappa_h < |k_{pq}| \end{cases} =$$

$$\frac{1}{2}\left[-\frac{2}{a^2+1}\left[\ln\left(a^2 + |a|\sqrt{a^2-1}\right)(a^2+1) + i\sqrt{a^2+1}\tan^{-1}\left(\frac{i|a|\sqrt{a^2+1}}{\sqrt{a^2-1}}\right) \right.\right.$$

$$\left.- \ln\left(x_m + \sqrt{x_m^2 - a^2}\right)(a^2+1) - i\sqrt{a^2+1}\tan^{-1}\left(\frac{ix_m\sqrt{a^2+1}}{\sqrt{x_m^2-a^2}}\right)\right]$$

$$+ i \, \text{sign}(k_{pq}) \frac{\pi(\sqrt{a^2+1}-1)}{\sqrt{a^2+1}}\Bigg], \tag{B.1}$$

where $a^2 = k_{qp}^2/\hat{\kappa}_j^2$, $x = \kappa_h^2/\hat{\kappa}_j^2$, $x_m = \kappa_{max}^2/\hat{\kappa}_j^2$. In the adiabatic limit, the simple result is

$$\int_0^{\kappa_{max}} \frac{\kappa_h d\kappa_h}{(\kappa_h^2 + \hat{\kappa}_j^2)} = \frac{1}{2}\ln\left(\frac{\hat{\kappa}_j^2 + \kappa_{max}^2}{\hat{\kappa}_j^2}\right). \tag{B.2}$$

For horizontal separations and in the quadratic approximation, the general integral is not solvable analytically. However, in the adiabatic approximation the substitution $x = \kappa_h^2/\hat{\kappa}_j^2$ can again be made, which gives the result

$$\int_0^{\kappa_{max}} \frac{\kappa_h^3 \hat{\kappa}_j d\kappa_h}{(\kappa_h^2 + \hat{\kappa}_j^2)^2} = \frac{\hat{\kappa}_j}{2}\left[\ln\left(\frac{\hat{\kappa}_j^2 + \kappa_{max}^2}{\hat{\kappa}_j^2}\right) - \frac{\kappa_{max}^2}{\hat{\kappa}_j^2 + \kappa_{max}^2} \right]. \tag{B.3}$$

Appendix C Random Surface Gravity Waves

The transport theory formalism developed here allows treatment of mode scattering by small random sea surface height fluctuations (Thorsos et al., 2010; Raghukumar and Colosi, 2015), which can be another important source of ocean sound field stochasticity. For a small vertical sea surface displacement, $h(r)$, the linearized coupling matrix for Eq. 8.2 is given by (Thorsos et al., 2010)

$$\rho_{mn}^{SW}(r) = -\frac{h(r)}{2\sqrt{k_n k_m}}\left[\frac{1}{\rho_0}\frac{d\phi_m}{dz}\frac{d\phi_n}{dz} \right]_{z=0}, \tag{C.1}$$

where the functions in the square bracket are evaluated at the unperturbed sea surface $z = 0$. This coupling matrix, which is linear in the surface displacement, is only accurate for small sea states, and this vertical slice transport theory approach does not account for out-of-plane scattering. Denoting the internal-wave coupling matrix as $\rho_{mn}^{IW}(r)$, the total coupling matrix is given by $\rho_{mn}(r) = \rho_{mn}^{SW} + \rho_{mn}^{IW}$. Because surface-wave and internal-wave fluctuations are expected to be uncorrelated, the correlation function of ρ_{mn} has the form

$$\Delta_{mn,qp}(\delta r) = \langle \rho_{mn}^{SW}(r)\rho_{qp}^{SW}(r+\delta r)\rangle + \langle \rho_{mn}^{IW}(r)\rho_{qp}^{IW}(r+\delta r)\rangle, \tag{C.2}$$

and thus the scattering matrix is the sum of two parts, one due to surface waves and another due to internal waves. If it is further assumed that the small surface waves are isotropic in the horizontal, then the scattering matrix can be written as

$$I_{mn,qp}^{SW} = \frac{1}{4\sqrt{k_m k_n k_p k_q}} \left[\frac{1}{\rho^2}\frac{d\phi_m}{dz}\frac{d\phi_n}{dz}\frac{d\phi_q}{dz}\frac{d\phi_p}{dz}\right]_{z=0} \int_0^\infty d\kappa_h F^{SW}(\kappa_h),$$

$$\begin{cases} \dfrac{1}{\sqrt{\kappa_h^2 - k_{pq}^2}}, & 0 \le |k_{pq}| < \kappa_h \\[2ex] \dfrac{i\,\text{sign}(k_{pq})}{\sqrt{k_{pq}^2 - \kappa_h^2}}, & 0 \le \kappa_h < |k_{pq}| \end{cases},$$

$$\tag{C.3}$$

where $F^{SW}(\kappa_h)$ is the surface-wave spectrum. A useful model for the surface spectrum is the Pierson-Moskovitz (PM) spectrum given by

$$F^{SW}(\kappa_h) = \frac{\alpha}{2\kappa_h^3}\exp\left[-\frac{\kappa_L^2}{\kappa_h^2}\right], \tag{C.4}$$

with $\alpha = 8.1 \times 10^{-3}$, $\kappa_L^2 = \beta g^2/U^4$, $\beta = 0.74$, $g = 9.8$ (m/s^2), and U is the wind speed in (m/s). This spectrum is normalized such that

$$\langle h^2\rangle = \int_0^\infty d\kappa_h F^{SW}(\kappa_h) = \frac{\alpha}{4\kappa_L^2} = \frac{\alpha U^4}{4\beta g^2}, \tag{C.5}$$

where it is seen that the surface hight variance scales as wind speed to the fourth power. The wavenumber integral in Eq. C.3 using the PM spectrum cannot be solved analytically, and thus a numerical approach is needed. Specific examples of the combined effects of surface and internal waves on high-frequency shallow-water propagation are given in Raghukumar and Colosi (2015).

References

Abramowitz, M., and Stegun, I.A. 1964. *Handbook of Mathematical Functions with Formulas, Graphs, and Mathematical Tables*. Dover Publications, New York.

Ainslie, M. 2010. *Principles of Sonar Performance Modeling*. Springer, London.

Alford, M.H., Lien, R.C., Simmons, H., Klymak, J., Ramp, S., Yang, Y.J., Tang, D., and Chang, M.H. 2010. Speed and evolution of nonlinear internal waves transiting the South China Sea. *J. Phys. Oceanogr.*, **40**(6), 1338–1355.

Andrew, R., Howe, B., and Mercer, J. 2005. Transverse horizontal spatial coherence of deep arrivals at megameter ranges. *J. Acoust. Soc. Am.*, **117**, 1511–1526.

Andrews, L.C., and Phillips, R.L. 2005. *Laser Beam Propagation through Random Media*. Vol. 10. SPIE Press, Bellingham, WA.

Antonov, J.I., Locarnini, R.A., Boyer, T.P., Mishonov, A.V., and Garcia, H.E. 2006. World Ocean Atlas 2005. Salinity. edited by S. Levitus (U.S.Government Printing Office, Washington, DC), *NOAA Atlas NESDIS62*, **2**, 182.

Apel, J., Badiey, M., Chiu, C.S., Finnette, S., Headrick, R., Kemp, J., Lynch, J., Newhall, A., Orr, M., Pasewark, B., Teilbuerger, D., Turgut, A., von der Heydt, K., and Wolf, S. 1997. An overview of the 1995 SWARM shallow water internal wave acoustic scattering experiment. *IEEE J. Oceanic Eng.*, **22**, 465–500.

Apel, J.R., Ostrovsky, L.A., Stepanyants, Y.A., and Lynch, J.F. 2007. Internal solitons in the ocean and their effect on underwater sound. *J. Acoust. Soc. Am.*, **121**, 695–722.

APL-UW. 2008. APL-UW High Frequency Ocean Environmental Acoustic Models Handbook. *APL-UW TR9407, AEAS 9501*, Applied Physics Laboratory, University of Washington.

Arnold, V.I. 1989. *Mathematical Methods of Classical Mechanics*. Springer-Verlag, New York.

Baym, G. 1973. *Lectures on Quantum Mechanics*. Benjamin/Cummings, Menlo Park.

Bell, T.H. 1974. Processing vertical internal wave spectra. *J. Phys. Oc.*, **4**, 669–670.

Beron-Vera, F.J., and Brown, M.G. 2003. Ray stability in weakly range dependent sound channels. *J. Acoust. Soc. Am.*, **114**, 123–130.

Beron-Vera, F.J., and Brown, M.G. 2004. Travel time stability in weakly range-dependent sound channels. *J. Acoust. Soc. Am.*, **115**, 1068–1077.

Beron-Vera, F.J., and Brown, M.G. 2009. Underwater acoustic beam dynamics. *J. Acoust. Soc. Am.*, **126**, 80–91.

Beron-Vera, F.J., Brown, M.G., Colosi, J.A., Tomsovic, S., Virovlyansky, A.L., Wolfson, M.A., and Zaslavsky, G.M. 2003. Ray dynamics in a long-range acoustic propagation experiment. *J. Acoust. Soc. Am.*, **114**, 1226–1241.

Born, M., and Wolf, E. 1999. *Principles of Optics: Electromagnetic Theory of Propagation*, Interference and Diffraction of Light, 7th edition Cambridge University Press, Cambridge.

Boyd, T.J., Luther, D.S., Knox, R.A., and Hendershott, M.C. 1993. High-frequency internal waves in the strongly sheared currents of the upper Equatorial Pacific: Observations and a simple spectral model. *J. Geophys. Res.*, **98**, 18089–18107.

Brekhovskikh, L.M., and Lysanov, Yu. 1991. *Fundamentals of Ocean Acoustics*, 2nd edition Springer-Verlag, Berlin.

Briscoe, M.G. 1975. Preliminary results from the tri-moored internal wave experiment IWEX. *J. Geophys. Res.*, **80**, 3877–3884.

Brown, M.G. 1982. Application of the WKBJ Green's function to acoustic propagation in horizontally stratified oceans. *J. Acoust. Soc. Am.*, **71**(6), 1427–1432.

Brown, M.G. 1994. A Maslov-Chapman wavefield representation for wide-angle one-way propagation. *Geophys. J. Int.*, **116**(3), 513–526.

Brown, M.G. 1998. Phase space structure and fractal trajectories in 1 1/2 degree of freedom Hamiltonian systems whose time dependence is quasi-periodic. *Nonlin. Proc. Geophys.*, **5**, 69–74.

Brown, M.G., Beron-Vera, F.J., Rypina, I., and Udovydchenkov, I.A. 2005. Rays, modes, wavefield structure and wavefield stability. *J. Acoust. Soc. Am.*, **117**, 1607–1610.

Brown, M.G., Colosi, J.A., Virovlyansky, A.L., Zaslavsky, G.M., Tomsovic, S., and Wolfson, M.A. 2003. Ray dynamics in ocean acoustics. *J. Acoust. Soc. Am.*, **113**, 2533–2547.

Brown, M.G., Tappert, T.F., and Goni, G. 1991. An investigation of sound ray dymanics in the ocean volume using an area-preserving map. *Wave Motion*, **14**, 93–99.

Buckingham, M.J. 2005. Compressional and shear wave properties of marine sediments: Comparisons between theory and data. *J. Acoust. Soc. Am.*, **117**, 137–152.

Carey, W.M. 1998. The determination of signal coherence length based on signal coherence and gain measurements in deep and shallow water. *J. Acoust. Soc. Am.*, **104**, 831–837.

Casati, G., and Chirikov, B.V. 1995. *Quantum Chaos: Between Order and Disorder: A Selection of Papers*. Cambridge University Press, New York.

Chandrayadula, T.K., Colosi, J.A., Worcester, P.F., Dzieciuch, M.A., Mercer, J.A., Andrew, R.K., and Howe, B.M. 2013. Observations and transport theory analysis of low frequency, acoustic mode propagation in the Eastern North Pacific Ocean. *J. Acoust. Soc. Am.*, **134**, 3144–3160.

Chapman, C. 2004. *Fundamentals of Seismic Wave Propagation*. Cambridge University Press, Cambridge.

Chernov, L.A. 1975. *Wave Propagation in Random Media*. Nauka, Moscow (In Russian).

Chu, P.C., and Hsieh, C.-P. 2007. Change of multifractal thermal characteristics in the western Philippine Sea upper layer during internal wave-soliton propagation. *J. of Oceangr.*, **63**, 927–939.

Clay, C.S., and Medwin, H. 1977. *Acoustical Oceanography: Principles and Applications*. John Wiley & Sons, New York.

Clifford, S. 1978. The classical theory of wave propagation in a turbulent medium. In: Strohbehn, J.W. (ed), *Laser Beam Propagation in the Atmosphere*. Springer-Verlag, Berlin.

Codona, J., Creamer, D., Flatté, S.M., Frehlich, R., and Henyey, F. 1985. Average arrival time of wave pulses through continuous random media. *Phys. Rev. Lett.*, **55**, 9–12.

Codona, J., Creamer, D., Flatté, S.M., Frehlich, R., and Henyey, F. 1986a. Solution for the fourth moment of waves propagating in random media. *Radio Sci.*, **21**, 929–948.

Codona, J.L., Creamer, D.B., Flatté, S.M., Frehlich, R.G., and Henyey, F.S. 1986b. Moment-equation and path-integral techniques for wave propagation in random media. *J. Math. Phys.*, **27**(1), 171–177.

Colladon, J.D., and Sturm, J.C.F. 1827. Memoir on the compression of liquids. *Ann. Chim. Phys.*, **2**(36), 225–257.

Colosi, J.A. 1999. A review of recent results on ocean acoustic wave propagation in random media: Basin scales. *IEEE J. Oc. Eng.*, **24**, 138–155.

Colosi, J.A. 2001. A scintillating problem: Basin scale acoustic propagation through a fluctuating ocean. *Proc. Inst. Acoust.*, **23**, 37–53.

Colosi, J.A. 2006. Geometric sound propagation through an inhomogeneous and moving ocean: Scattering by small scale internal wave currents. *J. Acoust. Soc. Am.*, **119**, 705–708.

Colosi, J.A. 2008. Acoustic mode coupling induced by shallow water nonlinear internal waves: Sensitivity to environmental conditions and space-time scales of internal waves. *J. Acoust. Soc. Am.*, **124**, 1452–1464.

Colosi, J.A. 2013. On horizontal coherence estimates from path integral theory for sound propagation through random ocean sound-speed perturbations. *J. Acoust. Soc. Am.*, **134**, 3116–3118.

Colosi, J.A. 2015. A reformulation of the $\Lambda - \Phi$ diagram for the prediction of ocean acoustic fluctuation regimes. *J. Acoust. Soc. Am.*, **137**, 2485–2494.

Colosi, J.A., and Baggeroer, A.B. 2004. On the kinematics of broadband multipath scintillation and the approach to saturation. *J. Acoust. Soc. Am.*, **116**, 3515–3522.

Colosi, J.A., Baggeroer, A.B., Cornuelle, B.D., Dzieciuch, M.A., Munk, W.H., Worcester, P.F., Dushaw, B.D., Howe, B.M., Mercer, J.A., Spindel, R.C., Birdsall, T.G., Metzger, K., and Forbes, A.M.G. 2005. Analysis of multipath acoustic field variability and coherence for the finale of broadband basin-scale transmissions in the North Pacific Ocean. *J. Acoust. Soc. Am.*, **117**, 1538–1564.

Colosi, J.A., and Brown, M.G. 1998. Efficient numerical simulation of stochastic internal wave induced sound speed perturbation fields. *J. Acoust. Soc. Am.*, **103**, 2232–2235.

Colosi, J.A., Chandrayadula, T., Voronovich, A.G., and Ostashev, V.E. 2013a. Coupled mode transport theory for sound propagation through an ocean with random sound-speed perturbations: Coherence in deep water environments. *J. Acoust. Soc. Am.*, **134**, 3119–3133.

Colosi, J.A., Duda, T.F., Lin, T.T., Lynch, J., Newhall, A., and Cornuelle, B.C. 2012. Observations of sound speed fluctuations on the New Jersey continental shelf in the summer of 2006. *J. Acoust. Soc. Am.*, **131**, 1733–1748.

Colosi, J.A., Duda, T.F., and Morozov, A.K. 2011. Statistics of low-frequency normal-mode amplitudes in an ocean with random sound-speed perturbations: Shallow-water environments. *J. Acoust. Soc. Am.*, **131**, 1749–1761.

Colosi, J.A., and Flatté, S.M. 1996. Mode coupling by internal waves for multimegameter acoustic propagation in the ocean. *J. Acoust. Soc. Am.*, **100**, 3607–3620.

Colosi, J.A., Flatté, S.M., and Bracher, C. 1994. Internal wave effects on 1000-km oceanic acoustic pulse propagation: Simulation and comparison to experiment. *J. Acoust. Soc. Am.*, **96**, 452–468.

Colosi, J.A., and Morozov, A.K. 2009. Coupled mode theory for the mean intensity of sound propagation through a random waveguide. *J. Acoust. Soc. Am,*, **126**, 1026–1035.

Colosi, J.A., Scheer, E.K., Flatté, S.M., Cornuelle, B.D., Dzieciuch, M.A., Munk, W.H., Worcester, P.F., Howe, B.M., Mercer, J.A., Spindel, R.C., Metzger, K., Birdsall, T., and Baggeroer, A. 1999. Comparisons of measured and predicted acoustic fluctuations for a 3250-km propagation experiment in the eastern North Pacific Ocean. *J. Acoust. Soc. Am.*, **105**, 3202–3218.

Colosi, J.A., Tappert, F.D., and Dzieciuch, M.A. 2001. Further analysis of intensity fluctuations from a 3252-km acoustic propagation experiment in the eastern North Pacific Ocean. *J. Acoust. Soc. Am.*, **110**, 163–169.

Colosi, J.A., Uffelen, L.J. Van, Cornuelle, B.D., Dzieciuch, M.A., Worcester, P.F., Dushaw, B.D., and Ramp, S.R. 2013b. Observations of sound speed fluctuations in the western Philippine Sea in the spring of 2009. *J. Acoust. Soc. Am.*, **134**, 3185–3200.

Colosi, J.A., Xu, J., Cornuelle, B.D., Dzieciuch, M.A., Munk, W.H., Worcester, P.F., Howe, B.M., and Mercer, J.A. 2009. Intensity fluctuations and spectra for low frequency, short range sound transmission in the Eastern North Pacific Ocean: Comparisons to weak fluctuation theory. *J. Acoust. Soc. Am.,* **126**, 1069–1083.

Cornuelle, B.D., and Howe, B.M. 1987. High spatial resolution in vertical slice ocean acoustic tomography. *J. Geophys. Res.*, **92**, 11680–11692.

Creamer, D. 1996. Scintillating shallow water waveguides. *J. Acoust. Soc. Am.*, **99**, 2825–2838.

da Vinci, Leonardo. 1483. Manuscript B, Institut de France, Folio 6 recto.

D'Asaro, E.A. 1984. Wind forced internal waves in the North Pacific and Sargasso Sea. *J. Phys. Oceanogr.*, **14**, 781–794.

D'Asaro, E.A., and Morehead, M.D. 1991. Internal waves and velocity fine structure in the Arctic Ocean. *J. Geophys. Res.*, **96**, 12725–12738.

Dashen, R. 1979. Path integrals for waves in random media. *J. Math. Phys.*, **20**(5), 894–920.

Dashen, R., Flatté, S.M., and Reynolds, S. 1985. Path-integral treatment of acoustic mutual coherence functions for rays in a sound channel. *J. Acoust. Soc. Am.*, **77**, 1716–1722.

DelGrosso, V.A. 1974. New equation for the speed of sound in natural waters. *J. Acoust. Soc. Am.*, **56**, 1084–1091.

Desaubies, Y.J.F. 1978. On the scattering of sound by internal waves in the ocean. *J. Acoust. Soc. Am.*, **64**(5), 1460–1469.

Dozier, L.B. 1983. A coupled mode model for spatial coherence of bottom-interacting energy. In: Spofford, C.W., and Haynes, J.M. (eds), *Proceedings of the Stochastic Modeling Workshop*. ARL-University of Texas, Austin, TX.

Dozier, L.B., and Tappert, F.D. 1978a. Statistics of normal-mode amplitudes in a random ocean. I. Theory. *J. Acoust. Soc. Am.*, **63**, 353–365.

Dozier, L.B., and Tappert, F.D. 1978b. Statistics of normal-mode amplitudes in a random ocean. II. Computations. *J. Acoust. Soc. Am.*, **64**, 353–365.

Duda, T.F. 2005. Ocean sound channel ray path perturbations from internal wave shear and strain. *J. Acoust. Soc. Am.*, **118**, 2899–2903.

Duda, T.F., and Bowlin, J.B. 1994. Ray-acoustic caustic formation and timing effects from ocean sound speed relative curvature. *J. Acoust. Soc. Am.*, **96**, 1033–1046.

Duda, T.F., Collis, J.M., Lin, Y.T., Newhall, A.E., Lynch, J.F., and DeFerrari, H.A. 2012. Horizontal coherence of low-frequency fixed-path sound in a continental shelf region with internal-wave activity. *J. Acoust. Soc. Am.*, **131**, 1782–1797.

Duda, T.F., and Cox, C.S. 1989. Vertical wave number spectra of velocity and shear at small internal wave scales. *J. Geophys. Res.*, **94**, 939–950.

Duda, T.F., and Farmer, D.M. 1999. *The 1998 WHOI/IOS/ONR Internal Solitary Wave Workshop: Contributed Papers*. Woods Hole Oceanographic Institution Technical Report.

Duda, T.F., Flatté, S.M., Colosi, J.A., Cornuelle, B.D., Hildebrand, J.A., Hodgkiss, W.S., Worcester, P.F., Howe, B.M., Mercer, J.A., and Spindel, R.C. 1992. Measured wave-front fluctuations in 1000-km pulse propagation in the Pacific Ocean. *J. Acoust. Soc. Am.*, **92**, 939–955.

Duda, T.F., Lynch, J.F., Newhall, A.E., Wu, L., and Chiu, C.S. 2004. Fluctuations of 400 Hz sound intensity in the 2001 ASIAEX South China Sea experiment. *IEEE J. Oceanic Eng.*, **29**, 1264–1280.

Dushaw, B.D. 2008. Another look at the 1960 Perth to Bermuda long-range acoustic propagation experiment. *Geophys. Res. Lett.*, **35**(8), L08601.

Dushaw, B.D., Howe, B.M., Mercer, J.A., and Spindel, R.C. 1999. Multi-megameter range acoustic data obtained by bottom mounted hydrophone arrays for measurement of ocean temperature. *IEEE J. Oceanic Eng.*, **24**, 203–215.

Dushaw, B.D., and Worcester, P.F. 1998. Resonant diurnal internal tides in the North Atlantic. *Geophys. Res. Lett.*, **25**, 2189–2193.

Dushaw, B.D., Worcester, P.F., Cornuelle, B.D., Howe, B.M., and Luther, D.S. 1995. Baroclinic and barotropic tides in the central North-Pacific Ocean determined from long-range reciprocal acoustic transmissions. *J. Phys. Oceanogr.*, **25**, 631–647.

Dushaw, B.D., Worcester, P.F., and Dzieciuch, M.A. 2011. On the predictability of mode-1 internal tides. *Deep-Sea Res.*, **58**, 677–698.

Dyer, I. 1970. Statistics of sound propagation in the ocean. *J. Acoust. Soc. Am.*, **48**, 337–345.

Dyson, F.J. 1949. The radiation theories of Tomonaga, Schwinger, and Feynman. *Phys. Rev.*, **75**(3), 486–502.

Dyson, F., Munk, W., and Zetler, B. 1976. Interpretation of multi path scintillations Eleuthera to Bermuda in terms of internal waves and tides. *J. Acoust. Soc. Am.*, **59**, 1121–1133.

Dzieciuch, M.A. 2014. Signal processing and tracking of arrivals in ocean acoustic tomography. *J. Acoust. Soc. Am.*, **136**(5), 2512–2522.

Dzieciuch, M.A., Munk, W.H., and Rudnick, D. 2004. Propagation of sound through a spicy ocean, the sofar overture. *J. Acoust. Soc. Am.*, **116**, 1447–1462.

Dzieciuch, M.A., and Vera, M.D. 2006. Horizontal coherence of tracked arrivals in the North Pacific Acoustic Laboratory98 (NPAL98). *J. Acoust. Soc. Am.*, **120**, 3022.

Dzieciuch, M.A., Worcester, P.F., and Munk, W.H. 2001. Turning point filters: Analysis of sound propagation on a gyre-scale. *J. Acoust. Soc. Am.*, **110**, 135–149.

Eckart, C. 1960. *Hydrodynamics of Oceans and Atmospheres*. Pergamon Press, Oxford.

Eckart, C., and Carhart, R.R. 1950. Fluctuation of sound in the sea. In *Basic Problems in Underwater Acoustics* (pp. 63–122). Committee on Undersea Warfare, National Research Council.

Eckert, E.G., and Foster, T.D. 1990. Upper Ocean internal waves in the marginal ice zone of the northeastern Greenland Sea. *J. Geophys. Res.*, **95**, 9569–9574.

Ehrenfest, P. 1927. Bemerkung ber die angenŁherte Gltigkeit der klassischen Mechanik innerhalb der Quantenmechanic. *Zeitschrift Physik*, **45**, 455–457.

Eriksen, C.C. 1978. Measurements and models of fine structure, internal gravity waves, and wave breaking in the deep ocean. *J. Geophys. Res.*, **83**, 2989–2310.

Eriksen, C.C. 1980. Evidence for a continuous spectrum of equatorial waves in the Indian Ocean. *J. Geophys. Res.*, **85**, 3285–3303.

Eriksen, C.C. 1985. Some characteristics of internal gravity waves in the Equatorial Pacific. *J. Geophys. Res.*, **90**, 7243–7255.

Eriksen, C.C. 1998. Internal wave reflection and mixing at Fieberling Guyot. *J. Geophys. Res.*, **103**, 2977–2994.

Esswein, R., and Flatté, S.M. 1980. Calculation of strength and diffraction parameters in oceanic sound transmission. *J. Acoust. Soc. Am.*, **67**, 1523–1531.

Esswein, R., and Flatté, S.M. 1981. Calculation of the phase structure function density from oceanic internal waves. *J. Acoust. Soc. Am.*, **70**, 1387–1396.

Ewart, T.E. 1976. Acoustic fluctuations in the open ocean – A measurement using a fixed refracted path. *J. Acoust. Soc. Am.*, **60**, 46–60.

Ewart, T.E. 1989. A model of the intensity probability distribution for wave propagation in random media. *J. Acoust. Soc. Am.*, **86**, 1490–1498.

Ewart, T.E., and Percival, D.B. 1986. Forward scattered waves in random media – The probability distribution of intensity. *J. Acoust. Soc. Am.*, **60**, 1745–1753.

Ewart, T.E., and Reynolds, S.A. 1984. The mid-ocean acoustic transmission experiment – MATE. *J. Acoust. Soc. Am.*, **75**, 785–802.

Ewart, T.E., and Reynolds, S.A. 1990. Instrumentation to measure the depth/time fluctuations in acoustic pulses propagated through Arctic internal waves. *J. Atmos. Ocean Tech.*, **7**, 129–140.

Ewart, T.E., and Reynolds, S.A. 1993. Ocean acoustic propagation measurements and wave propagation in random media. In: Ishimaru, A., and Zavorotny, V.U. (eds), *Wave Propagation in Random Media (Scintillation)*. SPIE Press, Bellingham, WA.

Ewart, T.E., Reynolds, S.A., and Rouseff, D. 1998. Determining an ocean internal wave model using acoustic log-amplitude and phase: A Rytov inverse. *J. Acoust. Soc. Am.*, **104**, 146–156.

Ewing, M., and Worzel, J.L. 1948. Long-range sound transmission. *Geol. Soc. Am. Mem.*, **27**, 1–32.

Ferarri, R., and Rudnick, D.L. 2000. Thermohaline variability in the upper ocean. *J. Geophys. Res.*, **105**, 16857–16883.

Feynman, R., and Hibbs, A. 1965. *Quantum Mechanics and Path Integrals*. McGraw-Hill, New York.

Fisher, F.H., and Simmons, V.P. 1977. Sound absorption in sea water. *J. Acoust. Soc. Am.*, **62**, 558–564.

Flatté, S.M. 1983a. Principles of acoustic tomography of internal waves. *Proc. Oceans '83*, **29**, 372–377.

Flatté, S.M. 1983b. Wave propagation through random media: Contributions from ocean acoustics. *Proc. IEEE*, **71**, 1267–1294.

Flatté, S.M. 1986. The Schrödinger equation in classical physics. *Am. J. Phys.*, **54**, 1088–1095.

Flatté, S.M., Bernstein, D., and Dashen, R. 1983. Intensity moments by path integral techniques for wave propagation through random media, with application to sound in the ocean. *Phys. Fluids*, **26**, 1701–1713.

Flatté, S.M., Bracher, C., and Wang, G. 1994. Probability density functions of irradiance for waves in atmospheric turbulence calculated by numerical simulation. *J. Opt. Soc. Am.*, **11**, 2080–2092.

Flatté, S.M., and Colosi, J.A. 2008. Anisotropy of the wavefront distortion for acoustic pulse propagation through ocean sound speed fluctuations: A ray perspective. *IEEE J. Oceanic Eng.*, **6**, 477–488.

Flatté, S.M., Dashen, R., Munk, W., Watson, K., and Zachariasen, F. 1979. *Sound Transmission through a Fluctuating Ocean*. Cambridge University Press, Cambridge.

Flatté, S.M., Leung, R., and Lee, S.Y. 1980. Frequency spectra of acoustic fluctuations caused by oceanic internal waves and other fine structure. *J. Acoust. Soc. Am.*, **68**, 1773–1780.

Flatté, S.M., Reynolds, S.A., and Dashen, R. 1987a. Path-integral treatment of intensity behavior for rays in a sound channel. *J. Acoust. Soc. Am.*, **82**, 967–972.

Flatté, S.M., Reynolds, S., Dashen, R., Buehler, B., and Maciejewski, P. 1987b. AFAR measurements of intensity and intensity moments. *J. Acoust. Soc. Am.*, **82**, 973–981.

Flatté, S.M., and Rovner, G. 2000. Calculation of internal-wave induced fluctuations in ocean acoustic propagation. *J. Acoust. Soc. Am.*, **108**, 526–534.

Flatté, S.M., and Stoughton, R. 1988. Predictions of internal wave effects on ocean acoustic coherence, travel time variance, and intensity moments for very long range propagation. *J. Acoust. Soc. Am.*, **84**, 1414–1424.

Flatté, S.M., and Stoughton, R.B. 1986. Theory of acoustic measurement of internal wave strength as a function of depth, horizontal position, and time. *J. Geophys. Res-Oceans (1978–2012)*, **91**(C6), 7709–7720.

Flatté, S.M., and Tappert, F.D. 1975. Calculation of the effects of internal waves on oceanic sound transmission. *J. Acoust. Soc. Am.*, **58**, 1151–1159.

Flatté, S.M., and Vera, M. 2003. Comparison between ocean acoustic fluctuations in parabolic equation simulations and estimates from integral approximations. *J. Acoust. Soc. Am.*, **114**, 697–706.

Fredricks, A., Colosi, J.A., Lynch, J.F., Gawarkiewicz, G., Chiu, C.S., and Abbot, P. 2005. Analysis of multi path scintillations from long range acoustic transmissions on the New England continental slope and shelf. *J. Acoust. Soc. Am.*, **117**, 1038–1057.

Frisk, G.V. 1994. *Ocean and Seabed Acoustics*. Prentice Hall, Englewood Cliffs, NJ.

Garrett, C.J., and Kunze, E. 2007. Internal tide generation in the deep ocean. *Annu. Rev. Fluid Mech.*, **39**, 57–87.

Garrett, C.J., and Munk, W.H. 1972. Space-time scales of internal waves. *Geophys. Fluid Dyn.*, **2**, 255–264.

Garrett, C.J., and Munk, W.H. 1975. Space-time scales of internal waves: A progress report. *J. Geophys. Res.*, **80**, 291–297.

Garrett, C.J.R. 1979. Mixing in the ocean interior. *Dyn. Atmos. Oceans*, **3**, 239–265.

Giannoni, M.J., Voros, A., and Zinn-Justin, J. 1991. *Chaos and Quantum Physics:Les Houches Session LII, 1989*. Elsevier Science, Amsterdam.

Gill, A.E. 1982. *Atmosphere-Ocean Dynamics*. Vol. 30. Academic Press, San Diego.

Godin, O.A. 2007. Restless rays, steady wave fronts. *J. Acoust. Soc. Am.*, **122**, 3353–3363.

Goldstein, H. 1980. *Classical Mechanics*. Addison-Wesley, Reading, MA.

Gould, W.J., Schmitz, W.J., and Wunsch, C. 1974. Preliminary field results for a mid-ocean dynamics experiment (MODE-0). In: *Deep Sea Research and Oceanographic Abstracts*, Vol. 21 (pp. 911–931). Elsevier, Amsterdam.

Gregg, M.C. 1977. A comparison of fine structure spectra in the main thermocline. *J. Phys. Oceanogr.*, **7**, 33–40.

Gutzwiller, M. 1990. *Chaos in Classical and Quantum Mechanics*. Springer-Verlag, New York.

Hamilton, E.L. 1980. Geoacoustic modeling of the seafloor. *J. Acoust. Soc. Am.*, **68**, 1313–1340.

Hamilton, E.L. 1987. Acoustic properties of sediments. In: Lara-Saenz, A., Ranz-Guerra, C., and Carbo-Fite, C. (eds), *Acoustics and Ocean Bottom*. C.S.I.C, Madrid, Spain.

Hamilton, G. 1977. Time variations of sound speed over long paths in the ocean. In: *International Workshop on Low-Frequency Propagation and Noise* (pp. 7–30). Department of the Navy.

Headrick, R.H., Lynch, J., Kemp, J., Newhall, A., von der Heydt, K., Apel, J., Badiey, M., Chiu, C.S., Finnette, S., Orr, M., Pasewark, B., Turgut, A., Wolf, S., and Teilbuerger, D. 1999. Acoustic normal mode fluctuation statistics in the 1995 SWARM internal wave scattering experiment. *J. Acoust. Soc. Am.*, **107**, 201–221.

Heaney, K.D., Kuperman, W.A., and McDonald, B.E. 1991. Perth–Bermuda sound propagation (1960): Adiabatic mode interpretation. *J. Acoust. Soc. Am.*, **90**, 2586–2594.

Henyey, F., and Ewart, T.E. 2006. Validity of the markov approximation in ocean acoustics. *J. Acoust. Soc. Am.*, **119**, 220–231.

Henyey, F., and Macaskill, C. 1996. Sound through the internal wave field. In: Adler, R.J., Müller, P., and Rozovskii, B.L. (eds), *Stochastic Modeling in Physical Oceanography*. Birkhauser Press, Boston.

Henyey, F.S. 1997. A quick introduction to path integrals. In: *'Aha Hulilo'a winter workshop: Monte Carlo Simulations in Oceanography*. University of Hawaii, School of Ocean and Earth Science and Technology.

Hotchkiss, F.S., and Wunsch, C. 1982. Internal waves in Hudson canyon with possible geological implications. *Deep-Sea Res.*, **29**, 415–442.

Howe, B.M., Worcester, P.F., and Spindel, R.C. 1987. Ocean acoustic tomography: Mesoscale velocity. *J. Geophys. Res-Oceans (1978–2012)*, **92**(C4), 3785–3805.

Huygens, C. 1690. *Traité de la Lumiere*. Published in Leiden Netherlands, also see 2012 re-print in the Forgotten Books' Classic Reprint Series.

Isaacson, E., and Keller, H.B. 1966. *Analysis of Numerical Methods*. John Wiley & Sons, New York.

Ishimaru, A. 1978. *Wave Propagation and Scattering in Random Media*, Vol. 2. Academic Press, New York.

Jensen, F.B., Kuperman, W.A., Porter, M.B., and Schmidt, H. 1994. *Computational Ocean Acoustics*. Springer-Verlag, New York.

Katznelson, B., and Pereselkov, S. 2000. Low-frequency horizontal acoustic refraction caused by internal wave solitons in a shallow sea. *Acoust. Phys.*, **46**, 684–691.

Katznelson, B., Petnikov, V., and Lynch, J. 2012. *Fundamentals of Shallow Water Acoustics*. Springer Science+Business Media, New York.

Kennedy, R.M. 1969. Phase and amplitude fluctuations in propagating through a layered ocean. *J. Acoust. Soc. Am.*, **46**(3B), 737–745.

Kunze, E., Rosenfeld, L.K., Carter, G.S., and Gregg, M.C. 2002. Internal waves in the Monterey submarine canyon. *J. Phys. Oc.*, **32**, 1890–1914.

Landau, L.D., and Lifshitz, E.M. 1975. *Classical Theory of Fields*. Pergamon Press, Oxford.

Landau, L.D., and Lifshitz, E.M. 1976. *Mechanics*. Pergamon Press, Oxford.

Landau, L.D., and Lifshitz, E.M. 1980. *Statistical Physics*, Third Edition, Part 1. Pergamon, Oxford.

Latora, V., and Baranger, M. 1999. Kolmogorov–Sinai entropy rate versus physical entropy. *Phys. Rev. Lett.*, **82**, 520–523.

Levine, M.D. 1990. Internal waves under the Arctic pack ice during AIWEX: The coherence structure. *J. Geophys. Res.*, **95**, 7347–7357.

Levine, M.D. 2002. A modification of the Garrett-Munk internal wave spectrum. *J. Phys. Oc.*, **32**, 3166–3181.

Levine, M.D., and Irish, J.D. 1981. A statistical description of temperature fine structure in the presence of internal waves. *J. Phys. Oceanogr.*, **11**, 676–691.

Levine, M.D., Irish, J.D., Ewart, J.D., and Reynolds, S.A. 1986. Simultaneous spatial and temporal measurements of the internal wave field during MATE. *J. Goephys. Res.*, **91**, 9709–9719.

Levine, M.D., Paulson, C.A., and Morison, J.H. 1987. Observations of internal gravity waves under the Arctic pack ice. *J. Geophys. Res.*, **92**, 779–782.

Lichtenberg, A.J., and Lieberman, M.A. 1983. *Regular and Stochastic Motion*. Applied Mathematical Sciences, Vol. 38. Springer-Verlag, New York.

Lien, R.-C., and Müller, P. 1992. Normal-mode decomposition of small-scale oceanic motions. *J. Phys. Oceanogr.*, **22**, 1583–1595.

Lighthill, J. 1978. *Waves in Fluids*. Cambridge University Press, Cambridge.

Lin, Y.T., Duda, T.F., and Lynch, J.F. 2009. Acoustic mode radiation from the termination of a truncated nonlinear internal gravity wave duct in a shallow ocean area. *J. Acoust. Soc. Am.*, **126**, 1752–1765.

Locarnini, R.A., Mishonov, A.V., Antonov, J.I., Boyer, T.P., and Garcia, H.E. 2006. World Ocean Atlas 2005. Temperature, edited by S. Levitus (U.S.Government Printing Office, Washington, DC), *NOAA Atlas NESDIS62*, **1**, 182.

Lovett, J.R. 1980. Geographic variation of low frequency sound absorption in the Atlantic, Indian, and Pacific Oceans. *J. Acoust. Soc. Am.*, **67**, 338–340.

Lynch, J.F., Colosi, J.A., Gawarkiewicz, G.G., Duda, T.F., Pierce, A.D., Badiey, M., Katsnelson, B.G., Miller, J.E., Siegmann, W., Ching-Sang, C., and Newhall, A. 2006. Consideration of fine-scale coastal oceanography and 3-D acoustics effects for the ESME sound exposure model. *IEEE J. Oceanic Eng.*, **31**, 33–48.

Lynch, J.F., Jin, G., Pawlowicz, R., Ray, D., Plueddemann, A.J., Chiu, C.S., Miller, J.H., Bourke, R.H., Parsons, A.R., and Muench, R. 1996. Acoustic travel-time perturbations due to shallow-water internal waves and internal tides in the Barents Sea Polar Front: Theory and experiment. *J. Acoust. Soc. Am.*, **99**(2), 803–821.

Lynch, J.F., Lin, Y.T., Duda, T.F., and Newhall, A.E. 2010. Acoustic ducting, reflection, refraction, and dispersion by curved nonlinear internal waves in shallow water. *IEEE J. Oceanic Eng.*, **35**, 12–28.

Macaskill, C., and Ewart, T.E. 1984. The probability distribution of intensity for acoustic propagation in a randomly varying ocean. *J. Acoust. Soc. Am.*, **76**(5), 1466–1473.

Macaskill, C., and Ewart, T.E. 1996. Numerical solution of the fourth moment equation for acoustic intensity correlations and comparison with the mid-ocean acoustic transmission experiment. *J. Acoust. Soc. Am.*, **99**, 1419–1430.

MacKenzie, K.V. 1981. Nine-term equation for sound speed in the ocean. *J. Acoust. Soc. Am.*, **70**, 807–812.

Mandelbrot, B.B. 1982. *The Fractal Geometry of Nature*. W. H. Freeman, New York.

Medwin, H., and Clay, C.S. 1997. *Fundamentals of Acoustical Oceanography*. Academic Press, San Diego.

Merzbacher, E. 1961. *Quantum Mechanics*. John Wiley & Sons, New York.

Michalevsky, P.N., Gavrilov, A.N., and Baggeroer, A.B. 1999. The trans-arctic acoustic propagation experiment and climate monitoring in the arctic. *IEEE J. Oc. Eng.*, **24**, 183–201.

Mignerey, P.C., and Orr, M.H. 2004. Observations of match-field autocorrelation time in the South China Sea. *IEEE J. Oceanic Eng.*, **29**, 1280–1291.

Milder, D.M. 1969. Ray and wave invariants for SOFAR channel propagation. *J. Acoust. Soc. Am.*, **46**, 1259–1263.

MODE-Group, et al. 1978. The mid-ocean dynamics experiment. *Deep-Sea Res.*, **25**(10), 859–910.

Morozov, A.K., and Colosi, J.A. 2004. Entropy and scintillation analysis for acoustical beam propagation through ocean internal waves. *J. Acoust. Soc. Am.*, **117**, 1611–1623.

Müller, P. 1976. On the diffusion of momentum and mass by internal gravity waves. *J. Fluid Mech.*, **77**(10), 789–823.

Müller, P., Holloway, G., Henyey, F., and Pomphrey, N. 1986. Nonlinear interactions amount internal gravity waves. *Rev. Geo. Phys.*, **24**(3), 493–536.

Müller, P., Olbers, D.J., and Willebrand, J. 1978. The IWEX spectrum. *J. Geophys. Res.*, **83**, 479–500.

Munk, W.H. 1966. Abyssal recipes. *Deep-Sea Res.*, **13**, 107–130.

Munk, W.H. 1974. Sound channel in an exponentially stratified ocean, with application to SOFAR. *J. Acoust. Soc. Am.*, **55**, 220–226.

Munk, W.H. 1981. Internal waves and small scale processes. In: Warren, B., and Wunsch, C. (eds), *The Evolution of Physical Oceanography*. MIT Press, Cambridge, MA.

Munk, W.H. 1998. Once again: Once again – tidal friction. *Prog. Ocean.*, **40**, 7–35.

Munk, W.H., and Forbes, A.M.G. 1989. Global warming: An acoustic measure? *J. Phys. Oc.*, **19**, 1765–1778.

Munk, W., Spindel, R., Baggeroer, A., and Birdsall, T. 1994. The heard island feasibility test. *J. Acoust. Soc. Am.*, **96**(4), 2330–2342.

Munk, W.H., Worcester, P.F., and Wunsch, C. 1995. *Ocean Acoustic Tomography*. Cambridge University Press, Cambridge.

Munk, W.H., Worcester, P.F., and Zachariasen, F. 1981. Scattering of sound by internal wave currents: The relation to vertical momentum flux. *J. Phys. Oc.*, **11**, 442–454.

Munk, W., and Wunsch, C. 1979. Ocean acoustic tomography: A scheme for large scale monitoring. *Deep-Sea Res.*, **26**(2), 123–161.

Munk, W.H., and Wunsch, C. 1983. Ocean acoustic tomography: Rays and modes. *Rev. Geophys. Space Phys.*, **21**, 777–793.

Munk, W.H., and Wunsch, C. 1998. Abyssal Recipes II. *Deep-Sea Res. Pt I*, **45**, 1977–2010.

Munk, W.H., and Zachariasen, F. 1976. Sound propagation through a fluctuating stratified ocean: Theory and observation. *J. Acoust. Soc. Am.*, **59**, 818–838.

Nichols, R.H., and Young, H.J. 1968. Fluctuations in low-frequency acoustic propagation in the ocean. *J. Acoust. Soc. Am.*, **43**(4), 716–722.

Ostashev, V., and Wilson, K. 2015. *Acoustics in Moving and Inhomogeneous Media*. CRC Press, Boca Raton, FL.

Pedlosky, J. 1987. *Geophysical Fluid Dynamics*. Springer, New York.

Penland, C. 1985. Acoustic normal mode propagation through a three dimensional internal wave field. *J. Acoust. Soc. Am.*, **78**, 1356–1365.

Phillips, O.M. 1977. *The Dynamics of the Upper Ocean*, 2nd edition. Cambridge University Press, Cambridge.

Pierce, A.D. 1994. *Acoustics: An introduction to Its Physical Principles and Application*. American Institute of Physics, Melville, NY.

Pinkel, R. 1983. Doppler sonar observations of internal wave: Wave-field structure. *J. Phys. Oceanogr.*, **13**, 804–815.

Pinkel, R. 1984. Dopper sonar observations of internal waves: The wavenumber - frequency spectrum. *J. Phys. Oceanogr.*, **14**, 1249–1270.

Pinkel, R. 2000. Internal solitary waves in the warm pool of the western equatorial Pacific. *J. Phys. Oceanogr.*, **30**, 2906–2926.

Pinkel, R. 2008. Advection, phase distortion, and the frequency spectrum of finescale fields in the sea. *J. Phys. Oceanogr.*, **38**, 291–313.

Pinkel, R. 2014. Vortical and internal wave shear and strain. *J. Phys. Oceanogr.*, **44**, 2070–2092.

Piperakis, G.S., Skarsoulis, E.K., and Makrakis, G.N. 2006. Rytov approximation of tomographic receptions in weakly range-dependent ocean environments. *J. Acoust. Soc. Am.*, **120**, 120–134.

Plueddemann, A.J. 1992. Internal wave observations from the Arctic environmental drifting buoy. *J. Geophys. Res.*, **97**, 12619–12638.

Polzin, K.L., and Lvov, Y.V. 2011. Toward regional characterizations of the oceanic internal wave field. *Rev. Geo. Phys.*, 1–61.

Pringle, J.M. 1999. Observations of high-frequency internal waves in the Coastal Ocean Dynamics region. *J. Geophys. Res.*, **104**, 5263–5281.

Raghukumar, K., and Colosi, J.A. 2014. High frequency normal mode statistics in a shallow water waveguide: I. The effect of random linear internal waves. *J. Acoust. Soc. Am.*, **136**, 66–79.

Raghukumar, K., and Colosi, J.A. 2015. High frequency normal mode statistics in a shallow water waveguide: II. The combined effect of random linear surface and internal waves. *J. Acoust. Soc. Am.*, **137**, 2950–2961.

Rainville, L., and Pinkel, R. 2006. Propagation of low mode internal waves through the ocean. *J. Phys. Oceanogr.*, **36**, 1220–1236.

Ramp, S.R., Tang, T.Y., Duda, T.F., Lynch, J.F., Liu, A.K., Chiu, C-S., Bahr, F.L, Kim, H-R., and Yang, Y-J. 2004. Internal solitons in the northeastern South China Sea. Part I: Sources and deep water propagation. *IEEE J. Oceanic Eng.*, **29**(4), 1157–1181.

Ray, R.D., and Mitchum, G.T. 1996. Surface manifestations of internal tides generated near Hawaii. *Geophys. Res. Lett.*, **23**, 2101–2104.

Revelle, R.H. 1974. On starting a university.

Reynolds, S., Flatté, S.M., Dashen, R., and Maciejewski, P. 1985. AFAR measurements of acoustic mutual coherence functions of time and frequency. *J. Acoust. Soc. Am.*, **77**, 1723–1731.

Rouseff, D., Turgut, A., Wolf, S., Finnette, S., Orr, M., Pasewark, B., Apel, J., Badiey, M., Chiu, C.S., Headrick, R., Kemp, J., Lynch, J., Kemp, J., Newhall, A., von der Heydt, K., and Teilbuerger, D. 2002. Coherence of acoustic mode propagation through shallow water internal waves. *J. Acoust. Soc. Am.*, **111**, 1655–1666.

Rytov, S. 1937. Wave and geometrical optics. *Comptes Rendus (Doklady) de l Acad. des Sciences, USSR*, **18**, 263–300.

Rytov, S.M., Kravtsov, Y.A., and Tatarskii, V.I. 1989. *Principles of Statistical Radiophysics 4 Wave Propagation Through Random Media*. Springer-Verlag, Berlin.

Sakurai, J.J. 1985. *Modern Quantum Mechanics*. Addison-Westley, Reading, MA.

Sato, H., Fehler, M.C., and Maeda, T. 2012. *Seismic Wave Propagation and Scattering in the Heterogeneous Earth*. Vol. 496. Springer, Berlin.

Schulman, L.S. 1981. *Techniques and Applications of Path Integration*. John Wiley & Sons, New York.

Shankar, R. 1994. *Principles of Quantum Mechanics*, 2nd edition Plenum Press, New York.

Shannon, C.E. 1948. A mathematical theory of communication. *Bell Syst. Tech. J.*, **27**, 379–423.

Sherman, J.T., and Pinkel, R. 1991. Estimates of the vertical wavenumber-frequency spectra of vertical shear and strain. *J. Phys. Oceanogr.*, **21**, 292–303.

Simmen, J., Flatté, S.M., and Wang, G.Y. 1997. Wavefront folding, chaos, and diffraction for sound propagation through ocean internal waves. *J. Acoust. Soc. Am.*, **102**, 239–255.

Skarsoulis, E.K., and Cornuelle, B.D. 2004. Travel-time sensitivity kernels in ocean acoustic tomography. *J. Acoust. Soc. Am.*, **116**, 227–238.

Smith, K.B., Brown, M.G., and Tappert, F.D. 1992a. Acoustic ray chaos induced by mesoscale ocean structure. *J. Acoust. Soc. Am.*, **91**(4), 1950–1959.

Smith, K.B., Brown, M.G., and Tappert, F.D. 1992b. Ray chaos in underwater acoustics. *J. Acoust. Soc. Am.*, **91**(4), 1939–1949.

Spiesberger, J.L., and Worcester, P.F. 1981. Fluctuations of resolved acoustic multipaths at long range in the ocean. *J. Acoust. Soc. Am.*, **70**, 565–577.

Steinberg, J.C., and Birdsall, T.G. 1966. Underwater sound propagation in the Straits of Florida. *J. Acoust. Soc. Am.*, **39**(2), 301–315.

Stoughton, R.B., Flatté, S.M., and Howe, B.M. 1986. Acoustic measurement of internal wave rms displacement and rms horizontal current off Bermuda in late 1983. *J. Geophys. Res.*, **91**, 7721–7732.

Tabor, M. 1989. *Chaos and Integrability in Nonlinear Dynamics*. John Wiley & Sons, New York.

Tang, D., Moum, J.N., Lynch, J.F., Abbot, P.A., Chapman, R., Dahl, P.H., Duda, T.F., Gawarkiewicz, G.G., Glenn, S.M., Goff, J.A., Graber, H.C., Kemp, J.N., Maffei, A.R., Nash, J.D., and Newhall, A.E. 2007. Shallow Water 06: A joint acoustic propagation/nonlinear internal wave physics experiment. *Oceanography*, **20**, 156–167.

Tang, X., Tappert, F.D., and Creamer, D. B. 2006. Simulations of large acoustic scintillations in the Straits of Florida. *J. Acoust. Soc. Am.*, **120**(6), 3539–3552.

Tappert, F.D. 1974. Parabolic equation method in underwater acoustics. *J. Acoust. Soc. Am.*, **55, S34**.

Tappert, F.D. 2003. Theory of explosive beam spreading due to ray chaos. *J. Acoust. Soc. Am.*, **114**, 2775–2781.

Tappert, F.D., and Brown, M.G. 1996. Asymptotic phase errors in parabolic approximations to the one-way Helmholtz equation. *J. Acoust. Soc. Am.*, **99**, 1405–1413.

Tappert, F.D., and Hardin, R.H. 1973. In: *A Synopsis of the AESD Workshop on Acoustic Modeling by non-Ray Techniques*. Office of Naval Research, AESD TN-73-05.

Tappert, F.D., and Tang, X. 1996. Ray chaos and eigenrays. *J. Acoust. Soc. Am.*, **99**, 185–195.

Tatarskii, V.I. 1971. *The Effects of the Turbulent Atmosphere on Wave Propagation*. Israel Program for Scientific Translation: Jerusalem, Israel.

Thorpe, S.A. 1975. The excitation, dissipation, and interaction of internal waves in the deep ocean. *J. Goephys. Res.*, **80**, 328–338.

Thorpe, W.H. 1967. Analytic description of the low frequency attenuation coefficient. *J. Acoust. Soc. Am.*, **42**, 270.

Thorsos, E.I., Henyey, F.S., Elam, W.T., Hefner, B.T., Reynolds, S.A., and Yang, J. 2010. Transport theory for shallow water propagation with rough boundaries. *AIP Conference Proceedings*, **1272**, 99–105.

Tielburger, D., Finnette, S., and Wolf, S. 1997. Acoustic propagation through an internal wave field in a shallow water waveguide. *J. Acoust. Soc. Am.*, **101**, 789–808.

Topinka, M.A., and Westervelt, R.M. 2003. Imaging electron flow. *Phys. Today*, **56**, 47–52.

Turner, J.S. 1979. *Buoyancy Effects in Fluids*. Cambridge University Press, Cambridge.

Udovydchenkov, I.A., and Brown, M.G. 2008. Modal group time spreads in weakly range-dependent deep ocean environments. *J. Acoust. Soc. Am.*, **123**, 41–50.

Udovydchenkov, I.A., Brown, M.G., Duda, T.F., A., Mercer J.A., Andrew, R.K., Worcester, P.F., Dzieciuch, M.A., Howe, B.M., and Colosi, J.A. 2012. Modal analysis of the range evolution of broadband wavefields in the North Pacific Ocean: Low mode numbers. *J. Acoust. Soc. Am.*, **131**, 4409–4427.

Udovydchenkov, I.A., Brown, M.G., Duda, T.F., Mercer, J.A., Andrew, R.K., Worcester, P.F., Dzieciuch, M.A., Howe, B.M. and Colosi, J.A. 2013. Weakly dispersive modal pulse propagation in the North Pacific Ocean. *J. Acoust. Soc. Am.*, **134**, 3386–3394.

Urick, R.J. 1979. *Sound Propagation in the Sea*. Defense Advanced Research Agency, Los Altos, CA.

Uscinski, B.J. 1982. Intensity fluctuations in a multiple scattering medium. Solution of the fourth moment equation. *Proc. R. Soc. London Ser. A*, **380**, 137–169.

Van Kampen, N.G. 1981. *Stochastic Processes in Physics and Chemistry*. North-Holland, New York.

Van Uffelen, L.J., Worcester, P.F., Dzieciuch, M.A., and Rudnick, D. 2009. The vertical structure of shadow-zone arrivals at long range in the ocean. *J. Acoust. Soc. Am.*, **125**, 3569–3588.

Vera, M.D. 2007. Comparison of ocean acoustic horizontal coherence predicted by path-integral approximations and parabolic equation simulation results. *J. Acoust. Soc. Am.*, **121**, 166–174.

Virovlyansky, A.L. 2003. Ray travel times at long ranges in acoustic waveguides. *J. Acoust. Soc. Am.*, **113**, 2523–2532.

Virovlyansky, A.L. 2014. Ray-based description of mode coupling by sound speed fluctuations in the ocean. *J. Acoust. Soc. Am.*, **137**, 2137.

Virovlyansky, A.L., and Zaslavsky, G.M. 2000. Evaluation of the smoothed interference pattern under conditions of ray chaos. *Chaos*, **10**, 211–223.

Voronovich, A.G., and Ostashev, V.E. 2006. Low frequency sound scattering by internal waves in the ocean. *J. Acoust. Soc. Am.*, **119**, 1406–1419.

Voronovich, A.G., Ostashev, V.E., Colosi, J.A., Cornuelle, B.D., Dushaw, B.D., Dzieciuch, M.A., Howe, B.M., Mercer, J.A., Munk, W.H., Spindel, R.C., and Worcester, P.F. 2005. Horizontal refraction of acoustic signals retrieved from North Pacific Acoustic Laboratory billboard array data. *J. Acoust. Soc. Am.*, **117**, 1527–1537.

Wage, K.E., Baggeroer, A.B., and Preisig, J. 2003. Modal analysis of broadband acoustic receptions at 3515-km range in the North Pacific using short-time Fourier techniques. *J. Acout. Soc. Am.*, **113**, 801–817.

Wage, K.E., Dzieciuch, M.A., Worcester, P.F., Howe, B.M., and Mercer, J.A. 2005. Mode coherence at megameter ranges in the North Pacific Ocean. *J. Acout. Soc. Am.*, **117**, 1565–1581.

Weinburg, H., and Burridge, R. 1974. Horizontal ray theory for ocean acoustics. *J. Acoust. Soc. Am.*, **55**, 63–79.

Wheelon, A.D. 2003. *Electromagnetic Scintillation*, Vols. 1–3. Cambridge University Press, New York.

White, A.W., Andrew, R.K., Mercer, J.A., Worcester, P.F., Dzieciuch, M.A., and Colosi, J.A. 2013. Wavefront intensity statistics for 284-Hz broadband transmissions to 107-km range in the Philippine Sea: Observations and modeling. *J. Acoust. Soc. Am.*, **134**, 3347–3358.

Wijesekera, H., Padman, L., Dillon, T., Levine, M., and Paulson, C. 1993. The application of internal-wave dissipation models to a region of strong mixing. *J. Phys. Oceanogr.*, **23**, 269–286.

Wilson, W.D. 1960. Equation for the speed of sound in sea water. *J. Acoust. Soc. Am.*, **32**(10), 1357–1357.

Wolfson, M.A., and Tappert, F.D. 2000. Study of horizontal multipaths and ray chaos due to ocean mesoscale structure. *J. Acout. Soc. Am.*, **107**, 154–162.

Wolfson, M.A., and Tomsovic, S. 2001. On the stability of long-range sound propagation through a structured ocean. *J. Acoust. Soc. Am.*, **109**, 2693–2703.

Wood, A.B. 1930. *A Textbook of Sound*. George Bell and Sons, London.

Worcester, P.F. 1977. Reciprocal acoustic transmission in a midocean environment. *J. Acoust. Soc. Am.*, **62**, 895–905.

Worcester, P.F., Cornuelle, B.D., Dzieciuch, M.A., Munk, W.H., Colosi, J.A., Howe, B.M., Mercer, J.A., Spindel, R.C., Metzger, K., Birdsall, T., and Baggeroer, A. 1999. A test of basin-scale acoustic thermometry using a large-aperture vertical array at 3250-km range in the eastern North Pacific Ocean. *J. Acoust. Soc. Am.*, **105**, 3185–3201.

Worcester, P.F., Cornuelle, B.D., Hildebrand, J.A., Hodgkiss, W.S., Duda, T.F., Boyd, J., Howe, B.M., Mercer, J.A., and Spindel, R.C. 1994. A comparison of measured and predicted acoustic arrival patterns in travel time depth coordinates at 1000-km range. *J. Acoust. Soc. Am.*, **95**, 3118–3128.

Worcester, P.F., Dzieciuch, M.A., Mercer, J.A., Andrew, R.K., Dushaw, B.D., Baggeroer, A.B., Heaney, K.D., D'Spain, G., Colosi, J.A., and Stephen, R.A. 2013. The North Pacific Acoustic Laboratory deep-water acoustic propagation experiments in the Philippine Sea. *J. Acoust. Soc. Am.*, **134**(4), 3359–3375.

Worcester, P.F., Howe, B.M., Mercer, J.A., and Dzieciuch, M.A. 2000. A comparison of long-range acoustic propagation at ultra-low (28Hz) and very low (84 Hz) frequencies. In: Talanov, V.I. (ed), *Proceedings of the US-Russia Workshop on Experimental Underwater Acoustics* (93–104). Institute of Applied Physics, Russian Academy of Sciences, Nizhny Novgorod.

Worcester, P.F, and Spindel, R.C. 2005. North Pacific acoustic laboratory. *J. Acoust. Soc. Am.*, **117**(3), 1499–1510.

Worcester, P.F., Williams, G.O., and Flatté, S.M. 1981. Fluctuations of resolved acoustic multipaths at short range in the ocean. *J. Acoust. Soc. Am.*, **70**, 825–840.

Wunsch, C. 1976. Geographic variability of the internal wave field: A search for sources and sinks. *J. Phys. Oceanogr.*, **6**, 471–485.

Wunsch, C., and Hendry, R. 1972. Array measurements of the bottom boundary layer and the internal wave field on the continental slope. *Geophys. Fluid Dyn.*, **4**, 101–145.

Wunsch, C., and Webb, S. 1979. The climatology of the deep ocean internal wave field. *J. Phys. Oceanogr.*, **9**, 235–243.

Young, W.R., Rhines, P.B., and Garrett, C.J.R. 1982. Shear-flow dispersion, internal waves and horizontal mixing in the ocean. *J. Phys. Oceanogr.*, **12**, 515–527.

Zaslavsky, G.M. 1980. Stochasticity in quantum systems. *Phys. Rev.*, **80**, 157–250.

Zhou, J.X., Zhang, X.Z., and Rogers, P.H. 1991. Resonant interaction of sound wave with internal solitons in the coastal zone. *J. Acoust. Soc. Am.*, **90**(4), 2042–2054.

Index